NCCER's Instructor Res

MW01538147

This website is passcode-protected and is only meant for instructor use. The access code below provides you with access to the following:

- Test Generator software and module exams
- Lesson plans
- PowerPoints®
- Performance Profile sheets

Your access code is good for all levels of the specified craft and edition it accompanies.

To get started, follow the steps below:

1. **Visit www.nccerirc.com.**

2. **Select the appropriate craft area.**

3. **Locate your book cover/edition and click "Login"**

4. **If you are a first-time user, select "Register" and agree to the privacy statement. Select "Yes" if you have ever created a Pearson account. Select "No" if you need to set up a new one.**

5. **Enter your access code provided below and follow the prompts.**

The Access Code is valid for one-time redemption upon intial registration. You may then log-in using your username and password every time you return to this craft area.

Note: Additional resources may be added to this site as they become available.

Pearson Customer Technical Support
For assistance with the NCCER IRC, you can access our 24-hour Tech Support Site at http://247pearsoned.custhelp.com

Access Code Millwright

PESMGH-QUAIL-CONCH-ORGAN-SLANT-RISES

STORE IN A SECURE AREA!

PEARSON

Millwright

Level Five

Annotated Instructor's Guide
Third Edition

PEARSON
Prentice Hall

nccer

Upper Saddle River,
New Jersey
Columbus, Ohio

National Center for Construction Education and Research

President: Don Whyte
Director of Curriculum Revision and Development: Daniele Stacey
Millwright Project Manager: Tania Domenech/Patty Bird
Production Manager: Tim Davis
Quality Assurance Coordinator: Debie Ness
Editors: Rob Richardson and Matt Tischler
Desktop Publishing Coordinator: James McKay
Production Assistant: Laura Wright

NCCER would like to acknowledge the contract service provider for this curriculum:
Topaz Publications, Liverpool, New York.

This information is general in nature and intended for training purposes only. Actual performance of activities described in this manual requires compliance with all applicable operating, service, maintenance, and safety procedures under the direction of qualified personnel. References in this manual to patented or proprietary devices do not constitute a recommendation of their use.

PEARSON
Prentice
Hall

nccer

Contren® Learning Series

10 9 8 7 6 5 4 3
ISBN 0-13-609961-0
ISBN 13 978-0-13-609961-1

PREFACE

TO THE INSTRUCTOR

In training millwrights, you are creating a workforce that can enjoy a career that demands complex mechanical knowledge and sharp analytical ability in a variety of industrial work environments. Since its humble beginnings in the construction of wood mills, the millwright trade has expanded to include work in metal and machinery of ever-increasing technology and precision. Millwrights install, align, and troubleshoot machinery in factories, power plants (particularly the precision machinery required in nuclear power plants), and other industrial sites. They install conveyor systems, connect machinery to power supplies and piping, direct hoisting and setting of machines, and adjust the moving and stationary parts of machines to certain specifications. Millwrights are extremely skilled at mathematics and interpreting blueprints and specs to set machines at perfect measurements, sometimes working with clearances no bigger than thousandths of an inch.

Millwrights are a specialized and elite group, as there are only about 60,000 millwrights employed in the United States today. However, over the next decade, there will be a demand for a third more (U.S. Bureau of Labor Statistics). As the population grows, especially in developing countries, increased demands for energy and travel in particular will require more millwrights working in power plants, refineries, and factories. Trained and experienced millwrights enjoy a comfortable salary and the chance for different avenues of professional development. Millwrights can progress upwards in their trade, undergoing full apprenticeship, becoming supervisors, and/or obtaining higher education; millwrights may also opt for careers in related professions, such as machinist, equipment engineer, or aircraft assembler.

NCCER wishes you the best as you instruct future millwrights. With solid millwright training and experience, there are countless avenues open to trainees in an industrial sector that shows no signs of slowing down.

We invite you to visit the NCCER website at www.nccer.org for the latest releases, training information, newsletter, and much more. You can also reference the Contren® product catalog online at www.crafttraining.com. Your feedback is welcome. You may email your comments to curriculum@nccer.org or send general comments and inquiries to info@nccer.org.

CONTREN® LEARNING SERIES

The National Center for Construction Education and Research (NCCER) is a not-for-profit 501(c)(3) education foundation established in 1995 by the world's largest and most progressive construction companies and national construction associations. It was founded to address the severe workforce shortage facing the industry and to develop a standardized training process and curricula. Today, NCCER is supported by hundreds of leading construction and maintenance companies, manufacturers, and national associations. The Contren® Learning Series was developed by NCCER in partnership with Pearson Education, Inc., the world's largest educational publisher.

Some features of NCCER's Contren® Learning Series are as follows:

- An industry-proven record of success
- Curricula developed by the industry for the industry
- National standardization, providing portability of learned job skills and educational credits
- Compliance with the Office of Apprenticeship requirements for related classroom training (*CFR 29:29*)
- Well-illustrated, up-to-date, and practical information

NCCER also maintains a National Registry that provides transcripts, certificates, and wallet cards to individuals who have successfully completed modules of NCCER's Contren® Learning Series. *Training programs must be delivered by an NCCER Accredited Training Sponsor in order to receive these credentials.*

Contents

Contren® Curricula

NCCER's training programs comprise over 50 construction, maintenance, and pipeline areas and include skills assessments, safety training, and management education.

Boilermaking
Cabinetmaking
Carpentry
Concrete Finishing
Construction Craft Laborer
Construction Technology
Core Curriculum:
  Introductory Craft Skills
Drywall
Electrical
Electronic Systems Technician
Heating, Ventilating, and
  Air Conditioning
Heavy Equipment Operations
Highway/Heavy Construction
Hydroblasting
Industrial Coating and Lining
  Application Specialist
Industrial Maintenance
  Electrical and Instrumentation
  Technician
Industrial Maintenance Mechanic
Instrumentation
Insulating
Ironworking
Masonry
Millwright
Mobile Crane Operations
Painting
Painting, Industrial
Pipefitting
Pipelayer
Plumbing
Reinforcing Ironwork
Rigging
Scaffolding
Sheet Metal
Site Layout
Sprinkler Fitting
Welding

Pipeline
Control Center Operations,
  Liquid
Corrosion Control
Electrical and Instrumentation
Field Operations, Liquid
Field Operations, Gas
Maintenance
Mechanical

Safety
Field Safety
Safety Orientation
Safety Technology

Management
Introductory Skills for the
  Crew Leader
Project Management
Project Supervision

Spanish Translations
Albañilería Nivel Uno
Andamios
Currículo Básico
  Habilidades Introductorias
  del Oficio
Instalación de Rociadores
  Nivel Uno
Orientación de Seguridad
Principios Básicos de Maniobras
Seguridad de Campo

Supplemental Titles
Applied Construction Math
Careers in Construction
Your Role in the Green
  Environment

Acknowledgments

This curriculum was revised as a result of the
farsightedness and leadership of the following sponsors:

ABC SW Pelican Training Center and Performance Contractors, Inc.
Bierlein Companies
Cianbro Institute
Constellation Energy
"LTC" River Parishes Campus
QCI/GE Energy Services
Sunoco, Inc.
Zachry Construction Corporation

This curriculum would not exist were it not for the dedication and unselfish energy of those volunteers who served on the Authoring Team. A sincere thanks is extended to the following:

Terry Auger
Gerald Kenyon
Ed LePage

Richard Platt
Gregory Spooner
Bill Wall
John Ziegler

NCCER PARTNERING ASSOCIATIONS

American Fire Sprinkler Association
Associated Builders and Contractors, Inc.
Associated General Contractors of America
Association for Career and Technical Education
Association for Skilled and Technical Sciences
Carolinas AGC, Inc.
Carolinas Electrical Contractors Association
Center for the Improvement of Construction
   Management and Processes
Construction Industry Institute
Construction Users Roundtable
Design Build Institute of America
Green Advantage
Merit Contractors Association of Canada
Merit Shop Training, Inc.
Metal Building Manufacturers Association
NACE International
National Association of Manufacturers

National Association of Minority Contractors
National Association of Women in Construction
National Insulation Association
National Ready Mixed Concrete Association
National Systems Contractors Association
National Technical Honor Society
National Utility Contractors Association
NAWIC Education Foundation
North American Crane Bureau
North American Technician Excellence
Painting & Decorating Contractors of America
Portland Cement Association
SkillsUSA
Steel Erectors Association of America
U.S. Army Corps of Engineers
University of Florida
Women Construction Owners & Executives, USA

Product Supplements

Windows/Macintosh-Based
TestGen

ISBN	0-13-609087-7
ISBN 13	978-0-13-609087-8

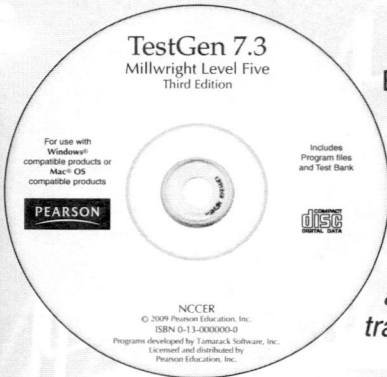

Ensure test security with NCCER's computerized testing software. This software allows instructors to scramble the module exam questions and answer keys in order to print multiple versions of the same test, customize tests to suit the needs of their training units*, add questions, or easily create a final exam.

Due to NCCER's Accreditation Guidelines, instructors may not delete existing questions from exams. Doing so may seriously jeopardize either the accreditation status or the training program sponsor or any recognition of trainees, instructors, and trainers through the NCCER National Registry.

Transparency
Masters

ISBN	0-13-610671-4
ISBN 13	978-0-13-610671-5

Spend more time training and less time at the copier. In response to instructor feedback NCCER offers loose, reproducible copies of the overhead transparencies referenced in the Instructor's Guides. The transparency masters package includes most of the Trainee Guide graphics, enlarged for projection and printed on loose sheets* for easy copying onto transparency film using your photocopier.

* Transparency masters are provided on regular, loose sheets of paper, not acetates.

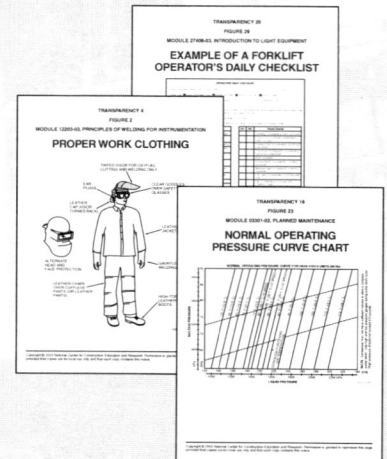

Microsoft®
PowerPoint® Presentation Slides

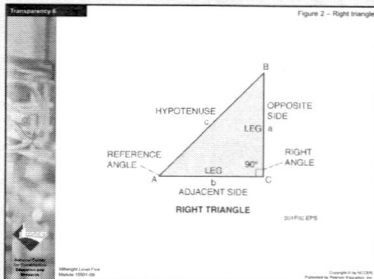

This slide show presentation will be a powerful addition to your classroom discussions! NCCER's Microsoft® PowerPoint® CD features all the drawings and photos from the Traineee Guide, in color and sequenced as they are called out in the Annotated Instructor's Guide annotations. Please contact your Pearson Sales Representative for availability.

ISBN	0-13-609088-5
ISBN 13	978-0-13-609088-5

To order these supplements, contact Pearson Education, Inc. at

800-922-0579

Reverse Alignment

NCCER STANDARDIZED CRAFT TRAINING PROGRAM

The National Center for Construction Education and Research (NCCER) provides a standardized national program of accredited craft training. Key features of the program include instructor certification, competency-based training, and performance testing. The program provides trainees, instructors, and companies with a standard form of recognition through a National Craft Training Registry. The program is described in full in the *Guidelines for Accreditation*, published by NCCER. For more information on standardized craft training, contact the NCCER by writing us at 3600 NW 43rd St., Bldg. G, Gainesville, FL 32606; calling 352-334-0911; or emailing info@nccer.org. More information may be found at our website, www.nccer.org.

HOW TO USE THIS ANNOTATED INSTRUCTOR'S GUIDE

Each page presents two sections of information. The larger section displays each page exactly as it appears in the Trainee Module. The narrow column ties suggested trainee and instructor actions to each page and provides icons (detailed below) to call your attention to material, safety, audiovisual, or testing requirements. The bottom of each page includes space for your notes.

The **Audiovisual** icon indicates an appropriate time to show a transparency or other audiovisual aid.

The **Classroom** icon prompts you to define a term, stress a point, ask trainees to explain a concept, or give examples.

The **Demonstration** icon directs you to show trainees how to perform tasks.

The **Examination** icon tells you to administer the written module examination.

The **Homework** icon is placed where you may wish to assign reading for the next session, assign a project, or advise trainees to prepare for an examination.

The **Laboratory** icon is used when trainees are to practice performing tasks.

The **Materials** icon is a reminder for you to gather materials needed for classes, labs, and testing.

The **Performance Testing** icon tells you to administer a performance test or a portion thereof.

The **Safety** icon is used to emphasize safety issues. It is often keyed to *Caution* and *Warning!* statements in the Trainee Module.

The **Teaching Tip** icon indicates additional guidance is available, such as how to conduct an exercise, get the most educational value from a field trip, or encourage class participation. Teaching Tips may expand on a feature (*Think About It*, *Did You Know?*) or provide *Quick Quizzes* or similar exercises. You will be referred to the Teaching Tips section at the back of the module if there is additional material.

The **Combination** icon indicates that the laboratory listed corresponds with a performance task. If desired, you can note the proficiency of the trainees during the laboratory, and use it to satisfy performance testing requirements.

PREPARATION

Before teaching this module, you should review the Objectives, Performance Tasks, Materials and Equipment List, and Module Outline. Be sure to allow ample time to prepare your own training or lesson plan and gather all required materials and equipment.

Reverse Alignment
Annotated Instructor's Guide

MODULE OVERVIEW

This module covers setting up reverse dial indicator jigs and performing reverse dial alignment using both the chart and mathematical methods. Basic information about shaft alignment and coupling stress is also presented.

PREREQUISITES

Prior to training with this module, it is recommended that the trainee shall have successfully completed *Core Curriculum*; *Millwright Level One*; *Millwright Level Two*; *Millwright Level Three*; and *Millwright Level Four*.

OBJECTIVES

Upon completion of this module, the trainee will be able to do the following:

1. Explain how machinery can be misaligned.
2. Explain the conditions that can cause misalignment.
3. Measure shaft and coupling runout, using a dial indicator.
4. Set up complex reverse dial indicator jigs.
5. Measure indicator sag using complex reverse dial indicator jigs.
6. Perform reverse dial indicator alignment, using a graphical alignment chart.
7. Perform reverse dial indicator alignment, using the mathematical equation.

PERFORMANCE TASKS

Under the supervision of the instructor, the trainee should be able to do the following:

1. Measure shaft runout, using a dial indicator jig.
2. Set up a complex reverse alignment jig.
3. Measure indicator sag, using a complex reverse dial indicator jig.
4. Perform reverse alignment, using the alignment demonstration rig and the graphical chart.
5. Perform reverse alignment, using the alignment demonstration rig and the mathematical equation.

MATERIALS AND EQUIPMENT LIST

Overhead projector and screen
Transparencies
Blank acetate sheets
Transparency pens
Whiteboard/chalkboard
Markers/chalk
Pencils and scratch paper
Dial indicator on a base
Complex reverse dial indicator jig
Dial indicators

Alignment demonstration rig(s)
Alignment simulators or equipment to be aligned
Graph paper
Calculators
Reverse dial indicator plotting guide
Graphical alignment chart
Copies of Quick Quizzes*
Module Examinations**
Performance Profile Sheets**

* Located at the back of this module.

**Located in the Test Booklet.

ADDITIONAL RESOURCES

This module is intended to present thorough resources for task training. The following reference works are suggested for both instructors and motivated trainees interested in further study. These are optional materials for continued education rather than for task training.

A Millwright's Guide to Motor/Pump Alignment, 2nd ed. Tommy B. Harlon. New York, NY: Industrial Press, 2008.

The Optalign Training Book. Galen Evans and Pedro Casanova. Miami, FL: Ludeca, Inc.

TEACHING TIME FOR THIS MODULE

An outline for use in developing your lesson plan is presented below. Note that each Roman numeral in the outline equates to one session of instruction. Each session has a suggested time period of 2½ hours. This includes 10 minutes at the beginning of each session for administrative tasks and one 10-minute break during the session. Approximately 30 hours are suggested to cover *Reverse Alignment*. You will need to adjust the time required for hands-on activity and testing based on your class size and resources. Because laboratories often correspond to Performance Tasks, the proficiency of the trainees may be noted during these exercises for Performance Testing purposes.

Topic	Planned Time
Session I. Introduction; Descriptive Terms and Conditions	
A. Introduction	_____
B. Descriptive Terms and Conditions	_____
C. Conditions	_____
1. Checking for Soft Foot, Rough Alignment, and Shaft Runout	_____
D. Laboratory	_____
Have trainees practice checking for shaft runout.	
Session II. Coupling Stress	
A. Coupling Stress	
B. Causes of Coupling Stress	_____
1. Incorrect Pipe Weldments	_____
2. Improper Placement of Pipe Hangers	_____
3. Defective Anchor Bolts	_____
4. Bad Bearings	_____
5. Improper Foundations	_____
Session III. Reverse Dial Indicator Jigs	
A. Reverse Dial Indicator Jigs	_____
B. Alignment Demonstration Rig	_____
C. Dial Indicators	_____
D. Measuring Shaft Runout	_____
E. Laboratory	_____
Have trainees measure shaft runout using a dial indicator jig. This laboratory corresponds to Performance Task 1.	
Session IV. Reverse Dial Indicator Alignment, Part One	
A. Setting Up Complex Reverse Dial Indicator Jigs	_____
1. Same-Side Mounting	_____
2. Opposite-Side Mounting	_____
3. Checking Indicator Sag	_____
B. Laboratory	_____
Have trainees set up a complex reverse dial indicator jig and check for indicator sag. This laboratory corresponds to Performance Tasks 2 and 3.	

Sessions V–VII. Reverse Dial Indicator Alignment, Part Two

   A. Performing Reverse Dial Indicator Alignment   _____

      1. Charting Alignment   _____

      2. Performing Alignment   _____

   B. Alignment Equation   _____

   C. Recording Alignment   _____

Sessions VIII–XI. Reverse Dial Indicator Alignment, Part Three

   A. Laboratory   _____

Have trainees perform reverse alignment using the alignment demonstration rig, graphical chart, and mathematical equation. This laboratory corresponds to Performance Tasks 4 and 5.

Session XII. Review and Testing

   A. Review   _____

   B. Module Examination   _____

      1. Trainees must score 70% or higher to receive recognition from NCCER.

      2. Record the testing results on Craft Training Report Form 200, and submit the results to the Training Program Sponsor.

   C. Performance Testing   _____

      1. Trainees must perform each task to the satisfaction of the instructor to receive recognition from NCCER. If applicable, proficiency noted during laboratory exercises can be used to satisfy the Performance Testing requirements.

      2. Record the testing results on Craft Training Report Form 200, and submit the results to the Training Program Sponsor

Millwright Level Five

15501-09

Reverse Alignment

Assign reading of Module 15501-09.

15501-09
Reverse Alignment

Topics to be presented in this module include:

Overview

Reverse alignment uses dial indicators to make shafts collinear. In this module, trainees will learn to determine the required movements in both angularity and offset mathematically and graphically. All of the changes can be calculated by either method before any movements are carried out, allowing for an efficient alignment procedure. Trainees will also be taught to set up a complex reverse alignment jig.

Instructor's Notes:

Objectives

When you have completed this module, you will be able to do the following:

1. Explain how machinery can be misaligned.
2. Explain the conditions that can cause misalignment.
3. Measure shaft and coupling runout, using a dial indicator.
4. Set up complex reverse dial indicator jigs.
5. Measure indicator sag using complex reverse dial indicator jigs.
6. Perform reverse dial indicator alignment, using a graphical alignment chart.
7. Perform reverse dial indicator alignment, using the mathematical equation.

Trade Terms

Collinear   Thermal growth
Machinery train

Required Trainee Materials

1. Pencil and paper
2. Appropriate personal protective equipment

Prerequisites

Before you begin this module, it is recommended that you successfully complete *Core Curriculum*; *Millwright Level One*; *Millwright Level Two*; *Millwright Level Three*; and *Millwright Level Four*.

This course map shows all of the modules in the fifth level of the Millwright curriculum. The suggested training order begins at the bottom and proceeds up. Skill levels increase as you advance on the course map. The local Training Program Sponsor may adjust the training order.

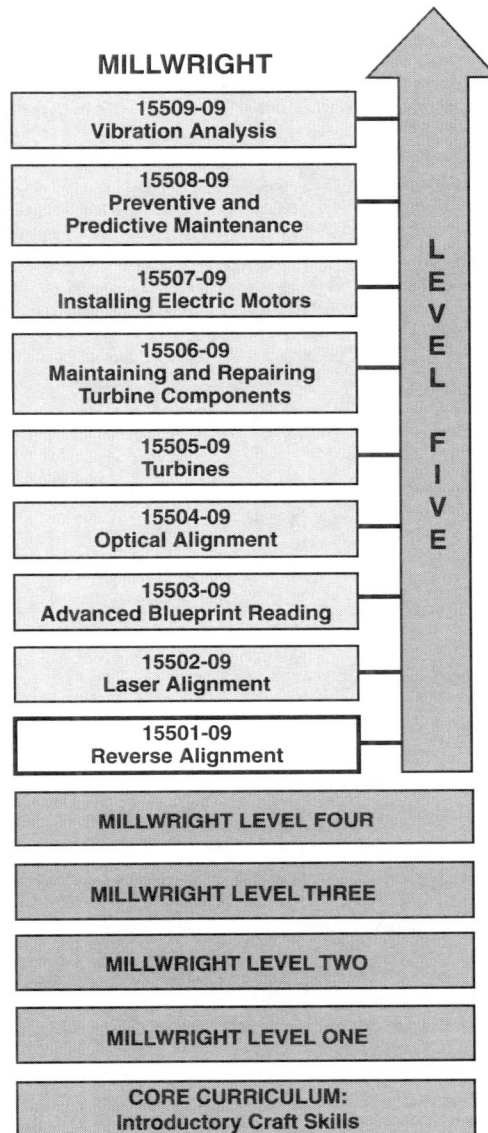

MILLWRIGHT

LEVEL FIVE
15509-09 Vibration Analysis
15508-09 Preventive and Predictive Maintenance
15507-09 Installing Electric Motors
15506-09 Maintaining and Repairing Turbine Components
15505-09 Turbines
15504-09 Optical Alignment
15503-09 Advanced Blueprint Reading
15502-09 Laser Alignment
15501-09 Reverse Alignment

| MILLWRIGHT LEVEL FOUR |
| MILLWRIGHT LEVEL THREE |
| MILLWRIGHT LEVEL TWO |
| MILLWRIGHT LEVEL ONE |
| CORE CURRICULUM: Introductory Craft Skills |

501CMAP.EPS

Ensure that you have everything required to teach the course. Check the Materials and Equipment List at the front of this module.

See the general Teaching Tip at the end of this module.

Explain that terms shown in bold are defined in the Glossary at the back of this module.

Show Transparency 1, Objectives, and Transparency 2, Performance Tasks. Review the goals of the module, and explain what will be expected of the trainees.

Review the modules covered in Level Five, and explain how this module fits in.

Introduce the reverse
dial indicator method
of shaft alignment.

Define collinear.

Review the list of tasks
to be completed prior
to performing an align-
ment.

Discuss the views of
a rotary shaft that
describes its position
in space.

Define offset and
angularity, and explain
how misalignment is
described.

Ensure that trainees
understand the terms
and conditions associ-
ated with misalign-
ment.

Show Transparency 3
(Figure 1). Describe the
types of misalign-
ments illustrated.

Show Transparency 4
(Figure 2). Discuss the
dimensional defini-
tions.

Review the conditions
that may interfere with
a successful align-
ment.

1.0.0 ◆ INTRODUCTION

The reverse dial indicator method of shaft alignment is one of many methods used to align the
rotational center lines of machines. Shaft alignment is defined as positioning two or more
machines so that their rotational center lines are
collinear at the coupling point under operating
conditions. The reverse dial indicator method is as
accurate as optical or laser alignment but requires
a high degree of skill, dependable fixtures, time,
and patience.

The benefits of nearly perfectly aligned shafts
are reduced costs, less energy used, and time
saved. The equipment runs longer and wastes less
energy and maintenance time when the shafts are
aligned properly.

Before performing alignment, complete the following steps:

Step 1 Inspect the base of the equipment for
level, center line, and distortion.

Step 2 Check for soft foot.

Step 3 Set coupling gap.

Step 4 Eliminate outside stress influences such
as pipe stress, conduit, etc.

Step 5 Perform rough alignment of equipment
within mechanical limits of the indicators
(0.001 inch).

2.0.0 ◆ DESCRIPTIVE TERMS AND CONDITIONS

The position of a rotating shaft center line in space
can be completely described by showing both vertical and horizontal views. These views are sometimes called the side view and top view and are
also called the elevation and plan view. Each view
displays offset and angularity. The combination of
offset and angularity is the most common occu-
rance (*Figure 1*).

To approach true shaft alignment, you must
fully understand the difference between offset and
angularity, and between vertical and horizontal.

Once these ideas are clear, you may then combine
the four variables, because in a real alignment situation, there is not one simple plane of misalignment. Since two views describe the location of a
shaft and each view can have offset and angularity,
four terms and numbers are used to describe the
misalignment. Misalignment can be described by
the following terms:

• Vertical offset (VO)
• Horizontal offset (HO)
• Vertical angularity (VA)
• Horizontal angularity (HA)

The basic dimensional definitions are vertical
to horizontal and front to back (*Figure 2*). The stationary machine, or STAT, refers to the machine
that does not move and stays firmly bolted. The
machine to be moved, or MTBM, is where all the
adjustments take place. Each term listed above
describes the position of the MTBM in relation to
the STAT, using the center line of the STAT as zero.
As in any form of alignment, the order of the procedure is: setup; vertical angularity and offset;
then horizontal angularity and offset.

The convention for determining positive and
negative is defined in three planes: rotational, vertical, horizontal, and combinations of these three.

2.1.0 Conditions

Existing conditions can prevent a successful precision alignment. A good habit is to check for all the
possible problems, such as soft foot, bent shafts,
grossly misaligned equipment, and coupling
stress. The machine base could also be weak,
twisted, or on a bad foundation, or the MTBM
could be bolt-bound, which means that the motor
is out of the adjustment range so that the bolts are
bent or difficult to loosen because of the side load.
The time spent looking for these functional problems can prevent having to do the alignment
twice. The most important external factors that
cause problems are thermal expansion, soft foot,
and failure to calculate indicator sag.

1.2 MILLWRIGHT ◆ LEVEL FIVE

Instructor's Notes:

_____

_____

_____

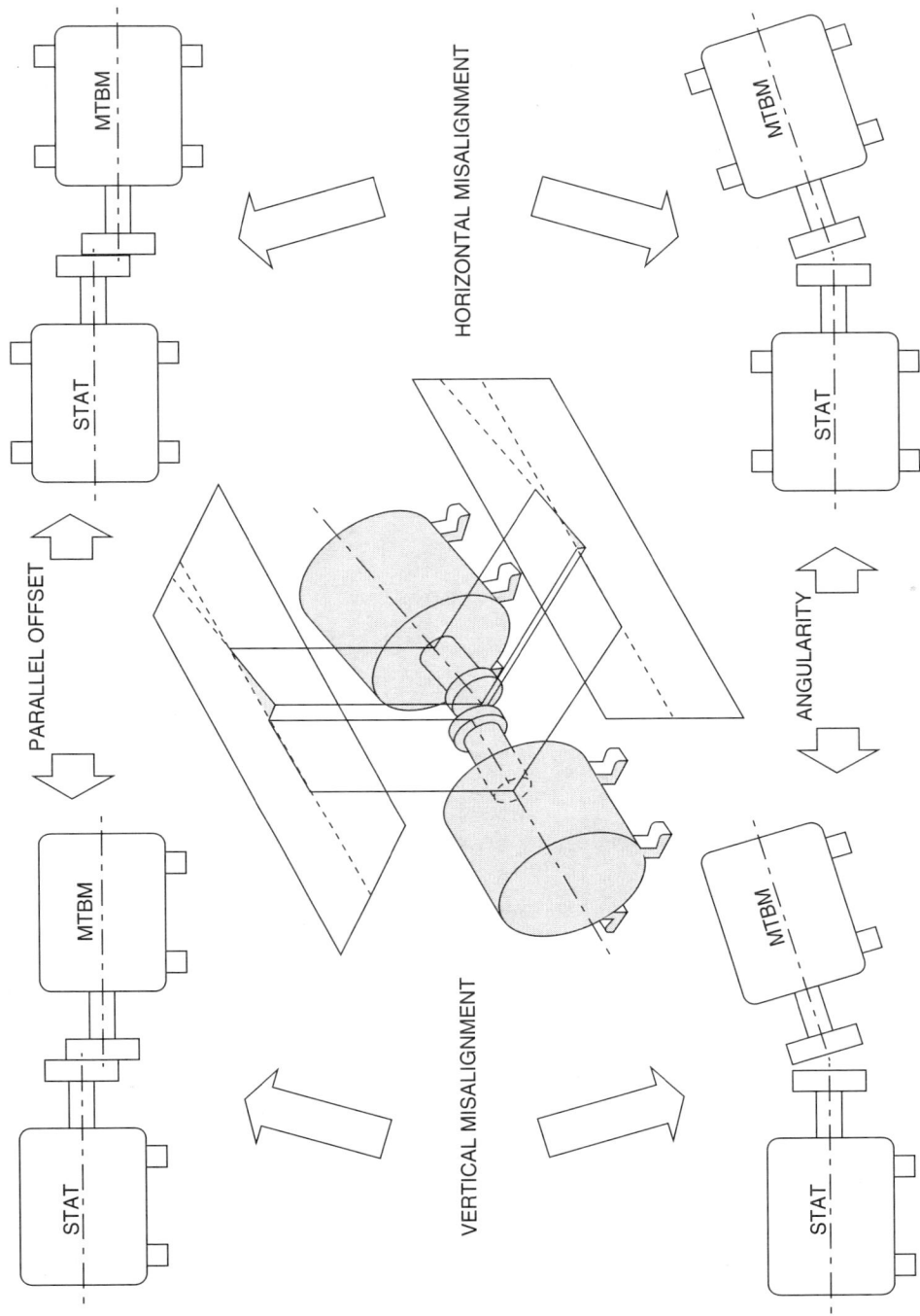

Figure 1 ◆ Combined misalignments.

501F01.EPS

Describe soft foot. Discuss the causes and effects of soft foot.

Show Transparencies 5–7 (Figures 3–5).

Explain how to perform a rough alignment.

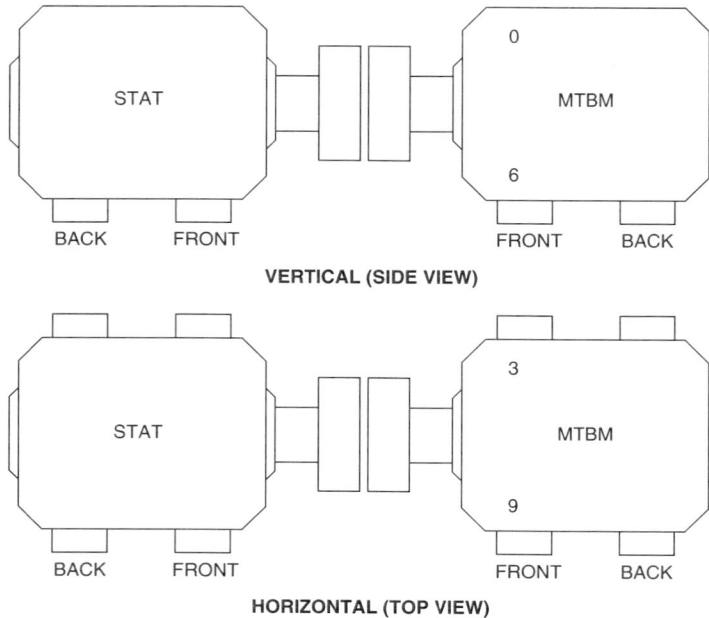

Figure 2 ◆ Dimensional definition.

2.1.1 Checking for Soft Foot

Soft foot is a condition in which the tightening or loosening of the bolt(s) of a single foot distorts the machine frame (*Figure 3*). If the looseness indicates a soft foot, the condition should be analyzed to determine its cause before any corrective action is taken. The effects of soft foot vary because of operating conditions and where the weakest conditions are. A machine with a soft foot that is bolted down firmly distorts and affects the operation of the **machinery train** in many ways (*Figure 4*). It also prevents proper alignment. The condition of the shims is critical to aligning the machine; therefore, damaged shims should be discarded and new shims installed before alignment.

One method for checking for soft foot uses a dial indicator mounted at one of the feet. To check for soft foot, zero the indicator, and loosen the hold-down bolt. If the indicator moves, there is a soft foot condition. Check the remaining feet, and write down all indicator movement at each foot. Insert the necessary shims to the soft foot, and recheck the condition. Indiscriminate shimming will often make things worse, and not all soft feet are the same. Soft feet are caused by a variety of reasons, some of which may not even be related to the machine itself. The following are some causes of soft foot:

• Parallel air gap
• Bent foot
• Squishy foot
• Induced soft foot
• Gaps without soft foot

These causes and the corrections for them are detailed later. Before aligning shafts using the reverse dial indicator method, the soft foot must be eliminated. *Figure 5* shows types of soft foot.

2.1.2 Rough Alignment

To perform a good reverse alignment, the shaft rotational center lines should be as close to each other as possible. Check the rough alignment using a good straightedge or feeler gauges to confirm that the position of the two machines is within range of final alignment.

Instructor's Notes:

Describe shaft runout
and explain how to
check for it.

0.008" SHAFT
MOVEMENT
DUE TO
SOFT FOOT

0.020" SOFT FOOT

BOLTED
TIGHT

|← 10" →|← 10" →|

0.016"
DUE TO
SOFT FOOT

0.008"

0.020"
SOFT
FOOT

|← 10" →|← 10" →|

MISALIGNMENT AT COUPLING

DUE TO SOFT FOOT

INTERNAL MISALIGNMENT

501F03.EPS

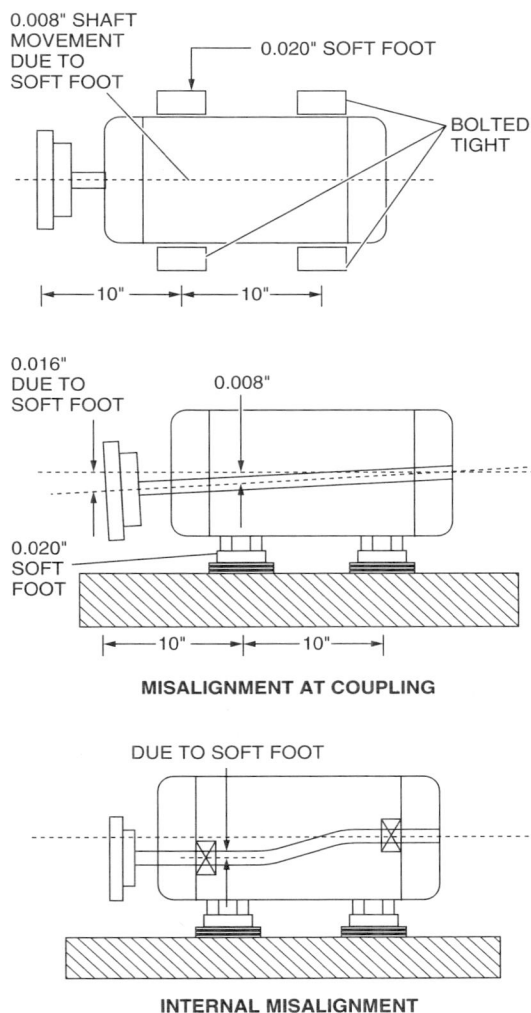

Figure 3 ◆ Soft foot.

2.1.3 Checking for Shaft Runout

A shaft usually has some runout, or eccentricity, not more than 0.002 inch. This could be because the shaft is bent, the bearings are worn, or the coupling is eccentric. If the runout is beyond permissible limits, this condition must be corrected before alignment. The indicator can be set up two ways to check for runout. An indicator on a magnetic base can be positioned to read the shaft after it is uncoupled. Rotate the shaft completely around; record the total reading difference. This is the shaft surface runout, but not necessarily the shaft center line runout. At operational speeds and loads, the true center line may change.

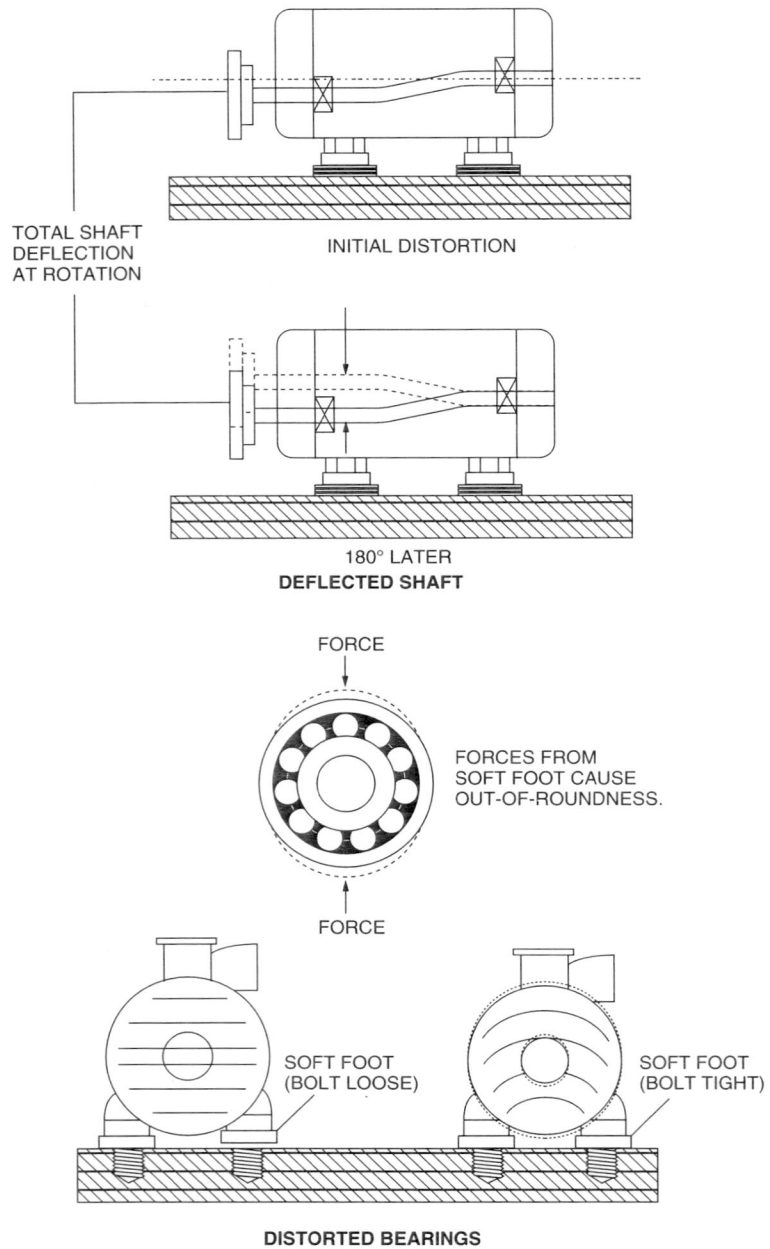

TOTAL SHAFT
DEFLECTION
AT ROTATION

INITIAL DISTORTION

180° LATER
DEFLECTED SHAFT

FORCE

FORCES FROM
SOFT FOOT CAUSE
OUT-OF-ROUNDNESS.

FORCE

SOFT FOOT
(BOLT LOOSE)

SOFT FOOT
(BOLT TIGHT)

DISTORTED BEARINGS

501F04.EPS

Figure 4 ◆ Effects of soft foot.

Instructor's Notes:

_____

_____

_____

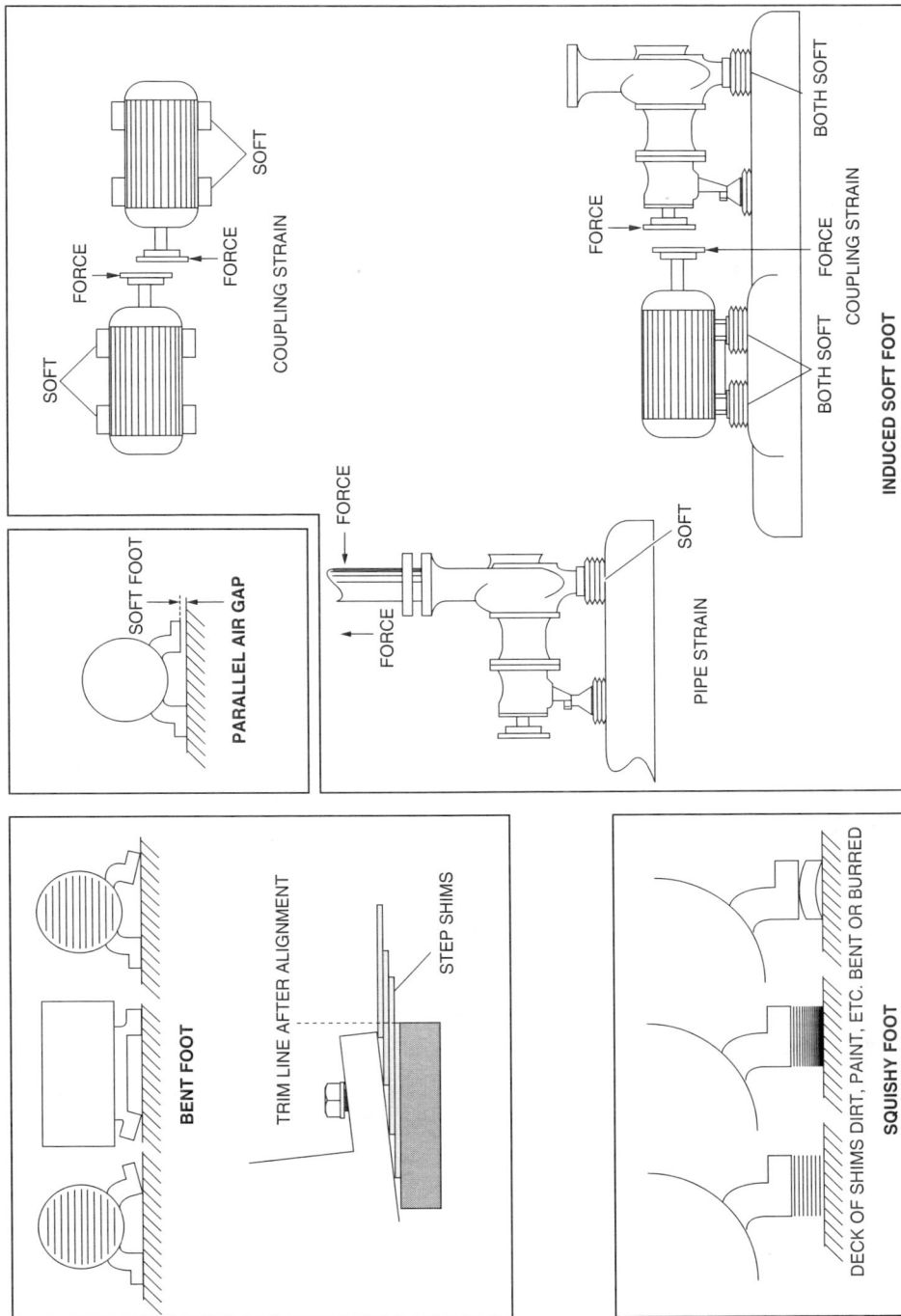

Figure 5 ◆ Types of soft foot.

501F05.EPS

Another method is to use the rim and face method on one of the unbolted couplings. Although the alignment can be successful with shaft or coupling runout within tolerances, it is good to check both sides and have a record of where and how much runout exists. *Figure 6* shows a runout check.

3.0.0 ◆ COUPLING STRESS

Coupling stress is any external condition or situation that places abnormal stress on a STAT or MTBM that can cause misalignment of the shaft center lines. Coupling stress can make it very difficult to align the shafts and has very harmful effects. It can cause vibration, overheating, excessive wear on the equipment, and can increase operating costs through power loss. Coupling stress must be eliminated before the shafts are aligned; therefore, it is important to understand the causes and possible solutions for coupling stress.

Coupling stress can be caused by one or more of many conditions. If, during alignment, it becomes apparent that aligning the shafts is not curing the problem, coupling stress is the next area to analyze. After curing the stress problem, set up the dial indicators and align the shafts properly. The following sections explain some of the common causes of coupling stress and some possible solutions. Common causes of coupling stress include:

- Incorrect pipe weldments
- Improper placement of pipe hangers

- Defective anchor bolts
- Bad bearings
- Improper foundations

3.1.0 Incorrect Pipe Weldments

Pipe that is incorrectly fitted and welded can cause coupling stress. If the pipe flange is welded to the pipe so that it is not perfectly square to the pump connection, bolting the flange can cause considerable stress on the pump, which is transferred to the coupling. If the flange is short of the pump connection, leaving a large gap, bolting the flange and pulling it up to the pump connection will also cause coupling stress. Therefore, the pipe flange should be accurately fitted to the pump, with just enough gap to install the gasket. To eliminate pipe flange angular misalignment, a tapered filler piece may be necessary. Excessive parallel offset cannot be cured with a filler piece, but it can be helped by offsetting several successive joints slightly, taking advantage of the bolt hole clearances.

3.2.0 Improper Placement of Pipe Hangers

Improper placement of pipe hangers can create excessive stress on a STAT and MTBM by forcing the weight of the pipe to be carried by the flange. Piping should be well fitted, firmly supported, and sufficiently flexible so that no more than 0.002 inch of soft foot movement occurs when the last

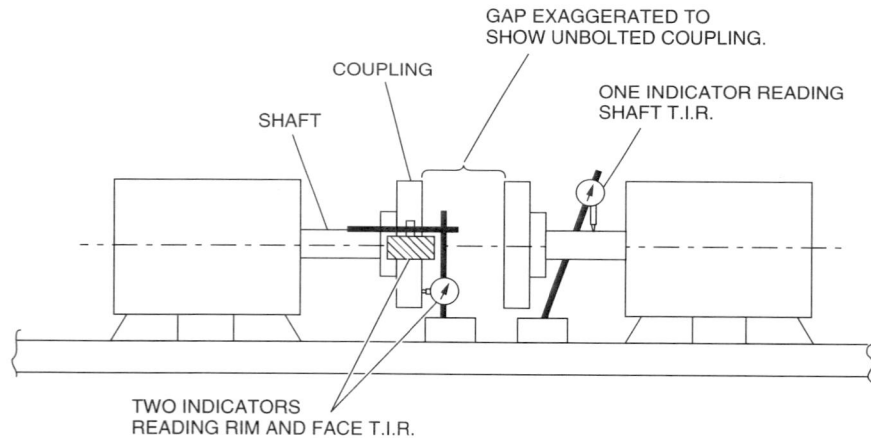

Figure 6 ◆ Runout check.

501F06.EPS

Instructor's Notes:

_____

_____

_____

pipe flanges are tightened. Pipe expansion or movement may cause misalignment and increased vibration. Pipe hangers should be placed and checked before the bolts are installed to ensure that no load is being carried by the flange or pump.

3.3.0 Defective Anchor Bolts

Anchor bolts that have been improperly installed can cause coupling stress by twisting the pump or motor baseplate. The machine base may cause these bolts to skew and bend, which puts a load on the machinery. Before installation, ensure that the base pads are level, flat, parallel, coplanar (on the same plane), and clean. The use of properly sized hold-down bolts with enough clearance to permit corrective movement during alignment is necessary. Anchor bolts should be installed straight and torqued properly. Too many, too few, or improperly spaced anchor bolts in a baseplate can also cause coupling stress.

3.4.0 Bad Bearings

Damaged or marginal bearings in machine units are a progressive cause of coupling stress and misalignment. As the bearings get worse, the shaft vibrates more and creates a wobbly rotational center line. This creates a complex set of alignment problems which, left uncorrected, can lead to catastrophic failure in the machine system.

3.5.0 Improper Foundations

An improper foundation can be an unlevel baseplate, an improperly sized concrete pad, or bad grouting. All these conditions can twist the machine frame and cause misalignment and coupling stress. The corrective action is to check for soft foot, and if close enough to allow it, make shims to level the equipment. If the foundation is very bad, it will have to be remade. A good rule of thumb calls for concrete weight equal to three times the machine weight for rotating machines and five times the weight for reciprocating machines.

4.0.0 ◆ REVERSE DIAL INDICATOR JIGS

The reverse dial indicator jig (*Figure 7*) is sturdier than the other types of jigs, and it is very accurate and easy to use. Two indicator jigs are used at the same time to perform reverse dial indicator alignment. A reverse dial indicator jig consists of a

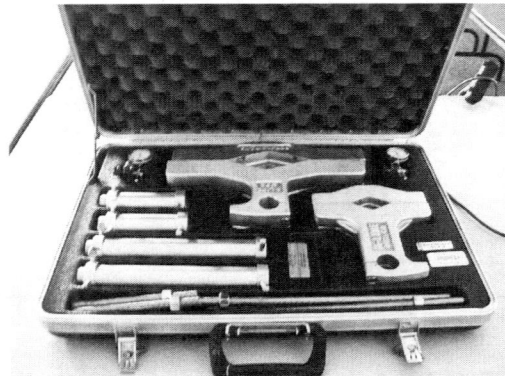

501F07.EPS

Figure 7 ◆ Reverse dial indicator jig components.

clamping device with bolts, a mounting pin, and an indicator support device. A dial indicator is mounted to these fixtures, and both units are mounted to the machine shafts. This type of jig is much more accurate over long distances because it is more rigid, takes both readings with one setup, and has less indicator sag than other jigs. The accuracy and repeatability of shaft alignment is important because of the savings in time and equipment.

4.1.0 Alignment Demonstration Rig

If possible, it is good to have a tabletop demonstration rig for alignment. This can be a simple unit that has all of the adjustments a pump and motor may have. Most maintenance shops have spare pillow block bearings, couplings, and plate that can be assembled to provide a hands-on teaching device. This device should be similar to the demonstration rig shown in *Figure 8*, or it may include specific needs of your teaching experience.

4.2.0 Dial Indicators

A dial indicator is a direct-reading instrument used to measure machined parts for accuracy or to measure surfaces of machinery to determine runout or accuracy of alignment. Dial indicators are also used when installing bearings and seals, setting up lathes and milling machines, and checking the concentricity of a diameter. A dial indicator consists of a graduated dial with an indicator hand and a contact point attached to a spring-loaded plunger. Any movement of the plunger causes the pointer on the dial to move. Dial indicators range in size from 1 to 4½ inches in

Describe the stresses caused by defective anchor bolts, bad bearings, and improper foundations.

See the Teaching Tip for Sections 2.0.0 and 3.0.0 at the end of this module.

Assign reading of Sections 4.0.0 - 4.3.0.

Ensure that you have everything required for teaching this session.

Describe reverse dial indicator jigs and explain how they are used.

Provide a reverse dial indicator jig for trainees to examine.

Show Transparency 9 (Figure 8). Describe the setup of this alignment demonstration rig.

Provide dial indicators for trainees to examine.

Explain how a dial indicator is read.

Show Transparency 10 (Figure 9). Compare the two types of dial indicators.

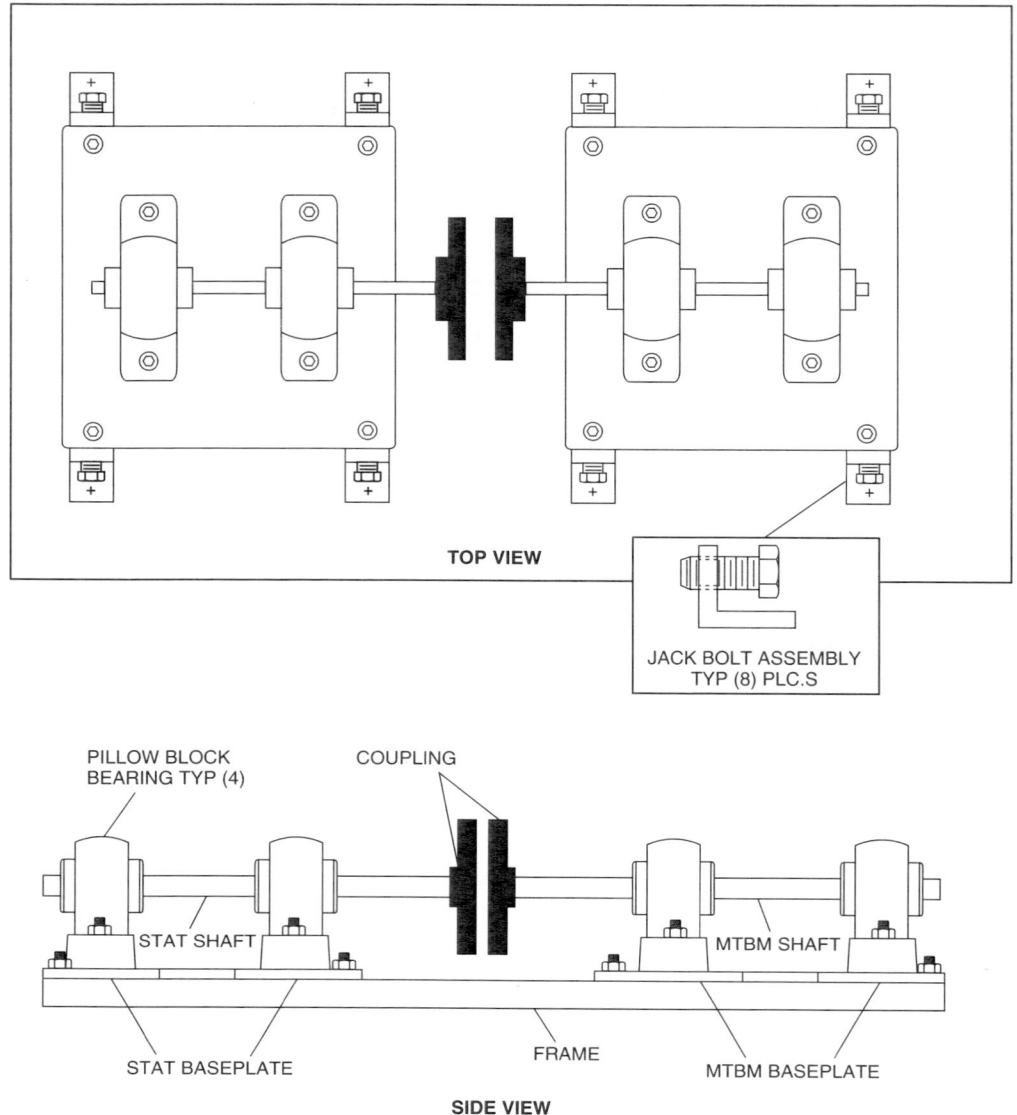

TOP VIEW

JACK BOLT ASSEMBLY
TYP (8) PLC.S

PILLOW BLOCK
BEARING TYP (4)   COUPLING

STAT SHAFT   MTBM SHAFT

STAT BASEPLATE   FRAME   MTBM BASEPLATE

SIDE VIEW

501F08.EPS

Figure 8 ◆ Alignment demonstration rig.

face diameter. The dials are usually divided into one of the following increments:

- Hundredths of a millimeter (0.01 millimeter)
- Two-thousandths of a millimeter (0.002 millimeter)
- Thousandths of an inch (0.001 inch)
- Ten-thousandths of an inch (0.0001 inch)

The two types of dial indicators are the balanced type and the continuous-reading type. The numbers on the balanced type start at zero and increase in both directions. The numbers on the continuous-reading type start at zero and continue around the dial clockwise. The dials can be reset to zero by rotating the entire dial face. *Figure 9* shows two types of dial indicators.

Instructor's Notes:

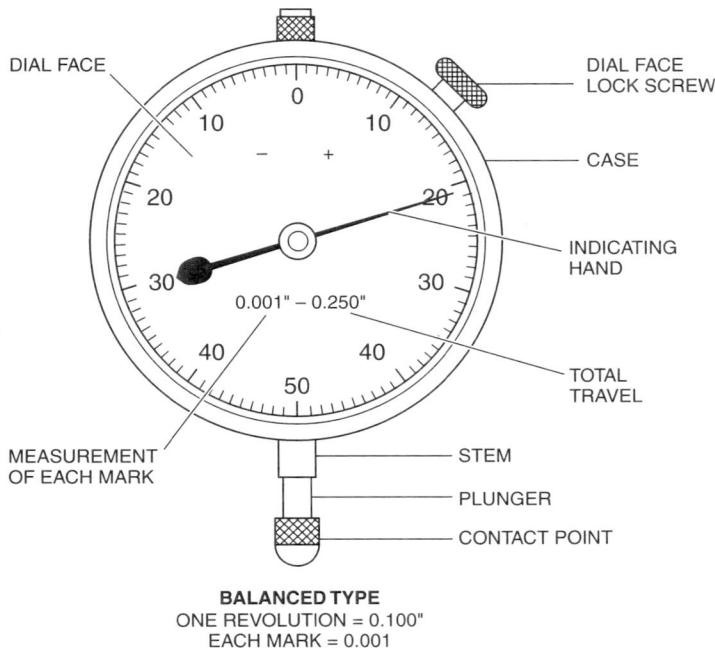

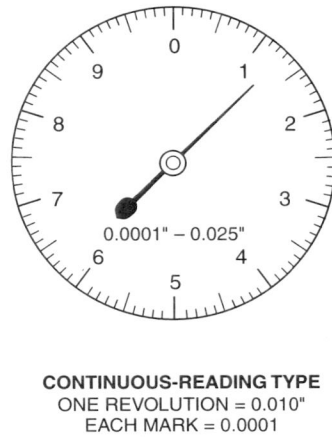

DIAL FACE

DIAL FACE
LOCK SCREW

CASE

INDICATING
HAND

0.001" – 0.250"

TOTAL
TRAVEL

MEASUREMENT
OF EACH MARK

STEM

PLUNGER

CONTACT POINT

BALANCED TYPE
ONE REVOLUTION = 0.100"
EACH MARK = 0.001

0.0001" – 0.025"

CONTINUOUS-READING TYPE
ONE REVOLUTION = 0.010"
EACH MARK = 0.0001

501F09.EPS

Figure 9 ◆ Dial indicators.

4.3.0 Measuring Shaft Runout Using a Dial Indicator and Base

Shaft runout can manifest the same symptoms as shaft misalignment. These can include premature bearing failure, loss of power, failed seals, or high operating temperatures. Before reverse dial indicator alignment is performed, the shaft runout should be checked. Follow these steps to measure shaft runout, using a dial indicator and base:

Step 1 Clean all surfaces that the dial indicator will touch.

Step 2 Determine where to attach the indicator base, and clamp the indicator base onto the work piece or machine, using a C-clamp or a magnetic base.

Step 3 Mount the indicator on the indicator base using the attachments on the indicator holding rod, and tighten all locknuts securely.

Step 4 Loosen the clamp on the base post, and push the indicator into the work piece by at least half the total travel of the indicator to preload the indicator.

Step 5 Tighten the locknuts of the base post clamp securely.

Step 6 Loosen the dial face lock screw, and rotate the face to read absolute zero to zero the indicator.

Step 7 Tighten the dial face lock screw. Be careful not to bottom the indicator plunger because this will damage the indicator. If the indicator is too close, reposition it with less preload.

Step 8 Check all other adjustments to ensure that they are tight and secure.

Step 9 Rotate the shaft to obtain a total runout.

> **NOTE**
> Record the most extreme position of measurement in both directions of travel.

Step 10 Position the indicator base and take new measurements to get at least three points of measurement for an average.

Explain why shaft runout should be checked before performing an alignment.

Walk through the steps to measure shaft runout using a dial indicator and base.

Have trainees measure shaft runout using a dial indicator jig. This laboratory corresponds to Performance Task 1.

Assign reading of Sections 5.0.0 - 5.1.0.

5.0.0 ◆ REVERSE DIAL INDICATOR ALIGNMENT

The alignment of shafts using the reverse dial indicator method is simple if you follow each step and avoid shortcuts or guessing. This method can be adapted to many situations. Close attention to **thermal growth**, elimination of soft foot, and factoring out the indicator sag helps ensure a successful alignment.

5.1.0 Setting Up Complex Reverse Dial Indicator Jigs

Setting up the reverse dial indicator jig is somewhat complicated and can be done two ways. Two jigs and two indicators are used. Because two jigs are used together, this is referred to as a complex reverse dial indicator jig. One jig is clamped around the driving coupling half, or MTBM, and the other is clamped around the driven coupling half, or STAT. The two jigs can be positioned so that the mounting pin of one jig is in line with the flat on top of the other jig or on the same side. *Figure 10* shows mounting the complex reverse dial indicator jigs on the same side.

Another method is to mount the indicators 180 degrees apart from each other. *Figure 11* shows mounting the complex reverse dial indicator jigs on the opposite side.

One indicator and its hardware are mounted on the MTBM coupling jig so that the indicator con-

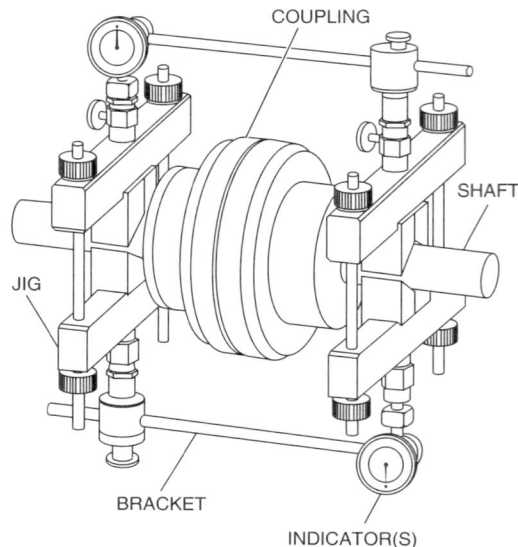

Figure 11 ◆ Opposite-side mounting.

tacts the flat on top of the STAT coupling jig. The other indicator and its hardware are mounted on the STAT coupling jig so that the indicator contacts the flat on top of the MTBM coupling jig.

Determining indicator sag is critical to any indicator alignment setup. Gravity causes the

Figure 10 ◆ Same-side mounting.

Instructor's Notes:

_____

_____

_____

extension bracket to droop by a very slight amount, so all brackets have some degree of sag. This amount could be as small as 0.001 inch or up to 0.050 inch, depending on the span and rigidity of the fixture. In either case, the error induced by this sagging is enough to make true center line alignment impossible.

The bracket sags toward the shaft when the shaft is in the 12-o'clock position. When rotated 180 degrees to the 6-o'clock position, that same amount of sag pulls away from the shaft, thereby doubling the amount of total movement and error that must be added to the alignment calculations.

To measure sag, mount the entire measurement fixture, including the brackets, bars, and indicators, on the machine shaft or off the machine on a piece of straight pipe about 2 inches in diameter. Adjust the fixture and position the brackets until they are the same distance apart as they will be when they are mounted on the machinery. With the indicators at the 12-o'clock position, zero the dials. Rotate the shaft until the indicators are at the 6-o'clock position, and record this value. This number is subtracted from the alignment total calculated from the vertical readings. *Figure 12* shows how to measure sag.

5.2.0 Performing Reverse Dial Indicator Alignment

Before starting reverse dial indicator alignment, lock out and tag the machinery to ensure that it will not start up during this procedure. After the reverse dial indicator jigs are mounted in the preferred position, the technique of reverse dial indicator alignment must be chosen.

Chart alignment involves taking the indicator readings and plotting them on an accurate graph to arrive at a corrective movement or shim thickness. The other method involves taking the indicator readings and inserting them into a mathematical formula. Both of these methods also require measuring the distances from feet to couplings. *Figure 13* shows distance measurements.

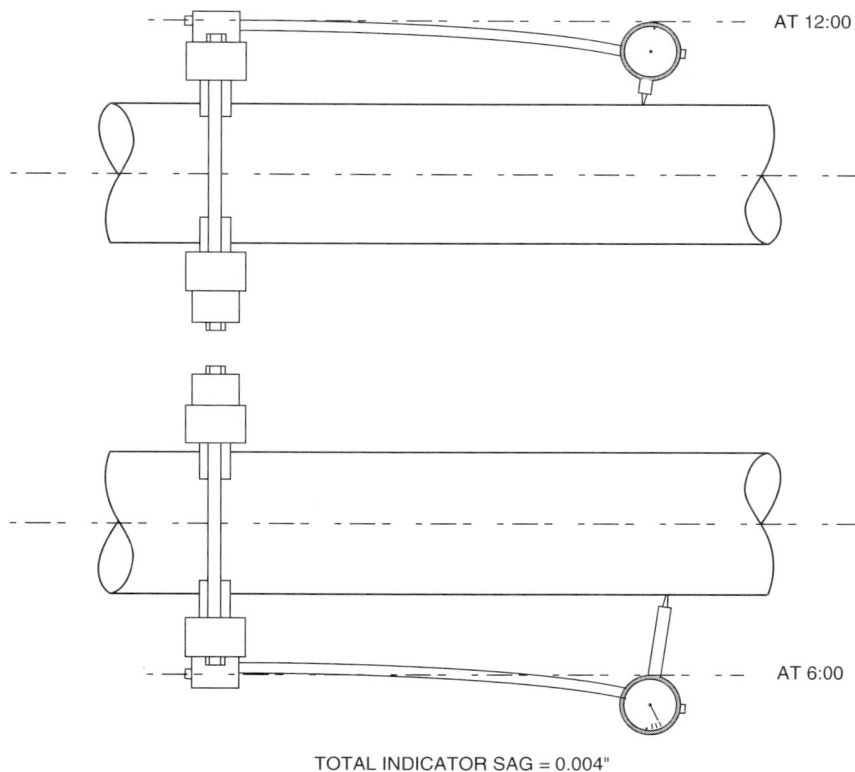

Figure 12 ◆ Indicator sag.

Explain why determining indicator sag is critical to an indicator alignment setup.

Show Transparency 12 (Figure 12). Describe how sag is measured.

Show trainees how to set up a complex reverse alignment jig and measure indicator sag using the jig.

Have trainees set up a complex reverse alignment jig and measure indicator sag using the jig. This laboratory corresponds to Performance Tasks 2 and 3.

Assign reading of Sections 5.2.0–5.2.4.

Ensure that you have everything required for teaching this session.

Describe how to perform reverse dial indicator alignment.

Explain how to measure the distances from feet to couplings.

Show Transparency 13 (Figure 13).

Describe the charting alignment method.

Show Transparency 14 (Figure 14). Demonstrate how to indicate dimensions on the chart.

Describe and/or demonstrate how to perform vertical alignment using the step-by-step procedure.

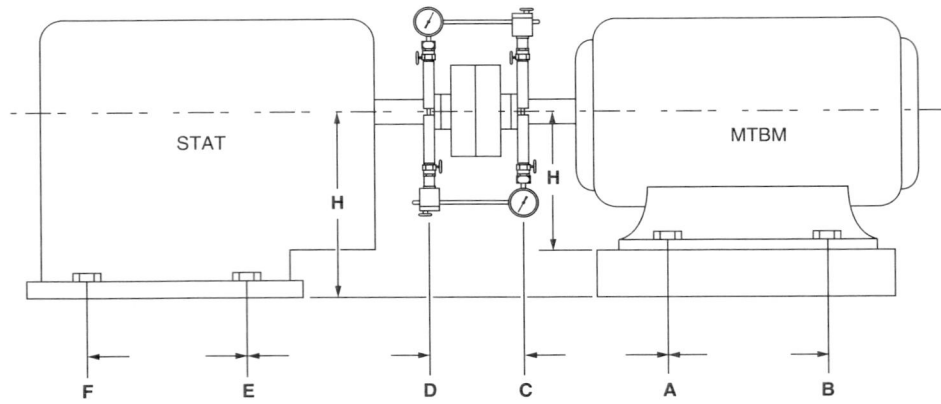

Figure 13 ◆ Distance measurements.

Both methods require practice and attention to detail. It is very easy to misread the direction of travel on the dial indicator and have a positive value recorded as a negative. Make it a habit to read every position at least twice. Always measure and note the indicator sag for each setup.

5.2.1 Charting Alignment

Charting alignment on a graph requires a chart that is divided into thousandths of an inch vertically for the indicator readings and into whole inches horizontally for the distance measurements. The plotting of points on this chart correlates to the shim thickness or distance to be moved. The chart can be customized as needed. *Figure 14* shows an alignment chart. A graph for this practice can be found in *Appendix A*.

Follow these steps to chart alignment:

Step 1 Measure and chart the dimensions from foot to foot and feet to couplings shown as distances A, B, C, D, E, and F in *Figure 13*.

Step 2 Measure the indicator sag, and write the value on the chart.

Step 3 Note on the chart whether the reading is a cold or hot indicator reading.

5.2.2 Performing Alignment

Follow these steps to perform vertical alignment.

Step 1 Indicate the direction at the tip of the arrow and the top of the graph paper as up.

Step 2 Set the indicator on top to read the positive value of indicator sag.

Step 3 Rotate the shafts 180 degrees, and note the direction of rotation and the final reading.

Step 4 Record the reading as total indicator reading (TIR) MTBM on the graph paper.

Step 5 Divide the TIR MTBM by 2, and plot this value on the left side of the graph.

Step 6 Set the indicator that is now on top to read the positive value of sag.

Step 7 Rotate the shafts 180 degrees. Note the direction of rotation and the final reading.

Step 8 Record the reading as TIR STAT on the graph.

Step 9 Divide the TIR STAT by 2, and plot this value at a distance equal to the span between indicators to the right of the previous dot.

Step 10 Connect the dots with a straight line. Use the numbers on the side of the graph to show the location of the feet on the MTBM.

Step 11 Make the proper alignment corrections.

Step 12 Continue to calculate and plot shims until the machines are aligned.

Step 13 Zero the indicator on the side of the coupling opposite from you.

Instructor's Notes:
_____
_____
_____

MISALIGNMENT TOLERANCE CALCULATIONS

Misalignment = $\dfrac{\text{½ T.I.R.}}{\text{SPAN BETWEEN INDICATORS}}$

Vertical Misalignment _____

Horizontal Misalignment _____

FIRST   SECOND
INDICATOR   INDICATOR

(MTBM) $\dfrac{+\ \text{CENTER}\ -}{-\ \text{LINE}\ +}$ (STAT)

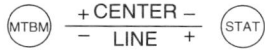

Side-to-Side Indicator Readings

_____ T.I.R.   _____ T.I.R.

_____ ½ T.I.R.   _____ ½ T.I.R.

Plot first indicator ⊖ on the zero line. Plot (+) plus reading above center line and (−) minus below center line.

Plot second indicator ⊖ on the inch line equal to the span between indicators. Plot (−) minus reading above center line and (+) plus below center line.

STATIONARY (+)
(−)

Top-to-Bottom Indicator Readings

+ − SAG   + − SAG

_____ T.I.R.   _____ T.I.R.

_____ ½ T.I.R.   _____ ½ T.I.R.

FIRST INDICATOR

STAT   MTBM

S   M

COUPLING

FRONT FOOT   BACK FOOT

SECOND INDICATOR

STAT (S)   MTBM (M)

H   H

F   E   D   C   A   B

INCHES

5 10 15 20 25 30 35 40 45 50 55 60 65 70

THOUSANDTHS
70 60 50 40 30 20 10 -S- 10 20 30 40 50 60 70
THOUSANDTHS

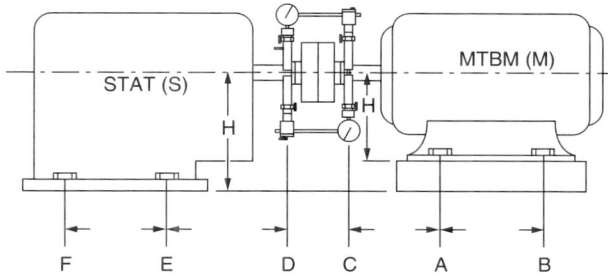

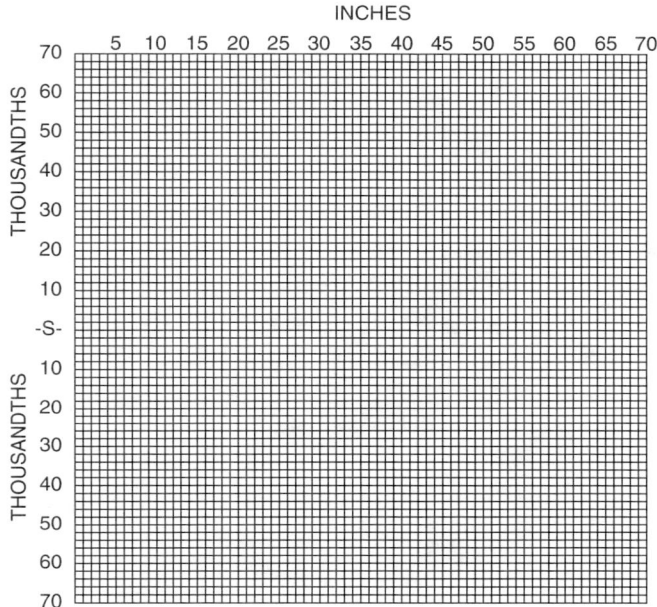

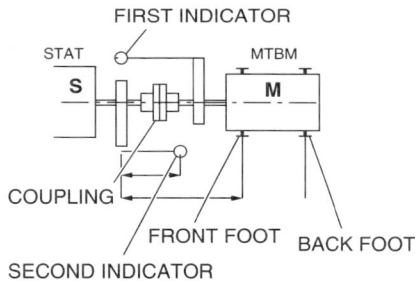

Vertical Adjustments (read from graph)

Front Foot _____ (−) Add (+) Remove Shim

Back Foot _____

Horizontal Adjustments (read from graph)

Front Foot _____ (−) Push (+) Pull

Back Foot _____

501F14.EPS

Figure 14 ◆ Graphical alignment chart.

Show Transparency 15 (Figure 15). Explain how the reverse dial indicator plotting guide is used.

Step 14 Rotate both shafts together in the correct direction of rotation until the indicator is on your side of the coupling.

Step 15 Record the final reading for MTBM as the TIR on the graph.

Step 16 Divide TIR MTBM by 2, and record the number on the left side of the graph paper.

> **NOTE**
> Plot the number above the STAT if it is positive and below if it is negative.

Step 17 Zero the indicator that is now on the side opposite you.

Step 18 Rotate the shafts in the correct direction of rotation until the indicator is on your side.

Step 19 Record the number as TIR STAT on the graph paper.

Step 20 Divide TIR STAT by 2, and plot this number at a distance to the right that is equal to the distance between indicators from the previously plotted number.

Step 21 Connect the two dots with a straight line. Use the numbers at the bottom of the graph to show the location of the feet on the MTBM.

The alignment can also be represented graphically on a reverse dial indicator plotting guide. *Figure 15* shows a reverse dial indicator plotting guide.

N = STAT R = MTBM

REVERSE INDICATOR PLOTTING GUIDE

REVERSE READING		PLOT LOCATION	ABSOLUTE VALUES
MTBM–STAT	STAT–MTBM	STAT D C MTBM	
+	–		R–N < N–R
+	–		R–N = N–R
+	–		R–N > N–R
+	+		
+	0		
–	+		R–N < N–R
–	+		R–N = N–R
–	+		R–N > N–R
–	–		
–	0		

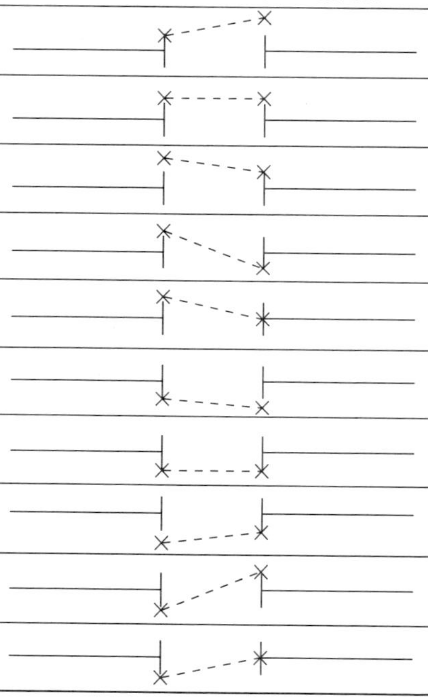

501F15.EPS

Figure 15 ◆ Reverse dial indicator plotting guide.

Instructor's Notes:

1.16 MILLWRIGHT ◆ LEVEL FIVE

5.2.3 Alignment Equation

Reverse dial indicator alignment can also be performed using the straight equation method. This method uses the same measurements obtained in the procedure in an earlier section to find the shim thickness and move dimensions.

The mathematical formula starts with the same indicator setup and readings taken at the four clock positions and the indicator sag measurement. The setup takes the indicated misalignment off the STAT to arrive at shim and move dimensions.

The equations that follow are derived from the rise over run calculations. The rise (or fall) is in thousandths of an inch, whereas the run is in inches. Nonetheless, the same formula for the ratio between the two can be used to express the relationship between the angles of the shafts of the STAT and MTBM. Having determined the ratio of rise to run, it is possible to calculate the adjustments required. The equation to calculate the ratio is:

[(TIR MTBM coupling ÷ 2) − (TIR STAT coupling ÷ 2)]
÷ distance between points

The answer is then used to calculate the rise by multiplying the numeric value of the ratio times the distance from the point on the STAT to the point where the measurement is to be calculated, and adding the result to the value of (TIR STAT ÷ 2).

If the TIR on the STAT coupling is −0.010, then the center of the shaft where the indicator is reading is half of 0.010, or 0.005, from being collinear.

If the TIR on the MTBM coupling is +0.016, then the second point is +0.008. If the gap between indicator tips is 10 inches, the run is 10 inches, and the rise is 0.008 − 0.005 = 0.003. Dividing 0.003 by 10, the answer is 0.0003. If the distance from the MTBM indicator to the nearest lug is 5 inches, then the rise would be 0.0003 × 15 = 0.0045, which would place the point at 0.005 + 0.0045 = 0.0095. If the distance between the lugs is 10 inches, then the rise would be 0.0003 × 25 + 0.005 = 0.0125. A graph of this calculation (*Figure 16*) allows you to calculate the shims immediately for each lug. Add shims to equal 0.0095 inches under the front lug, and 0.0125 inches of shim to the back lug, and the MTBM will be vertically aligned with the STAT.

The same sort of calculation is applied to correct horizontal angularity. The axis of the shaft becomes the run direction, and the deviation in the horizontal plane is the rise. The ratio between those two dimensions is then used to determine the distance and direction of horizontal movement to achieve collinearity.

5.2.4 Recording Alignment

The necessity of recording and maintaining records of all alignments is to ensure that a logical system can be productive and not cause more problems than it solves. Every alignment should be documented on a standard form that is distributed to all departments and that is stored in an area where it is accessible. Most companies now demand a record of the final alignment condition.

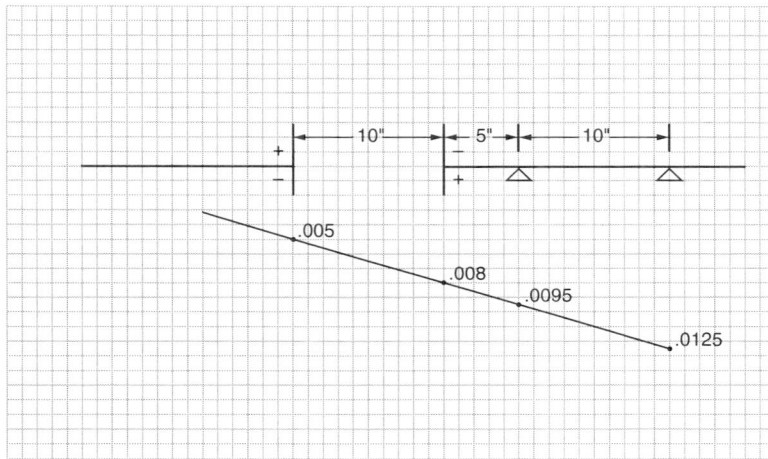

Figure 16 ◆ Graph of calculation.

Explain mathematical and graphical alignment procedures and recording requirements.

Show Transparency 16 (Figure 16).

Explain requirements for recording and maintaining alignment information.

See the Teaching Tip for Section 5.0.0 at the end of this module.

Ensure that you have everything required for teaching this session.

Show trainees how to perform reverse alignment using the alignment demonstration rig, the graphical chart, and the mathematical equation.

Have trainees perform reverse alignment using the alignment demonstration rig, the graphical chart, and the mathematical equation. This laboratory corresponds to Performance Tasks 4 and 5.

Have trainees complete the Review Questions, and go over the answers prior to administering the Module Examination.

Review Questions

1. The position of a rotating shaft center line in space can be described by showing the _____.
   a. speed of rotation
   b. direction of rotation
   c. horizontal and vertical views
   d. diagonal views

2. The basic dimensional definitions are vertical to horizontal and _____.
   a. front to back
   b. back to front
   c. left to right
   d. up to down

3. The most important external factors that cause problems are thermal expansion, soft foot, and failure to calculate _____.
   a. vertical offset
   b. horizontal offset
   c. indicator sag
   d. rotation speed

4. Soft foot is a condition in which the tightening and loosening of the bolt(s) of a single foot _____.
   a. softens the bolt
   b. straightens the frame
   c. distorts the frame
   d. aligns the shaft

5. A parallel air gap is a soft foot in which _____ when the other feet are bolted down.
   a. one foot is parallel to but above the base
   b. one foot is bent away from parallel
   c. the shims are too thick
   d. the shims are bent or burred

6. A shaft should not have more than 0.002 inch of runout.
   a. True
   b. False

7. The reverse alignment method of checking coupling runout in the field is _____.
   a. checking with a straightedge and feeler gauges
   b. turning with a lathe
   c. rim and face measurement on each half of an unbolted coupling
   d. rim and face measurement on bolted couplings

8. Before the shafts are aligned, _____ must be eliminated.
   a. vibration
   b. coupling stress
   c. bearings
   d. anchor bolts

9. Anchor bolts that have been improperly installed can _____ the baseplate.
   a. vibrate
   b. firmly hold
   c. twist
   d. thread

10. A rule of thumb for concrete baseplates calls for concrete weight equal to _____ times the machine weight for rotating machinery.
    a. two
    b. three
    c. four
    d. five

11. The numbers on a balanced type dial indicator start at _____ in both directions.
    a. one and decrease
    b. one and increase
    c. zero and decrease
    d. zero and increase

12. Which of the following can manifest the same symptoms as shaft misalignment?
    a. Motor burnout
    b. Shaft runout
    c. Indicator sag
    d. Loose bearings

Instructor's Notes:

_____
_____
_____

13. Dial indicators used for reverse alignment may be mounted either on the same side or _____.

    a. 90 degrees apart
    b. 45 degrees apart
    c. 180 degrees apart
    d. on top of one another

14. Graphical alignment requires a chart that is divided into _____ vertically and into _____ horizontally.

    a. millimeters; meters
    b. thousandths of an inch; whole inches
    c. quarter inches; whole inches
    d. centimeters; millimeters

15. In a vertical alignment, if the TIR on the MTBM coupling is +0.012, TIR on the STAT coupling is –0.006, the distance between the indicator points is 10 inches, the distance from the MTBM coupling to the front foot of the MTBM is 4 inches, and the distance between feet on the MTBM is 11 inches, _____ of shim is needed for the farthest foot on the MTBM.

    a. 0.0075
    b. 0.0015
    c. 0.075
    d. 0.0125

Summary

The reverse dial indicator alignment technique for optimizing equipment drivelines is very accurate and offers repeatable results when performed properly. The reverse dial indicator has equal accuracy to optical and laser alignment techniques and has the advantage of feel for the equipment. The attention to detail and time involved in a proper reverse dial indicator alignment provide the opportunity for the millwright to observe and note other conditions in the machine as well. The patience needed to perform a good reverse dial indicator alignment is rewarded by equipment that operates close to perfection.

Notes

Instructor's Notes:

_____

_____

_____

Trade Terms Introduced in This Module

Collinear: Two or more shaft center lines with no offset or angularity between them.

Machinery train: A group of separate machines connected to each other by a process or physically coupled to transfer power to each other.

Thermal growth: The expansion of materials with the rise in temperature.

Charting Alignment Practice Graph

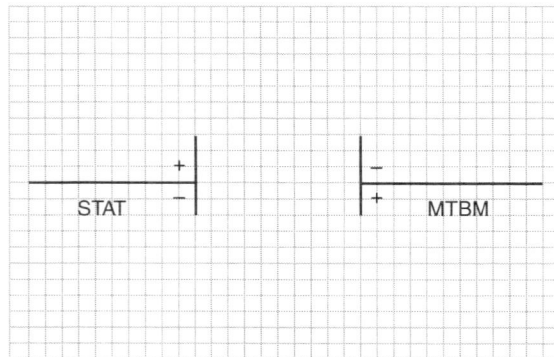

STAT   +   −

MTBM   −   +

501A01.EPS

Instructor's Notes:

_____

_____

_____

_____

Additional Resources

This module is intended to present thorough resources for task training. The following reference works are suggested for further study. These are optional materials for continued education rather than for task training.

A Millwright's Guide to Motor/Pump Alignment, 2nd Ed. Tommy B. Harlon, New York, NY: Industrial Press, 2008.

The Optalign Training Book. Galen Evans and Pedro Casanova. Miami, FL: Ludeca Inc.

Figure Credits

Cianbro Institute, 501F07, 501F10

MODULE 15501-09 — TEACHING TIPS

The following are suggested activities or instructional methods to help you teach the material in this module.

General   When you call on someone to answer a question, the rest of the class relaxes or even tunes out because they expect that the question and answer will take place only between you and the trainee you called on. Instead, use this technique to involve more trainees in answering questions and to keep them on their toes.

1. Ask trainees to define a term or explain a concept.
2. After one trainee has answered, ask a trainee seated nearby if the answer is right. Then ask whether a trainee in the back of the room agrees.
3. Ask trainees to explain why they think an answer is right or wrong.
4. Use the session to clear up incorrect ideas and encourage trainees to learn from their mistakes.

Sections 2.0.0 and 3.0.0   *Soft Foot and Coupling Stress*

This Quick Quiz will familiarize trainees with terms related to soft foot and coupling stress, which affect alignment. You will need photocopies of the quiz provided on the following page. Trainees will need pencils. If you allow trainees to use the Trainee Guide, decrease the amount of time you give them to complete the quiz, and remind them to bring their books to class.

1. Make a photocopy of the quiz for each trainee.
2. Give trainees between 5 and 10 minutes to complete the quiz.
3. Go over the answers to the quiz.
4. Ask trainees if they have questions.

Answers to Quick Quiz

1. c
2. d
3. e
4. b
5. a
6. h
7. g
8. f

Quick Quiz   *Soft Foot and Coupling Stress*

Identify the terms that are related to soft foot or coupling stress. In the blank provided, write the corresponding letter selected from the list below.

_____ 1. A condition in which one foot is at an angle to the plane of the other feet is called _____.

_____ 2. When a pipe flange is welded so that it has to be pulled to the flange on the STAT, there has been a(n) _____.

_____ 3. When one foot is parallel to but not in the same plane as the others, this condition is called _____.

_____ 4. When the machine base causes the anchor bolts to skew and bend, the reason is _____.

_____ 5. Dirt, paint, or burred or bent shims can cause _____.

_____ 6. A condition that can be caused by a baseplate that is not level, an improperly sized concrete pad, or bad grouting is a(n) _____.

_____ 7. If left uncorrected, catastrophic failure of the machine system can occur, caused by a complex set of problems that can be traced back to _____.

_____ 8. Excessive stress can be placed on a STAT and MTBM when the flange is forced to carry the weight of the pipe because of _____.

a. squishy foot
b. defective anchor bolts
c. bent foot
d. incorrect pipe weldment
e. parallel air gap
f. improper placement of pipe hangers
g. bad bearings
h. improper foundation

Section 5.0.0   *Reverse Dial Indicator Alignment*

Using several alignment simulators (or a motor and pump or gearbox), repeatedly misalign the MTBM and have the trainees measure, graph, and calculate the alignment. Be sure that they perform both the graphical and mathematical procedures, because of the specific performance task requirements, and because the two methods correct each other. Allow 2½ hours for this exercise. This laboratory corresponds to Performance Tasks 4 and 5.

MODULE 15501-09 — ANSWERS TO REVIEW QUESTIONS

	Answer	Section
1.	c	2.0.0
2.	a	2.0.0
3.	c	2.1.0
4.	c	2.1.1
5.	a	2.1.1; Figure 5
6.	a	2.1.3
7.	c	2.1.3
8.	b	3.0.0
9.	c	3.3.0
10.	b	3.5.0
11.	d	4.2.0
12.	b	4.3.0
13.	c	5.1.0
14.	b	5.2.1
15.	b*	5.2.3

* 15. $(0.012 \div 2) - (0.006 \div 2) \div 10 = 0.0003$
$10 + 4 + 11 = 25$
$25 \times 0.0003 = 0.0075$
$0.0075 - 0.006 = 0.0015$

NCCER makes every effort to keep these textbooks up-to-date and free of technical errors. We appreciate your help in this process. If you have an idea for improving this textbook, or if you find an error, a typographical mistake, or an inaccuracy in NCCER's Contren® textbooks, please write us, using this form or a photocopy. Be sure to include the exact module number, page number, a detailed description, and the correction, if applicable. Your input will be brought to the attention of the Technical Review Committee. Thank you for your assistance.

Instructors – If you found that additional materials were necessary in order to teach this module effectively, please let us know so that we may include them in the Equipment/Materials list in the Annotated Instructor's Guide.

Write: Product Development and Revision
National Center for Construction Education and Research
3600 NW 43rd St, Bldg G, Gainesville, FL 32606

Fax: 352-334-0932

E-mail: curriculum@nccer.org

Craft _____ Module Name _____

Copyright Date _____ Module Number _____ Page Number(s) _____

Description _____

_____

_____

_____

_____

(Optional) Correction _____

_____

_____

(Optional) Your Name and Address _____

_____

_____

_____

Laser Alignment

NCCER STANDARDIZED CRAFT TRAINING PROGRAM

The National Center for Construction Education and Research (NCCER) provides a standardized national program of accredited craft training. Key features of the program include instructor certification, competency-based training, and performance testing. The program provides trainees, instructors, and companies with a standard form of recognition through a National Craft Training Registry. The program is described in full in the *Guidelines for Accreditation*, published by NCCER. For more information on standardized craft training, contact the NCCER by writing us at 3600 NW 43rd St., Bldg. G, Gainesville, FL 32606; calling 352-334-0911; or emailing info@nccer.org. More information may be found at our website, www.nccer.org.

HOW TO USE THIS ANNOTATED INSTRUCTOR'S GUIDE

Each page presents two sections of information. The larger section displays each page exactly as it appears in the Trainee Module. The narrow column ties suggested trainee and instructor actions to each page and provides icons (detailed below) to call your attention to material, safety, audiovisual, or testing requirements. The bottom of each page includes space for your notes.

The **Audiovisual** icon indicates an appropriate time to show a transparency or other audiovisual aid.

The **Classroom** icon prompts you to define a term, stress a point, ask trainees to explain a concept, or give examples.

The **Demonstration** icon directs you to show trainees how to perform tasks.

The **Examination** icon tells you to administer the written module examination.

The **Homework** icon is placed where you may wish to assign reading for the next class, assign a project, or advise trainees to prepare for an examination.

The **Laboratory** icon is used when trainees are to practice performing tasks.

The **Materials** icon is a reminder for you to gather materials needed for classes, labs, and testing.

The **Performance Testing** icon tells you to administer a performance test or a portion thereof.

The **Safety** icon is used to emphasize safety issues. It is often keyed to *Caution* and *Warning!* statements in the Trainee Module.

The **Teaching Tip** icon indicates additional guidance is available, such as how to conduct an exercise, get the most educational value from a field trip, or encourage class participation. Teaching Tips may expand on a feature (*Think About It*, *Did You Know?*) or provide *Quick Quizzes* or similar exercises. You will be referred to the Teaching Tips section at the back of the module if there is additional material.

The **Combination** icon indicates that the laboratory listed corresponds with a performance task. If desired, you can note the proficiency of the trainees during the laboratory, and use it to satisfy performance testing requirements.

PREPARATION

Before teaching this module, you should review the Objectives, Performance Tasks, Materials and Equipment List, and Module Outline. Be sure to allow ample time to prepare your own training or lesson plan and gather all required materials and equipment.

MODULE OVERVIEW

This module covers the basic principles of lasers, as well as laser alignment, laser/detector operation, and troubleshooting lasers. This module also covers conditions such as soft foot and coupling stress.

PREREQUISITES

Prior to training with this module, it is recommended that the trainee shall have successfully completed *Core Curriculum; Millwright Level One; Millwright Level Two; Millwright Level Three; Millwright Level Four;* and *Millwright Level Five,* Module 15501-09.

OBJECTIVES

Upon completion of this module, the trainee will be able to do the following:

1. Explain lasers and laser alignment systems.
2. Operate a laser alignment system.
3. Explain soft foot, thermal growth, and coupling stress.
4. Troubleshoot repeatability and laser problems.

PERFORMANCE TASKS

Under the supervision of the instructor, the trainee should be able to do the following:

1. Identify the major components of a laser alignment system.
2. Perform a rough alignment.
3. Set up the laser alignment equipment.
4. Check the initial alignment.
5. Perform a vertical alignment using a laser.
6. Perform a horizontal alignment using a laser.

MATERIALS AND EQUIPMENT LIST

Overhead projector and screen

Transparencies

Blank acetate sheets

Transparency pens

Whiteboard/chalkboard

Markers/chalk

Pencils and scratch paper

Graph paper

Appropriate personal protective equipment

Alignment simulators or equipment to be aligned

Wrenches

Laser alignment equipment

Copies of the Quick Quizzes*

Module Examinations**

Performance Profile Sheets**

* Located at the back of this module.

** Located in the Test Booklet.

SAFETY CONSIDERATIONS

Ensure that the trainees are equipped with appropriate personal protective equipment and know how to use it properly. This module requires trainees to align machinery using laser alignment equipment. Ensure that all trainees are briefed on the appropriate shop safety procedures.

ADDITIONAL RESOURCES

This module is intended to present thorough resources for task training. The following reference work is suggested for both instructors and motivated trainees interested in further study. This is optional material for continued education rather than for task training.

The Optalign Training Book. Galen Evans and Pedro Casanova. Miami, FL: Ludeca, Inc.

TEACHING TIME FOR THIS MODULE

An outline for use in developing your lesson plan is presented below. Note that each Roman numeral in the outline equates to one session of instruction. Each session has a suggested time period of 2½ hours. This includes 10 minutes at the beginning of each session for administrative tasks and one 10-minute break during the session. Approximately 25 hours are suggested to cover *Laser Alignment*. You will need to adjust the time required for hands-on activity and testing based on your class size and resources. Because laboratories often correspond to Performance Tasks, the proficiency of the trainees may be noted during these exercises for Performance Testing purposes.

Topic	Planned Time
Session I. Introduction; Soft Foot; Thermal Growth; Coupling Stress	
A. Introduction	_____
B. Soft Foot	_____
1. Types of Soft Foot	_____
C. Thermal Growth	_____
D. Coupling Stress	_____
1. Causes of Coupling Stress	_____
Session II. Basic Laser Principles; Optalign® Laser Alignment	
A. Basic Laser Principles	_____
B. Laser Safety	_____
C. Optalign® Laser Alignment	_____
D. Descriptive Characteristics of Misalignment	_____
1. Optalign® System Capabilities/Limitations	_____
E. Laboratory	_____
Have trainees practice identifying the major components of a laser alignment system. This laboratory corresponds to Performance Task 1.	
Sessions III -V. Laser Detector Operation; Alignment Procedures, Part One	
A. Laser/Detector Operation	_____
B. Alignment Procedures	_____
C. Rough Alignment	_____
1. Laboratory	_____
Have trainees practice performing a rough alignment. This laboratory corresponds to Performance Task 2.	
D. Setting Up Laser Equipment; Initial Laser Alignment	_____
1. Laboratory	_____
Have trainees practice setting up the laser alignment equipment and checking the initial alignment. This laboratory corresponds to Performance Tasks 3 and 4.	

Sessions VI and VII. Laser Operation and Alignment Procedures, Part Two

   A. Aligning Machinery Trains   _____

   B. Laboratory   _____

       Have the trainees practice performing a horizontal alignment using a laser. This laboratory corresponds to Performance Task 6.

Session VIII. Laser Operation and Alignment Procedures, Part Three

   A. Determining Targets   _____

   B. Aligning Vertical Machines   _____

     1. Laboratory   _____

       Have the trainees practice performing a vertical alignment using a laser. This laboratory corresponds to Performance Task 5.

Session IX. Troubleshooting

   A. Machinery Defects   _____

   B. Incorrectly Installed Brackets   _____

   C. System Failure or Defect   _____

Session X. Review and Testing

   A. Review   _____

   B. Module Examination   _____

     1 Trainees must score 70% or higher to receive recognition from NCCER.

     2. Record the testing results on Craft Training Report Form 200, and submit the results to the Training Program Sponsor.

   C. Performance Testing   _____

     1. Trainees must perform each task to the satisfaction of the instructor to receive recognition from NCCER. If applicable, proficiency noted during laboratory exercises can be used to satisfy the Performance Testing requirements.

     2. Record the testing results on Craft Training Report Form 200, and submit the results to the Training Program Sponsor.

Millwright Level Five

15502-09

Laser Alignment

Assign reading of Module 15502-09.

15502-09
Laser Alignment

Topics to be presented in this module include:

Overview

In this module, the trainee will learn to use laser alignment equipment to align STAT and MTBM machines. The steps are based on Optalign® equipment, but are fairly similar to other machines. The skills for testing for soft foot and coupling stress are taught, as well as normal alignment procedures for a single pair of machines. Information is also presented on aligning a machine train, proceeding from one pair to the next.

Instructor's Notes:

_____

_____

_____

Objectives

When you have completed this module, you will be able to do the following:

1. Explain lasers and laser alignment systems.
2. Operate a laser alignment system.
3. Explain soft foot, thermal growth, and coupling stress.
4. Troubleshoot repeatability and laser problems.

Trade Terms

Coherent
Collimate
Laser

Required Trainee Materials

1. Pencil and paper
2. Appropriate personal protective equipment

Prerequisites

Before you begin this module, it is recommended that you successfully complete *Core Curriculum*; *Millwright Level One*; *Millwright Level Two*; *Millwright Level Three*; *Millwright Level Four*; and *Millwright Level Five*, Module 15501-09.

This course map shows all of the modules in the fifth level of the Millwright curriculum. The suggested training order begins at the bottom and proceeds up. Skill levels increase as you advance on the course map. The local Training Program Sponsor may adjust the training order.

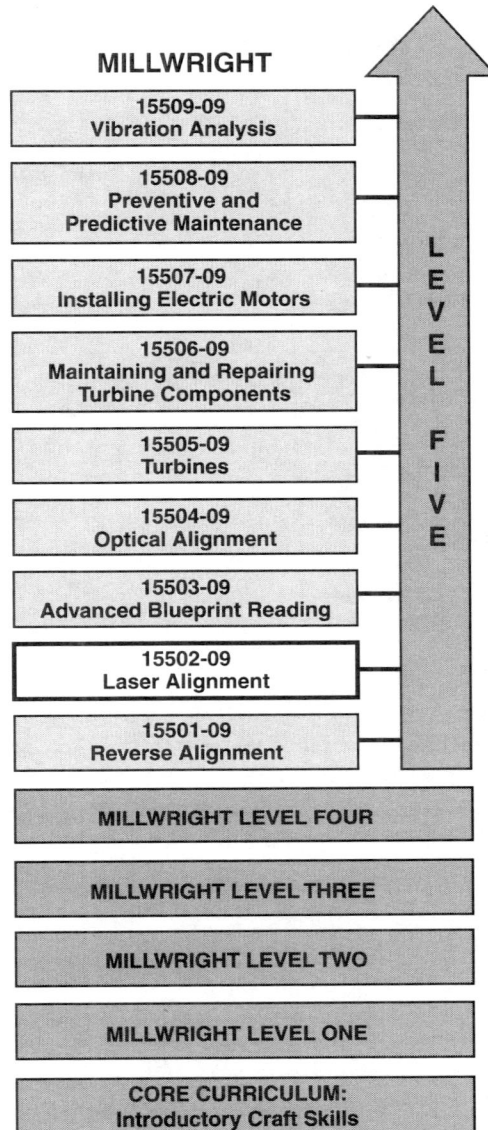

MILLWRIGHT

LEVEL FIVE

- **15509-09** Vibration Analysis
- **15508-09** Preventive and Predictive Maintenance
- **15507-09** Installing Electric Motors
- **15506-09** Maintaining and Repairing Turbine Components
- **15505-09** Turbines
- **15504-09** Optical Alignment
- **15503-09** Advanced Blueprint Reading
- **15502-09** Laser Alignment
- **15501-09** Reverse Alignment

MILLWRIGHT LEVEL FOUR

MILLWRIGHT LEVEL THREE

MILLWRIGHT LEVEL TWO

MILLWRIGHT LEVEL ONE

CORE CURRICULUM: Introductory Craft Skills

502CMAP.EPS

Ensure that you have everything required to teach the course. Check the Materials and Equipment List at the front of this module.

See the general Teaching Tip at the end of this module.

Explain that terms shown in bold are defined in the Glossary at the back of this module.

Show Transparency 1, Objectives, and Transparency 2, Performance Tasks. Review the goals of the module, and explain what will be expected of the trainees.

Review the modules covered in Level Five, and explain how this module fits in.

Define shaft alignment.

Identify the procedures that must be followed prior to alignment.

Define soft foot.

Show Transparencies 3 and 4 (Figures 1 and 2). Describe the effects of soft foot as shown in the diagrams.

1.0.0 ◆ INTRODUCTION

When machinery is installed, only the rotating shaft center lines of machines are aligned. The positions of the feet, couplings, shaft surfaces, machine housing, and bearings are not used to determine machine alignment. It is important to understand that alignment refers to the positions of two center lines of rotation and that the machined center line may or may not be the same as the rotational center line. Therefore, shaft alignment is defined as positioning two or more machines so that their rotational center lines are collinear at the coupling point under operating conditions.

The coupling point refers to the fact that vibration caused by misalignment originates at the point of power transmission, which is the coupling. It does not mean that the couplings are being aligned. The shaft center lines are being aligned, and the coupling center is the measuring point. Dynamic factors such as wear, thermal growth, dynamic loads, and support structure shifts after startup make it important to align machinery under operating conditions.

Although rough alignment can be made using straightedges, the only way to determine rotational center lines is to rotate the equipment. Alignment measurements taken without turning both shafts is aligning surfaces, not center lines of rotation. One of the fastest and most accurate methods for aligning machinery is using a **laser** alignment tool.

Before performing alignment, the millwright must complete the following steps:

Step 1   Inspect the base of the equipment for level, center line, and distortion.

Step 2   Check for soft foot.

Step 3   Set the coupling gap.

Step 4   Eliminate outside stress influences such as pipe stress, conduit, etc.

Step 5   Perform rough alignment of equipment within mechanical limits of the indicators (0.100 inch).

2.0.0 ◆ SOFT FOOT

Soft foot is a condition in which the tightening or loosening of the bolt(s) of a single foot distorts the machine frame. If the laser indicates a soft foot, the condition should be analyzed to determine its cause before any corrective action is taken. The effects of soft foot vary because of operating conditions and where the weakest conditions are. A machine with a soft foot that is bolted down firmly will distort and affect the operation of the machine train in many ways (*Figures 1* and 2).

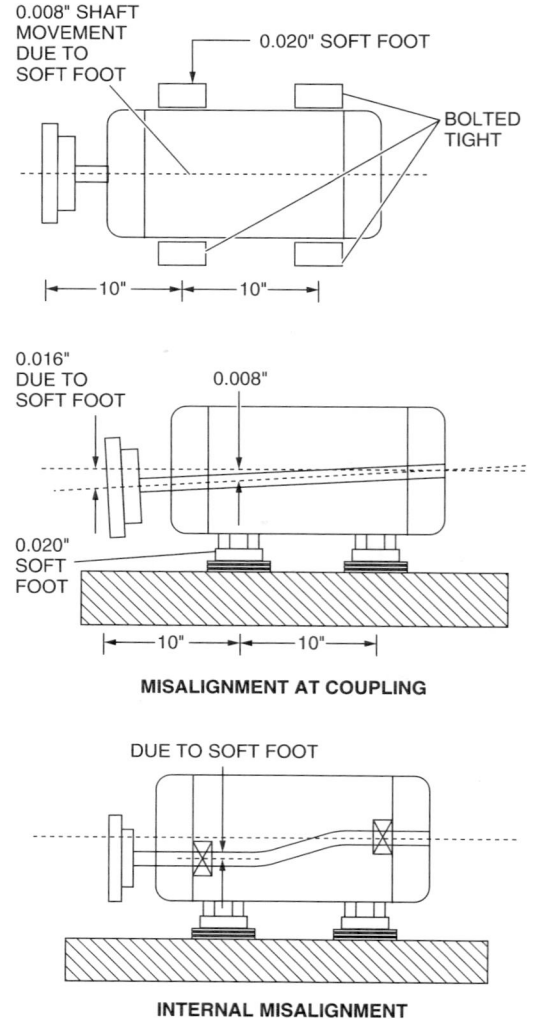

Figure 1 ◆ Soft foot.

Instructor's Notes:

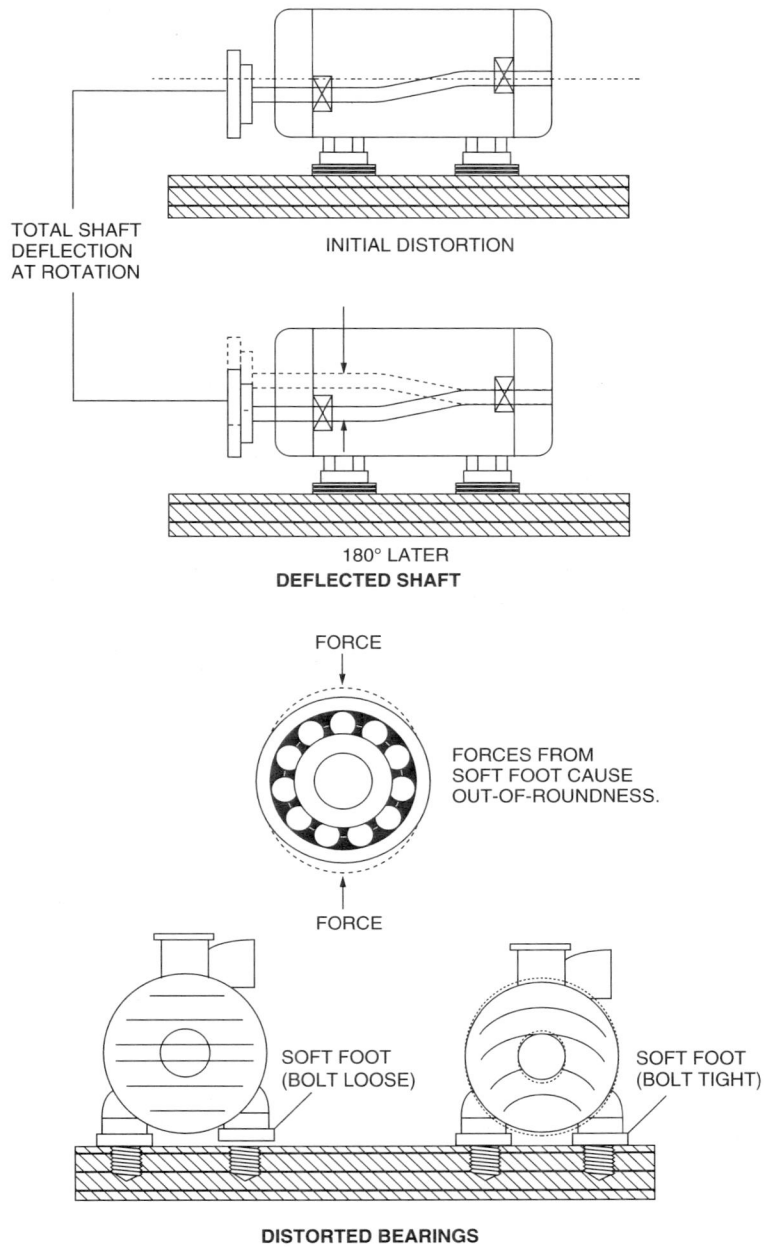

TOTAL SHAFT
DEFLECTION
AT ROTATION

INITIAL DISTORTION

180° LATER
DEFLECTED SHAFT

FORCE

FORCES FROM
SOFT FOOT CAUSE
OUT-OF-ROUNDNESS.

FORCE

SOFT FOOT
(BOLT LOOSE)

SOFT FOOT
(BOLT TIGHT)

DISTORTED BEARINGS

502F02.EPS

Figure 2 ◆ Effects of soft foot.

Discuss the symptoms of the various types of soft foot and the remedies for each type.

Show Transparency 5 (Figure 3).

Indiscriminate shimming will often make things worse. Soft feet are caused by a variety of reasons, some of which may not even be related to the machine itself. The following are some types of soft foot:

- Parallel air gap
- Bent foot
- Squishy foot
- Induced soft foot
- Gaps without soft foot

Soft foot can be difficult to find; it is therefore important to understand all the differences and causes of this condition. *Figure 3* shows types of soft foot.

2.1.0 Parallel Air Gap

Parallel air gap occurs when three mounting feet sit solid and flat but one foot does not touch. It is found as two diagonally opposed readings. Some causes may be that one leg is too short, one baseplate mounting pad is not level with the other three, or the shims under one foot are the incorrect size. The laser alignment computer shows which foot is short and how much it needs to be shimmed to correct it. Always shim one foot only instead of splitting the amount between the rocking feet. It is also advisable to confirm the reading, using a feeler gauge, before inserting and tightening the shims. Sometimes, the laser alignment computer shows that two diagonal feet have the same value. In this case, both feet should be shimmed equally.

2.2.0 Bent Foot

Bent foot occurs when the bottom of one foot is not parallel with the base. It appears as three readings the same, and one different. The feeler gauge readings show a slope from one corner to the other. The down corner may be touching, but when tightened, it acts as a lever and induces soft foot in the two opposing feet and sometimes in the fourth foot as well. This gives the machine the appearance of having three or four soft feet, but they disappear when the one bent foot is corrected.

The causes of bent foot can include damaged machinery, bent or poorly machined baseplates, severe vertical angularity misalignment, feet that have been welded, or a settling foundation. The only corrections other than replacing the bent machine are to mill or grind the bent foot to square or to build a step shim. If the foot is machined square, it will then be short and must

be corrected as a parallel air gap. To build a step shim may be more expedient, but observe the following procedure:

Step 1 Fill any gap that exists under the entire foot so that one corner or edge of the foot is touching the base.

Step 2 Measure the largest remaining gap, using the laser.

Step 3 Divide this gap by four or six, depending on the number of step shims, to obtain the step thickness.

Step 4 Select four or six shims of the step thickness, and insert them one at a time without lifting the machine until the fit is snug.

The purpose of using step shimming is to match the slope of the bent foot with a substantial wedge that will remain solid when the bolts are tightened. After rechecking the alignment using the laser, trim the outside of the step shims so that they are flush with the foot. Step shims are illustrated in *Figure 3*.

2.3.0 Squishy Foot

Squishy foot clearly shows as a soft foot measurement on the laser, but when measured using a feeler gauge, little or no gap is indicated. This is because there is dirt, grease, paint, or rust between the shims, because the shims are bent or burred, or because there are too many shims under the foot.

All shims have some imperfection and allow some compression when the bolts are tightened. This microscopic spring effect accumulates at the rate of 0.0003 inch per shim and increases as the shims are reused. As a standard rule, no more than four shims should be used under one foot.

The cleanliness of the base and foot area and condition and number of shims can eliminate squishy foot. Scrape, sand, brush, and wipe all surfaces under and around the feet. Always use new, clean shims and no more than four in a pack. Install the shim pack; tighten the mounting bolts, and recheck the alignment, using the laser.

2.4.0 Induced Soft Foot

Induced soft foot shows up as two soft feet on the same side or same end of the machine, and the feeler gauge indicates a gap that is parallel or nearly parallel. Secondary symptoms are that the condition does not improve, the condition

2.4 MILLWRIGHT ◆ LEVEL FIVE

Instructor's Notes:

_____

_____

_____

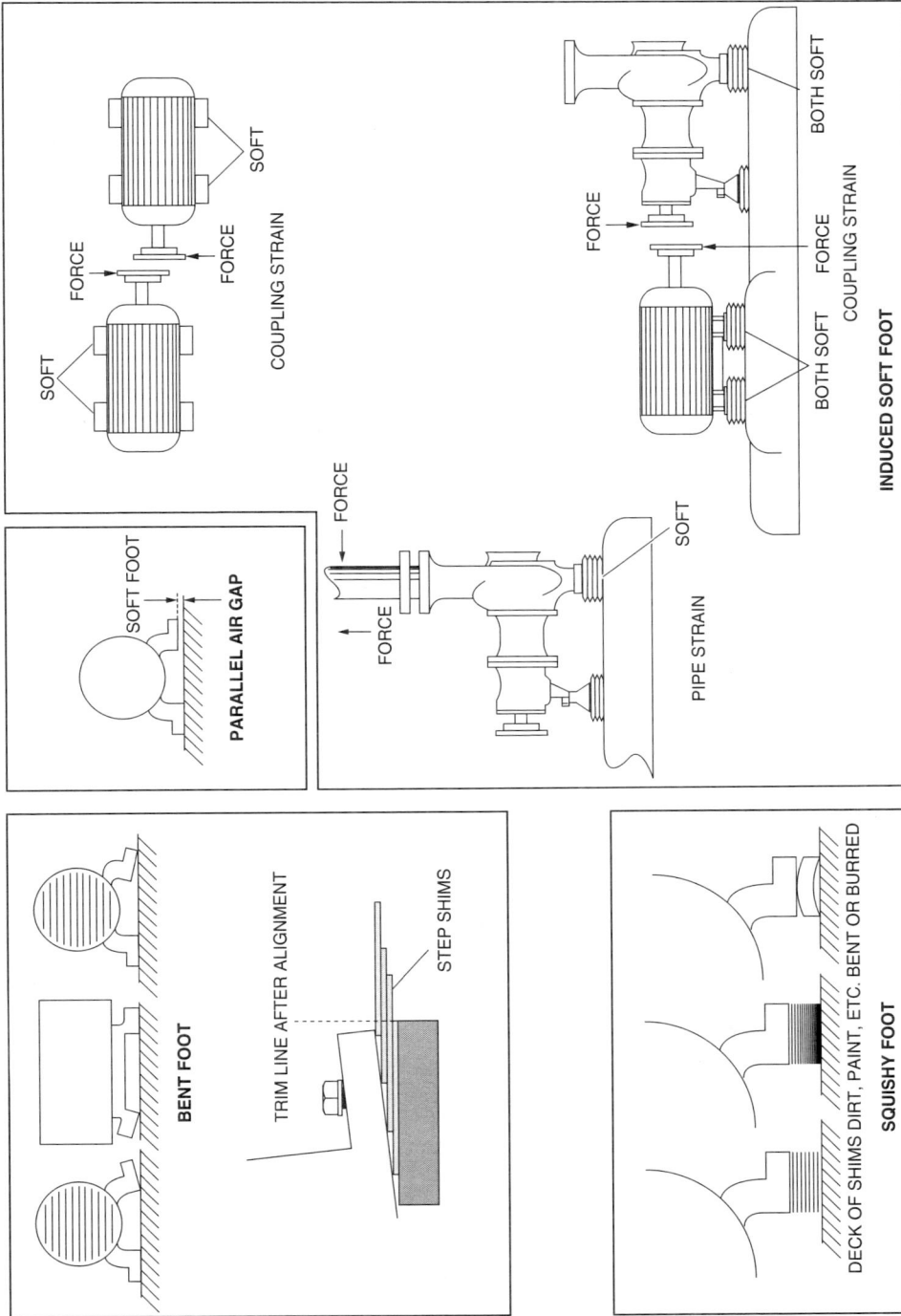

Figure 3 ◆ Types of soft foot.

Within the figure, the following labels appear:

- SOFT
- FORCE
- FORCE
- SOFT
- COUPLING STRAIN
- BOTH SOFT
- FORCE
- FORCE
- COUPLING STRAIN
- BOTH SOFT
- BOTH SOFT
- **INDUCED SOFT FOOT**
- FORCE
- FORCE
- SOFT
- PIPE STRAIN
- SOFT FOOT
- **PARALLEL AIR GAP**
- **BENT FOOT**
- TRIM LINE AFTER ALIGNMENT
- STEP SHIMS
- DECK OF SHIMS DIRT, PAINT, ETC. BENT OR BURRED
- **SQUISHY FOOT**
- 502F03.EPS

Discuss the characteristics of gaps without soft foot.

Explain thermal growth as it relates to alignment. Describe how thermal growth is measured.

Define coupling stress, and describe its harmful effects.

See the Teaching Tip for Section 2.0.0 at the end of this module.

becomes worse, or another foot becomes much worse after shimming the gap amount. These are caused by external forces on the machine. Coupling strain and pipe strain are the two most common causes of induced soft foot. If the coupling is difficult to bolt up, expect an induced soft foot until alignment is improved. Other external forces that can cause induced soft foot are as follows:

- Overhung machines or attachments
- Belt, gear, or chain loads
- Hoses or stressed conduit
- Structural bracing attached to the machine
- Jack bolts tight against the base

The solution for induced soft foot is to remove the external force. Use the laser to document measurements before and after the corrective action to confirm that the problem is diagnosed completely.

2.5.0 Gaps Without Soft Foot

Gaps without soft foot occur when there is a visible gap under the foot before tightening the bolt. However, when the foot is tightened down, there is no gap, but the laser does not show any soft foot or misalignment reading. The laser may indicate a relatively small soft foot, but the feeler gauges indicate a much larger gap. This may be caused by the base moving, the foot flexing without bending the machine, or the base being cracked, causing the foot to be loose. At this point, shimming to fill the gap is cosmetic; the logical solution is to replace the base or machine housing.

3.0.0 ◆ THERMAL GROWTH

Thermal growth generally refers to the expansion of materials with a rise in temperature. For alignment purposes, thermal growth refers to the movement of shaft center lines associated with the change in temperature from alignment to operating conditions. The change may be a rise in temperature, a drop in temperature, no change in the average temperature but a spot thermal spike in the machine, or some combination of change to average temperatures and gradients. These temperature changes can produce changes in alignment that may affect VO, HO, VA, HA, or any alignment combination.

Analysis of thermal growth can be as complex as the analyst desires; however, Optalign® can make the adjustment. A method for calculating thermal growth uses the movement of the shaft because of the temperature at the bearings or feet.

The formula requires knowing the change in temperature of the machine housing between the feet and the shaft bearings, the distance between the shaft center line and the feet, and the coefficient of thermal expansion for the machine housing. There are many chances for error in calculating, but usually the original equipment manufacturer's service and repair manual provides the correct information.

When feasible, it is far better to measure thermal growth than to calculate it. A common method of measuring thermal growth is to compare cold and hot alignment readings. This is done by first taking readings when the equipment is cold. Then, start the equipment, and let it run until the temperature is stable. Shut it down and immediately read the misalignment while the equipment is still hot. The hot readings minus the cold readings should equal the amount of thermal growth. The accuracy depends on taking the hot reading very quickly, and in some cases, this is impossible.

The cool-down rates and many design factors can hinder the timing and accuracy. The solution may be to continuously monitor the equipment movements. Optalign® makes a continual monitoring system as do other manufacturers. Continuous monitoring could also be accomplished with a combination of monitoring the bearing temperature and the load to chart a calculated growth after confirming the cold and hot misalignment, using the laser alignment system.

4.0.0 ◆ COUPLING STRESS

Coupling stress is any external condition or situation that places abnormal stress on a pump or motor which can cause misalignment of the shaft center lines. Coupling stress can make it very difficult to align the shafts and has very harmful effects. It can cause vibration, overheating, excessive wear on the equipment, and can increase operating costs through power loss. Coupling stress must be eliminated before the shafts are aligned; therefore it is important to understand the causes and possible solutions for coupling stress.

4.1.0 Causes of Coupling Stress

Coupling stress can be caused by one or more of many conditions. The following sections explain some of the common causes of coupling stress and some of the possible solutions. If during laser alignment it becomes apparent that aligning the shafts is not solving the problem, coupling stress would be the next area to analyze. After curing

---

2.6 MILLWRIGHT ◆ LEVEL FIVE

Instructor's Notes:

_____

_____

_____

the stress problem, reshoot the laser, and align the shafts properly. The following are some common causes of coupling stress:

- Incorrect pipe weldments
- Incorrect pipe hangers
- Defective anchor bolts
- Bad bearings
- Bad foundations

4.1.1 Incorrect Pipe Weldments

Incorrectly fitted and welded pipe can cause coupling stress. If the pipe flange is not welded perfectly square to the pump connection, bolting the flange can cause considerable stress to the pump, which is then transferred to the coupling and shaft. If the flange is short of the pump connection, it will leave a gap. When the flange is bolted to pull the flanges together, this causes stress in the pump, which affects alignment and coupling stress. Some gap is necessary for the gasket, but it is important that the two flanges be square. To eliminate pipe flange angular misalignment, a tapered filler piece may be necessary. Excessive parallel offset cannot be cured with a filler piece but can be helped by offsetting several successive joints slightly, taking advantage of the bolt hole clearances.

4.1.2 Incorrect Pipe Hangers

Improper placement of pipe hangers can create excessive stress on the pump and motor by forcing the weight of the pipe to be carried by the flange. Piping should be well fitted, firmly supported, and sufficiently flexible so that no more than 0.002 inch of soft foot movement occurs when the last pipe flanges are tightened. Pipe expansion or movement may cause misalignment and increased vibration. Pipe hangers should be placed and checked before they are bolted up to ensure that no load is being carried by the flange or the pump.

4.1.3 Defective Anchor Bolts

Improperly installed anchor bolts on any equipment that is coupled by a shaft can cause coupling stress by twisting either unit. The machine base may cause these bolts to skew and bend, which puts a load on the machinery. Before installation, the base pads should be checked to ensure that they are level, flat, parallel, coplanar, and clean. Using properly sized hold-down bolts with enough clearance to permit corrective movement during alignment is also necessary.

Anchor bolts should be installed straight and torqued properly. Improperly spaced anchor bolts or too few bolts can also cause twisting of machine bases.

4.1.4 Bad Bearings

Damaged or marginal bearings in machine units are a progressive cause of coupling stress and misalignment because as the bearing gets worse, the shaft vibrates more and creates a wobbly rotational center line. This creates a complex set of alignment problems. Left uncorrected, this can create catastrophic failure in the machine system.

4.1.5 Bad Foundations

A bad foundation under a machine system may be an unlevel baseplate, an improperly sized concrete pad, or bad grouting. All of these conditions can twist the machine frame and cause misalignment and coupling stress. The corrective action is to check for soft foot, and if close enough to allow it, make shims to level the equipment. If the foundation is very bad, it will have to be remade. A good rule of thumb calls for foundation concrete to weigh three times the machine weight for rotating machines and five times the weight for reciprocating machines.

5.0.0 ◆ BASIC LASER PRINCIPLES

A laser beam is an amplified and **collimated** beam of **coherent** light energy that travels perfectly straight in a vacuum and very nearly straight in the atmosphere. The cause of imperfect performance in the atmosphere is humidity or dust in the air, because each tiny drop of water acts as a prism that diffracts part of the laser beam. The emitter source in each laser and the amount of power applied to it determine the application for which it is used.

There are three classes of lasers, and each laser is classified according to its intensity. Most class I lasers are under 5 milliwatts and are used for optical alignment, reading bar codes, or measuring distance. The class II lasers have more power and can also be used for alignment and measuring, printing on objects, precision soldering, or cutting delicate tissue in microsurgery applications. The class III lasers include metal cutting, welding, and other processes in which precise control of intense power is necessary. Lasers are either continuous beam or pulsed beam and can be visible or invisible, depending on the emitter source.

Discuss each of the causes of coupling stress and the recommended remedies.

Assign reading of Sections 5.0.0–6.1.2.

Ensure that you have everything required for teaching this session.

Define a laser beam, and explain the basic principles of lasers.

Identify the three classes of lasers, and describe their characteristics and applications.

Stress the potential danger of lasers to the eyes and emphasize appropriate safety precautions.

Point out the hazards associated with the electrical energy required by lasers.

Show Transparency 6 (Figure 4). Describe the characteristics of optical alignment lasers.

Provide an overview of the Optalign® laser system referenced in this module.

Show Transparency 7 (Figure 5). Describe the components and operation of the Optalign® laser system.

Discuss misalignment terminology and the categories of offset and angularity.

Show Transparency 8 (Figure 6).

The optical alignment beams discussed in this module are very low power and are relatively safe but are limited in range and in their ability to penetrate smoke or mist. *Figure 4* shows the optical alignment laser.

5.1.0 Laser Safety

> **WARNING!**
> Lasers are capable of damaging your eyesight, even by reflection. Always follow all safety precautions when working with them.

Because of their intense, collimated energy beam, all lasers can damage the eye irreparably. Always wear polycarbonate safety glasses when working around lasers, and never look directly at a laser beam. Also, never point the beam at mirrors or other reflective surfaces because this could send the beam directly into an unprotected eye.

The Occupational Safety and Health Administration (OSHA) provides standards for laser use, as do some states. The laser manufacturer publishes these standards in each owner's manual or training guide. Some states require laser warning signs to be posted in any area in which a laser is being used.

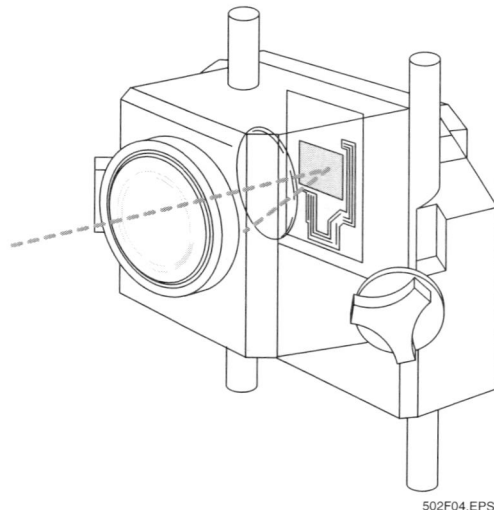

502F04.EPS

Figure 4 ◆ Alignment laser.

Another safety consideration is the power source because of the inherent inefficiency of a laser. To produce a small output of laser energy requires a large input of electrical energy; therefore, any laser with a substantial amount of power has a large source of electrical energy. The semiconductor type of laser used in alignment has no internally serviceable parts and should be serviced by the factory only.

6.0.0 ◆ OPTALIGN® LASER ALIGNMENT

There are several laser alignment systems available, and each has its advantages. The laser levels and transits used in site layout and surveying are similar in operation. This module uses information based on the Optalign® laser alignment system, which is a Class II, 5-milliwatt, semiconductor laser. The major components are the laser/detector, prism, computer, brackets, and level. The prism reflects the beam back to the detector or target, and the angularity and offset are calculated in the computer, which gives a numerical and graphic readout. *Figure 5* shows the Optalign® system.

6.1.0 Descriptive Characteristics

Correct and complete expression of misalignment is necessary before any meaningful tolerance or target discussion can take place. Actual misalignment is almost always a combination of angularity misalignment and offset misalignment. It is therefore necessary to describe any misalignment in a plane as a value of offset and a value of angularity.

The position of a rotating shaft center line in space can be completely described by showing both vertical and horizontal views. These views are sometimes called the side view and top view and are also called the elevation and plan view. Each view displays offset and angularity. The combination of offset and angularity is the most likely situation. *Figure 6* shows various misalignments.

Since two views describe the location of a shaft and each view can have offset and angularity, four terms and numbers are used to describe the misalignment. Misalignment can be described by the following terms:

- Vertical offset (VO)
- Horizontal offset (HO)
- Vertical angularity (VA)
- Horizontal angularity (HA)

Instructor's Notes:
_____
_____
_____

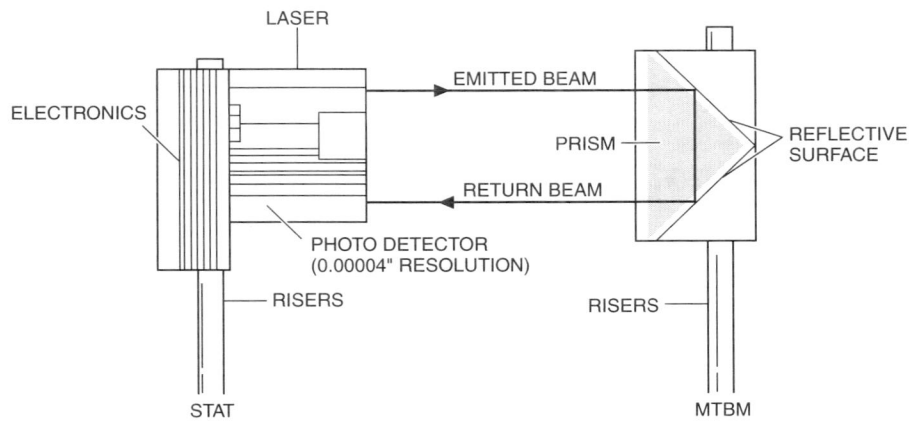

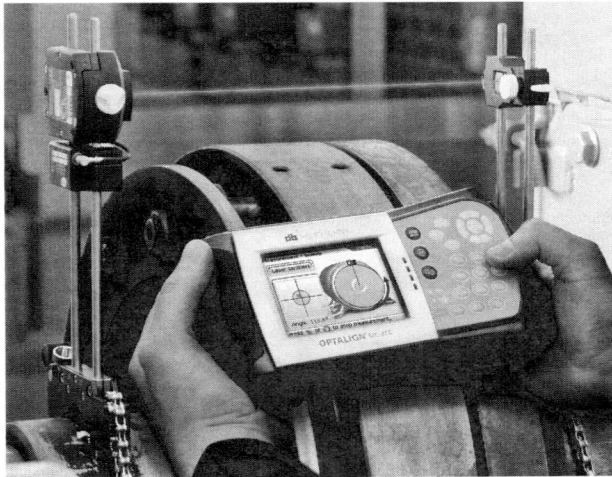

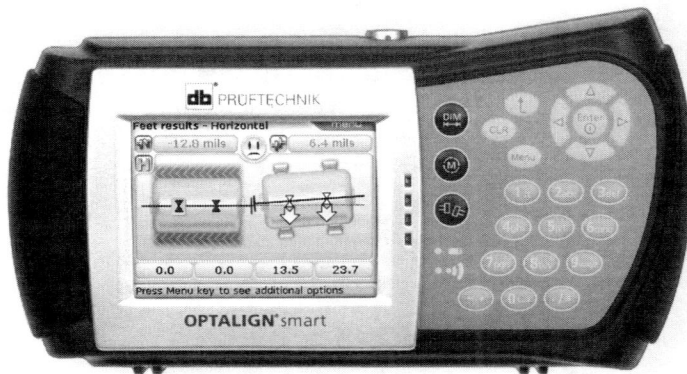

LASER

ELECTRONICS

EMITTED BEAM

PRISM

REFLECTIVE SURFACE

RETURN BEAM

PHOTO DETECTOR
(0.00004" RESOLUTION)

RISERS

RISERS

STAT

MTBM

502F05.EPS

Figure 5 ◆ Optalign® laser system.

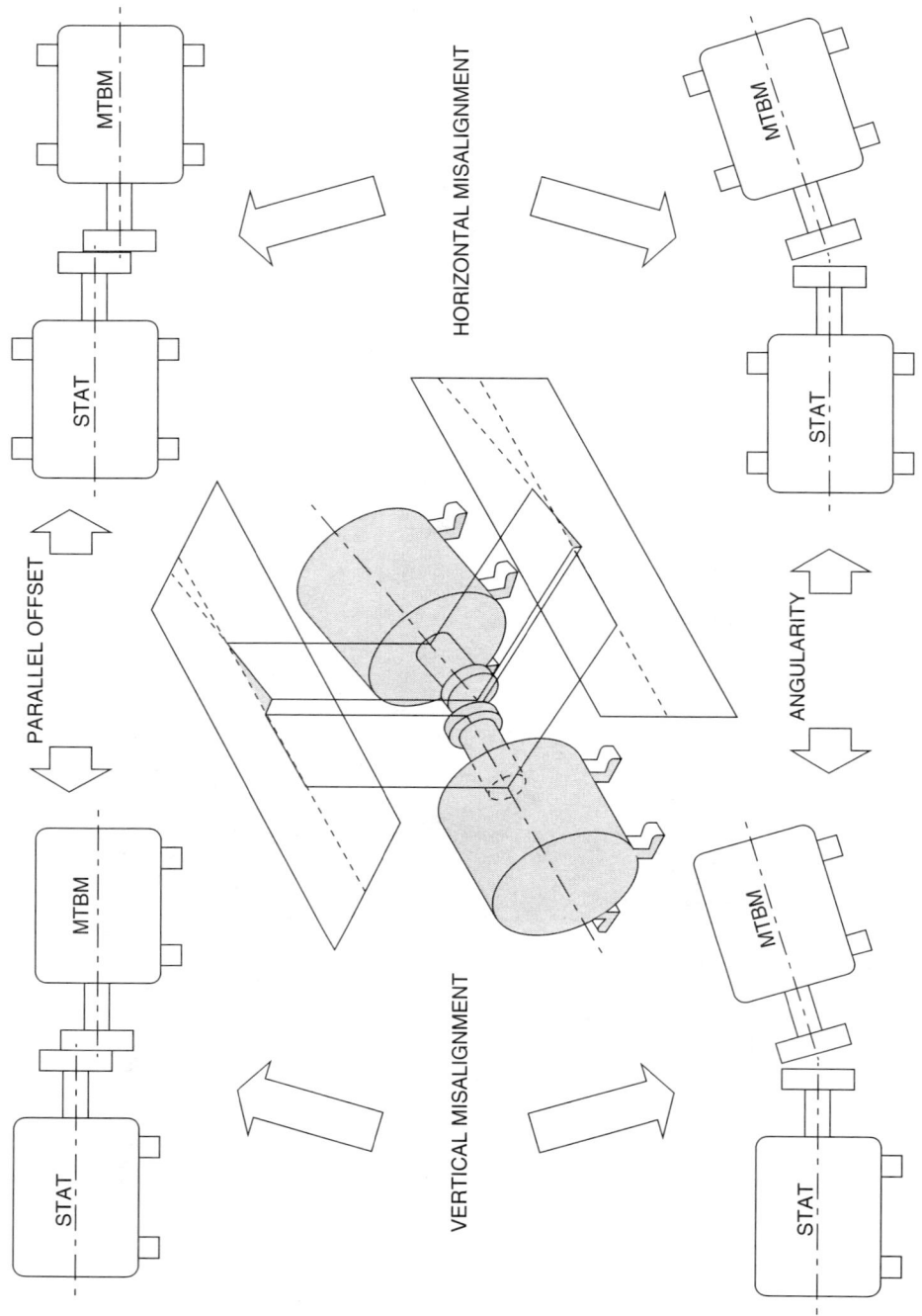

Figure 6 ◆ Combined misalignments.

Instructor's Notes:

The basic dimensional definitions are horizontal to vertical and front to back. The stationary machine, or STAT, refers to the machine that does not move and stays firmly bolted. The machine to be moved, or MTBM, is where all the adjustments take place. Each term listed above describes the position of the MTBM in relation to the STAT, using the center line of the STAT as zero. *Figure 7* shows the definition of the dimensions.

The convention for determining positive and negative is defined in three planes: rotational, vertical, horizontal, and combinations of these three. The rotational direction uses the clock to coordinate position. Clockwise is defined by looking along the shafts from the MTBM toward the STAT. The clock is used to describe which quadrant the offset is in and the number value based on Cartesian coordinates. From 12 o'clock to 3 o'clock, x and y are positive values; from 3 to 6, x is positive and y is negative; from 6 to 9, x and y are negative; and from 9 to 12, x is negative and y is positive. Positive offsets occur when the MTBM is high, or toward 3 o'clock, compared to the STAT. Negative offsets are when the MTBM is low, vertically, or toward 9 o'clock, horizontally, compared to the STAT. Angularity is defined by where the coupling gap is open. Open at 12 o'clock, or 0

degrees, is positive VA; open at 3 o'clock is positive HA. Likewise, negative VA means that the coupling gap is open at 6 o'clock; negative HA means that the coupling gap is open at 9 o'clock. *Figure 8* shows the laser detector target.

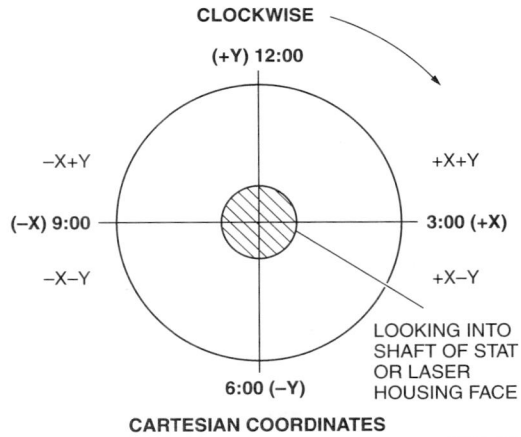

Figure 8 ◆ Laser detector target.

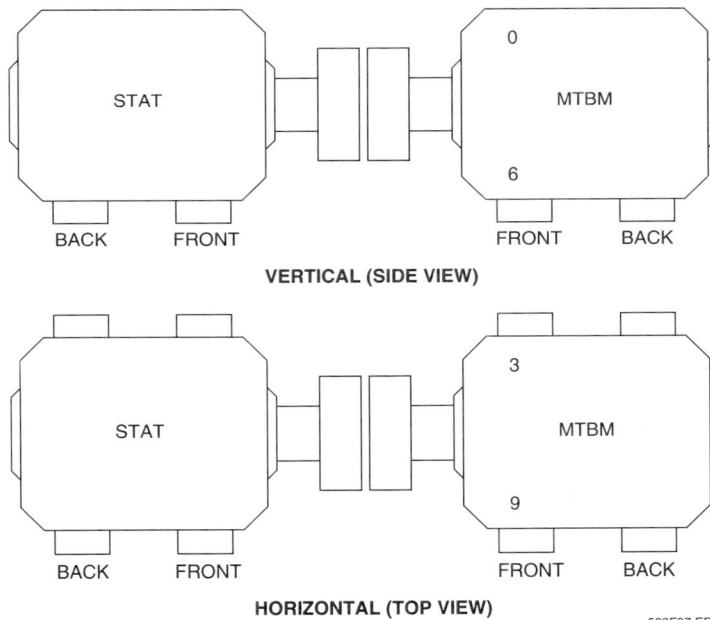

Figure 7 ◆ Dimensional definition.

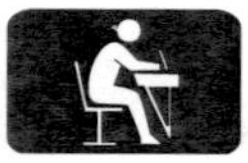

Explain why vibration generally has no effect on a laser.

Discuss the continuous sweep, clock, and set point modes of the Optalign® Plus™.

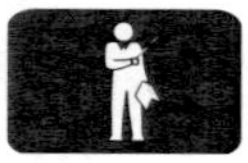

If possible, demonstrate these three modes using an Optalign® Plus™.

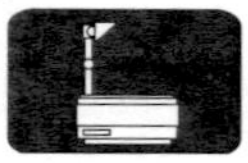

Show Transparency 11 (Figure 9). Identify the functions of the alignment keys.

6.1.1 Optalign® Laser System Capabilities

Normally, vibration has no effect on the laser because the beam travels so rapidly. Each measuring pulse travels at the speed of light across the span and back, which is 40 inches in 0.0000000034 of a second. During the transit time, even under severe vibration, the laser and prism will not move far. The beam deflection or lack of responsiveness due to uniform vibration is not a problem. If the STAT and MTBM are mounted on separate foundations or for some other reason are vibrating out of phase with each other, then it is possible that the readings will catch the machines in different positions with respect to each other and pick up an error. The maximum potential for error is approximately equal to the maximum amplitude of vibration from one machine to another.

The input to the computer consists of measurements made by the laser and distances keyed in by the operator. The computer performs the calculations that result in a shim thickness or move distance on the display. There are three modes for operation of the Optalign® Plus™. Continuous sweep mode is the most accurate and most commonly used mode. In this mode, turn the shaft and laser at least 70 degrees, stop, and press Results. This is only used when the shafts are coupled, but is the easiest and quickest method.

The second mode is the clock mode, and used either when the couplings are not connected, when the shafts are vertical, or when the feet are other than at 6 o'clock. The shaft is turned in at least three 45-degree components, pressing ENT at each point.

Set point mode requires at least three readings. First pick one point, press M (measure), then press ENT; turn to a second point, press ENT, then turn to a third point, and press ENT again. The three points must be at least 60 degrees apart. *Figure 9* shows the Optalign® Plus™ computer keys. Set point is usually used when major force is required to turn the shaft.

6.1.2 Laser Limitations

Some conditions can limit or stop a laser beam both in a plant and in the field. Heavy concentrations of dust, vapor, or other airborne contaminants can block or hinder the beam. The display

DIM **Dimensions**
Initiates entry of machine dimensions.

Results
Displays alignment condition at the coupling and at machine feet.

Soft foot
Checks whether all machine feet rest firmly on the foundation and helps determine corrections.

9 0 **Numbers**
For entry of numerical values.

ENT **Enter**
Confirms displayed value and proceeds to next value.

M **M = Measure**
Start alignment measurement by rotating the machine shaft.

'MOVE'
Guides the user interactively through machine positioning for alignment correction.

CLR **Clear**
Deletes displayed value to allow correction.

Arrows
For processing through program steps.

KEYS UNDER SLIDING DOOR

Alignment targets
Desired offsets and angularity can be entered to compensate for thermal growth. Correction values are then adjusted automatically.

Save/Open file
For storing and recalling machine setups and results in a file. Up to 99 files can be saved.

Vertical machines
Measurement at 3 or more clock positions gives flange offset and shimming values.

F **Function**
For special functions such as changing units, setting date and time, 6-feet machine setup, coupling type selection etc.

Printer
Directs output to the printer for producing a measurement report.

502F09.EPS

Figure 9 ◆ Optalign® Plus™ computer keys.

Instructor's Notes:

2.12 MILLWRIGHT ◆ LEVEL FIVE

shows an OFF message if the particles are extremely concentrated. The solution to this problem is to shoot the beam inside a length of PVC pipe or cardboard tube. Strong sunlight, infrared light, or welding flash directly striking the detector cause an error message to appear. Using a shade or hood or shooting the beam through a pipe will solve the problem. Theoretically, heat waves, steam, or temperature variations, if severe, can distort or vary the beam, but in reality, if you can see through the disturbance for the distance the beam has to travel, any beam variation is unlikely.

7.0.0 ◆ LASER/DETECTOR OPERATION

When the prism moves up and down, the return beam moves up or down twice the distance the prism moves. This is how the laser can measure offsets between two points. The offset measured is the offset of the prism relative to the laser. Because the prism is rotated in the horizontal plane, the detector can sense the up-and-down and side-to-side motion of the beam. *Figure 10* shows laser/detector movement.

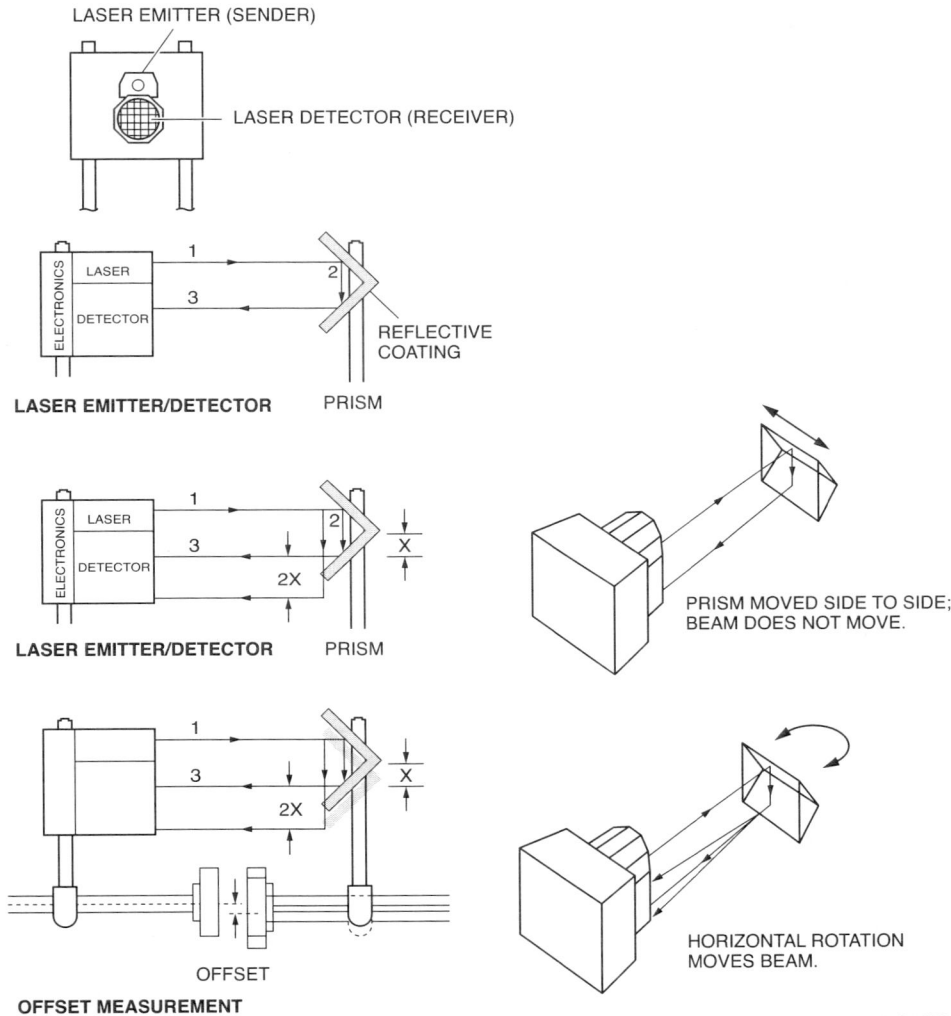

Figure 10 ◆ Laser/detector movement.

Discuss the conditions that can affect laser beam accuracy.

Describe methods used to compensate for these conditions.

Show trainees how to identify the major components of a laser alignment system.

Have trainees identify the major components of a laser alignment system. This laboratory corresponds to Performance Task 1.

Assign reading of Sections 7.0.0–7.1.3.

Ensure that you have everything required for teaching this session.

Explain laser/detector operation, including the function of the prisms.

Show Transparency 12 (Figure 10). Describe how the laser measures offsets.

Review the tasks associated with alignment procedures.

Show Transparency 13 (Table 1). Describe how tolerance tables are used.

Explain the procedures involved in rough alignment.

Show trainees how to perform a rough alignment.

Have trainees practice performing a rough alignment. This laboratory corresponds to Performance Task 2.

Describe the steps for setting up laser equipment.

Warn trainees that equipment should never be started up while an alignment is being performed.

In *Figure 10*, the angle the return beam moves is twice the angle the prism moves. If the beam is zeroed at the 12-o'clock position and then read at the 6-o'clock position, the horizontal beam movement is one leg of a 90-degree triangle. The distance from the laser to the prism is the other leg. The angle defined by these two legs is two times the actual angular misalignment between the shafts.

7.1.0 Alignment Procedures

The alignment procedures consist of performing a rough alignment, setting up the laser equipment, and performing an initial alignment. Specific alignment procedures, such as aligning machinery trains, measuring for shaft sag, and aligning vertical machines, are then performed depending on the needs of a particular job.

Tolerance tables (*Table 1*) set the standards to which the equipment must be aligned, based on the angularity limits of the machine. Tolerance tables are usually provided by the manufacturer of the equipment.

7.1.1 Rough Alignment

Before laser alignment is attempted, the STAT and MTBM must be aligned as closely as possible by hand. As a rough rule, the coupling offset should be less than ³⁄₃₂ inch and the angles between the coupling faces should be near parallel or within 10 mils per inch of coupling diameter. The most modern laser alignment equipment does not necessarily require even such rough alignment, and is capable of guiding the alignment all the way. Once the brackets are mounted and the lasers installed, the latest machines can go from very rough alignment to precision alignment, including adjustment for thermal growth, in a very few steps.

Ludeca, Inc., has three standard units, beginning with a machine (the Aligneo) that can align one pair of machines horizontally and vertically, providing the millwright with precise requirements for shimming and horizontal movement of the feet automatically, as well as soft foot adjustment. Ludeca also manufactures the Optalign®, as well as the Rotalign® Pro, which can align up to six machines, and the Rotalign® Ultra, which can align up to fourteen machines. The obvious speed advantage over conventional or reverse alignment with dial indicators makes these machines the equipment of choice for millwrightss that are required to align machinery very frequently. This ensures that the shafts are within range of the laser and that all the major causes of shaft misalignment have been corrected before setting up the laser.

7.1.2 Setting Up Laser Equipment

Follow these steps to set up the laser alignment equipment:

Step 1   Ensure that the machines are shut down, locked out, and tagged.

WARNING!
If the equipment is started up while an alignment is being performed, serious injury can occur.

Step 2   Mount the laser and prism on the shafts just aft of each coupling half.

Step 3   Clamp the brackets firmly in place and directly across from each other.

Step 4   Ensure that the beam hits the detector.

Table 1 Tolerance Tables

SHORT COUPLINGS							SPACER SHAFTS	
	Excellent			Acceptable			Excellent	Acceptable
	Offset	Angularity		Offset	Angularity			
RPM	Mils	Mils/in	Mils/10 in	Mils	Mils/inch	Mils/10 in	Mils/in of Spacer	Mils/in of Spacer
600	5.0	1.0	10.0	9.0	1.5	15.0	1.8	3.0
900	3.0	0.7	7.0	6.0	1.0	10.0	1.2	2.0
1200	2.5	0.5	5.0	4.0	0.8	8.0	0.9	1.5
1800	2.0	0.3	3.0	3.0	0.5	5.0	0.6	1.0
3600	1.0	0.2	2.0	1.5	0.3	3.0	0.3	0.5
7200	0.5	0.1	1.0	1.0	0.2	2.0	0.15	0.25

502T01.EPS

Instructor's Notes:
_____
_____
_____

The procedure that follows is the standard procedure for setting up the Optalign® Plus™ system. Some of this procedure may be the same for other systems. The standard brackets are of several types; however, the most common type is the chain bracket. Fasten the brackets to the shaft, and mount the posts. Use posts that are just tall enough that the laser will project over the coupling.

Having mounted the laser and prism, the actual operations are performed on the pendant keyboard. *Figure 11* shows the Optalign® Plus™ pendant. At the bottom of the keyboard is a slide cover that covers the six other special keys.

Mount the laser on the STAT and the prism on the MTBM. Attach the transducer cable to the laser, and clip the cable to the post of the laser. Attach the other end of the transducer cable to the right-hand socket on the top of the pendant. Turn the pendant on by moving the switch on top to the ON position. Once the laser is mounted properly, press the DIM switch. The first dimension is the distance from the arrow point on the laser to the hash mark on the prism. This dimension can be measured with a tape measure. When the numbers are entered, press "ENT." Second, enter the distance from the laser to the center of the coupling, and press "ENT" again. Third, enter the diameter of the coupling. The fourth dimension is the rpm of the motor. The next dimension is the distance from the laser to the front foot of the MTBM. Enter that dimension, and enter the distance from front foot to back foot of the MTBM. *Figure 12* illustrates this procedure.

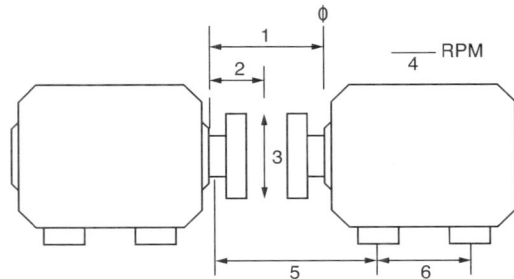

1. Laser to prism   ENT
   (Arrow point to hash mark "φ" on prism)
   (Center-to-center of support posts)
2. Laser to center of coupling   ENT
   (Center of flex planes)
   Note: See F73 if you have a spacer coupling.
3. Working diameter   10"   ENT
   (Coupling diameter)   (Default value)
4. RPM   ENT
5. Laser to front foot   ENT
6. Front foot to back foot   ENT

502F12.EPS

Figure 12 ◆ Laser alignment procedure.

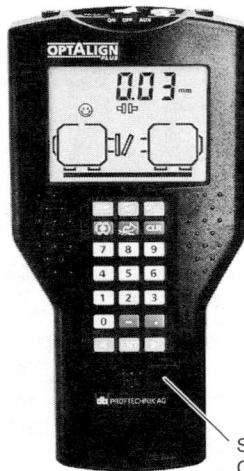

Describe how to mount the laser and prism.

Explain the layout of the Optalign® keyboard and the key functions. Describe how data is entered during laser alignment.

Show Transparency 14 (Figure 12).

Show trainees how to set up a laser and how to enter dimensions.

SLIDE COVER COVERS SPECIAL KEYS

502F11.EPS

Figure 11 ◆ Optalign® keyboard.

7.1.3 Initial Laser Alignment

After the laser and prism are in place and square to each other, an initial alignment procedure is performed. To check the initial alignment, enter the machinery and bracket dimensions in the computer using the DIM key (*Figure 13*).

After the dimensions have been entered, leave the cap on the prism until the laser beam is centered on the cap crosshairs. Press the Measure button. Remove the cap and adjust the thumbscrews on the prism until the coordinates are close to 0.0. "Turn" will appear on the screen. At this time, rotate the shaft at least 75 degrees.

Press the Results key until the coupling symbol appears. Press ENT or the forward and back arrow keys to work through the offset and angular misalignment (*Figure 14*). The results will appear in the following order: vertical coupling offset; vertical coupling angularity; horizontal coupling offset; and horizontal coupling angularity. To obtain the shimming and horizontal move corrections, press the Results key until the foot symbol appears, then use the ENT or arrow keys to read the moves required (*Figure 15*).

To take a set of readings before the move, after shimming is completed, the Move key requires that you choose one of the 45-degree shaft positions (45 degrees, 135 degrees, 225 degrees, or 315 degrees). Choose the direction of the move, either horizontal or vertical, with the right arrow key. When the correct choice appears, confirm the choice with the ENT key. If the beam is not at 0.0, adjust it. When "Entr" appears, press ENT, and move the machine into alignment. After each move, another set of readings must be taken. You would then follow these steps:

Step 1   Press the M key to zero the system.

Step 2   Rotate the system at least 60 degrees (*Figure 16*).

Step 3   Press the Results key on the Optalign® Plus or the Coupling key on the Rotalign® (*Figure 17*), and record the four values. The Optalign® can be connected to a PC to record automatically.

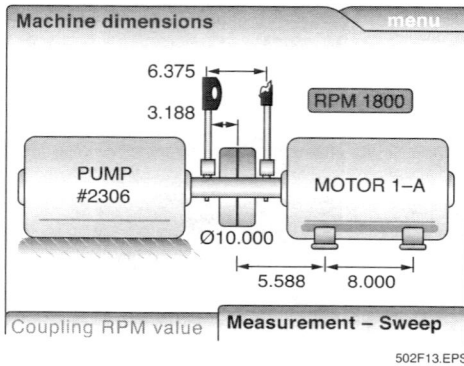

Figure 13 ◆ Entering dimensions.

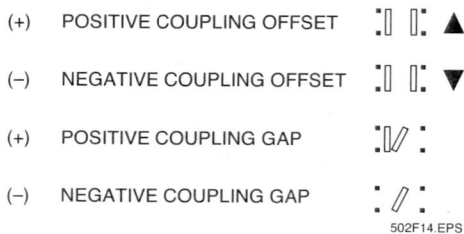

(+)   POSITIVE COUPLING OFFSET

(−)   NEGATIVE COUPLING OFFSET

(+)   POSITIVE COUPLING GAP

(−)   NEGATIVE COUPLING GAP

502F14.EPS

Figure 14 ◆ Coupling results.

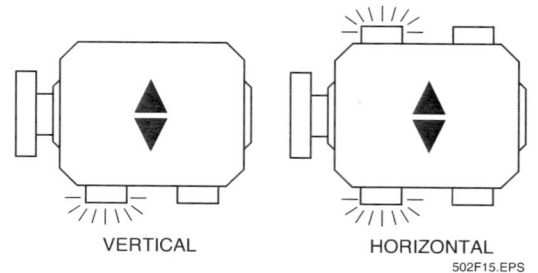

VERTICAL   HORIZONTAL

502F15.EPS

Figure 15 ◆ Move corrections.

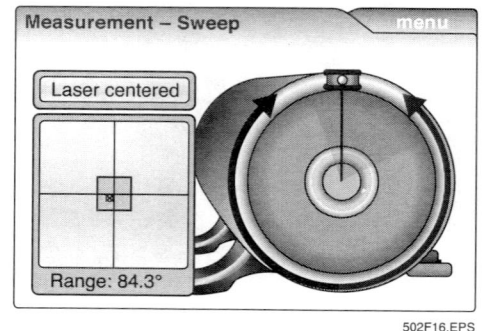

Figure 16 ◆ Measuring.

2.16 MILLWRIGHT ◆ LEVEL FIVE

Instructor's Notes:

Feet results – Horizontal

-12.8 mils   6.4 mils

0.0   0.0   13.5   23.7

502F17.EPS

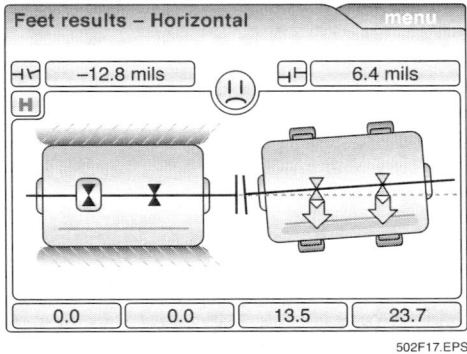

Figure 17 ◆ Reading results.

Step 4   Determine if the dimensions at all corners are within specification. If the dimensions are not within specification, move or shim the equipment the required amount. Always complete the shimming prior to making moves. After shimming, it is required to take another set of readings prior to moving the machine.

Step 5   Take another reading to confirm that the alignment is correct.

Soft foot is checked with the shaft at 90 degrees from vertical, that is, at 3 o'clock or 9 o'clock. Press the soft foot key and turn the shafts correctly. If needed, adjust the beam. Press ENT, press CLR, then loosen the bolt on the first foot, then press ENT again to enter the value, and tighten the bolt. Press the forward or back arrow key to move to the next foot.

7.1.4 Aligning Machinery Trains

Mark the shaft where brackets are to be mounted for trains. This alignment procedure is designed for more than two machines. The objective is to move the machines as little as possible. Draw a scale graphical plot of how much movement is available with each machine. This method avoids the problem of having to lower a machine with insufficient or zero shims and minimizes the chance of the machine becoming bolt-bound.

The vertical corrections are performed first and are rechecked before the horizontal adjustments are determined. This procedure is for zero-zero alignment, however, it can be used with intentional misalignment targets. Zero-zero alignment occurs when all the sight lines line up to no variation and the display shows zeroes at every placement.

If one machine is difficult to move, consider this machine as the STAT.

Follow these steps to align a machinery train:

Step 1   Draw the configuration of your machinery train on graph paper true to scale, and indicate the front and back feet of each machine.

Step 2   Use the machine on the far left as your base center line, and plot this as a horizontal line on your graph paper. The machine on the far left is machine A; the next one to the right is machine B.

Step 3   Mount the laser on machine A and the prism on machine B.

Step 4   Determine the position of machine B relative to machine A. This is setup AB.

Step 5   Plot on the graph paper the vertical corrections for the front and back feet of machine B relative to machine A.

Step 6   Draw a connecting line between these two points, and extend the line in both directions.

Step 7   Mount the laser on machine B and the prism on machine C.

Step 8   Determine the relative position of machine C to machine B. This is setup BC.

Step 9   Plot the vertical corrections for the front and back feet of machine C relative to machine B.

Step 10   Draw a connecting line between these two points, and extend the line in both directions.

Step 11   Repeat Steps 7 through 10 for each of the remaining machines. These are setups CD, DE, and so on.

Step 12   Draw a scale drawing of all the machines showing exactly how much movement is possible in all directions. This optimizes the bolt clearances and defines the adjustment limitations.

Step 13   Use the graph to determine the correct values for adjusting the machines to the ideal condition, and adjust the machines accordingly.

Step 14   Repeat Steps 1 through 11, while checking the misalignment at each coupling. Compare the vertical offsets and angularities with the permissible tolerances.

Show trainees how to set up the laser alignment equipment and check the initial alignment.

Have trainees practice setting up the laser alignment equipment and checking the initial alignment. This laboratory corresponds to Performance Tasks 3 and 4.

Assign reading of Section 7.1.4.

Ensure that you have everything required for teaching this session.

Explain the specific procedures for performing horizontal alignment using a laser.

> **NOTE**
>
> If the alignment is within tolerance, proceed to Step 15. If the alignment is not within tolerance, repeat Steps 12 and 13.

Step 15 Redraw the configuration of your setup of machinery on new graph paper true to scale, again indicating the front and back feet of each machine.

Step 16 Repeat Steps 2 through 11, but this time only plot the horizontal relationship between the machines.

Step 17 Draw a scale drawing of all the machines showing exactly how much movement is possible in all directions.

Step 18 Use the graph or rise/run calculations to determine the correct values for adjusting the machines to the ideal condition, and adjust the machines accordingly.

Step 19 Repeat Steps 2 through 11, while checking the misalignment at each coupling. Compare the horizontal offset and angularities with the permissible tolerances.

> **NOTE**
>
> If the tolerances are not correct, repeat Steps 17 through 19.

The Rotalign® lasers are capable of aligning trains of machinery on a single display screen.

7.1.5 Determining Targets

Alignment targets are defined as deliberate, initial misalignment specifications to compensate for differences in machine shaft positions between alignment conditions and operating conditions. Alignment targets are properly expressed as VO, HO, VA, and HA at a coupling center.

In order to specify targets, one of two data sets must be known. One is the machine growth characteristics supplied by the equipment manufacturer, which is based on a dimensional difference between average optimal running temperature and ambient temperature at rest. This is calculated based on given metal characteristics and is theoretical. The other data set is based on actual measurements made when the machine is cold and hot. These are more realistic, and simple to perform with the Optalign® system.

All known growth characteristics and values of these must be converted into offsets and angles at the coupling. These are entered into the Optalign® computer, which will then compensate for thermal growth when calculating the alignment.

Follow these steps to enter targets into the Optalign® computer.

Step 1 Determine the targets.

Step 2 Press the Target key.

Step 3 Enter the vertical offset target (VOt), vertical angularity target (VAt), horizontal offset target (HOt), and horizontal angularity target (HAt) one at a time, observing the proper sign, positive or negative.

Hot alignment readings must be taken as soon as possible after the shafts stop rotating, within 3 to 5 minutes. In order to achieve this, preparations must be made in advance and the readings must be taken by an experienced aligner. Although the term *hot readings* is commonly used, a more technically correct term is *operating readings* since the machines may heat or cool depending on the process.

Targets for cold alignment are cold conditions minus hot conditions. The actual cold temperature does not matter. However, alignment must be performed with the equipment stable at the cold temperature. If the machines are cooling or heating while the alignment is being done, the readings will probably be wrong.

7.1.6 Aligning Vertical Machines

Because the lateral alignment can change very easily when shimming, the millwright must first make shim corrections to correct angular misalignment and then measure the alignment once more before proceeding with horizontal move corrections to correct the offset misalignment.

Vertical machines are aligned using the special function F5. For this procedure, a 0-degree (12 o'clock) reference point is chosen. The angular positions are based on this reference. *Figure 18* shows aligning vertical machines.

The machine dimensions needed to align vertical machines are as follows:

- Distance from laser to prism
- Distance from laser to coupling center
- Greatest diameter (D_{max}) between two fastening bolts in the shim plane

Follow these steps to align vertical machines:

Step 1 Turn on the system. Press the Vertical Coupling key.

Instructor's Notes:

_____

_____

_____

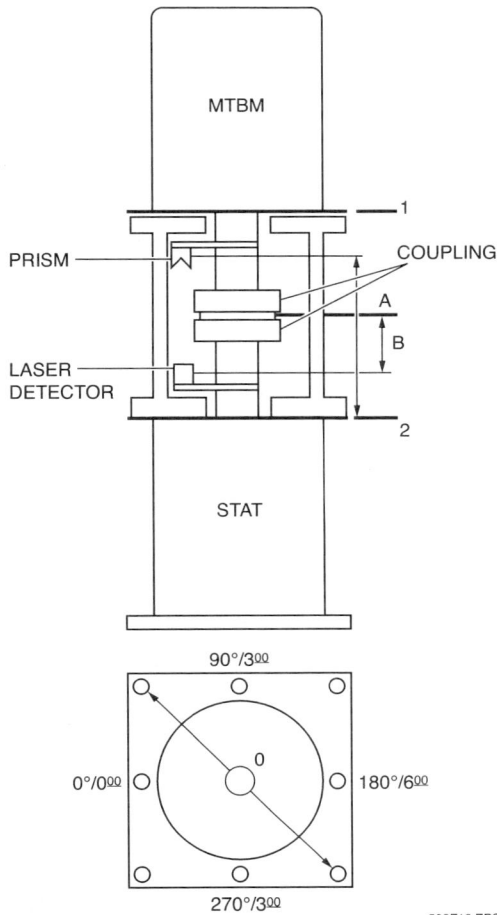

Figure 18 ◆ Aligning vertical machines.

502F18.EPS

Step 2 Measure the machine dimensions listed above and enter them under the DIM key.

Step 3 Enter the laser-to-prism distance.

Step 4 Enter the laser-to-coupling center distance.

Step 5 Enter the laser-to-front-foot distance as the distance from the laser to the coupling center.

Step 6 Enter the maximum bolt circle diameter and the angular position of each fastening bolt using the Arrow and ENT keys. Repeatedly pressing the ENT key causes the corrective value for each bolt to be displayed automatically.

Step 7 You will need to determine which fastener bolt is in the 0 or 12 position. Take measurements in at least three of the eight positions.

Step 8 Press the Result key to activate the vertical shimming function. If the + sign appears, insert shims; if the – sign appears, remove shims.

NOTE

The computer internally calculates the relative position of the MTBM to the STAT on the entire circumference where the greatest distance must be lowered. This amount is later added to the respective corrective value so that negative corrective values are eliminated. If lowering is desired because the entire machine would otherwise stand too high, zero must be entered as D_{max}.

The angle entered first is used as a default increment for the other angles. With careful selection of the 12 o'clock position, entry of each successive angle occurs automatically.

Step 9 Shim the machinery where the computer specifies.

Step 10 Take alignment measurements in at least three of the eight positions, and repeat the vertical shimming procedure until angularity is within tolerance. Once the results indicate that angularity is within tolerance, proceed with horizontal moves.

Step 11 Press the Results key twice to view the horizontal corrections of the MTBM.

As in all alignments with Optalign®, the coupling function can be used to enter coupling target specifications and determine compliance with tolerances.

8.0.0 ◆ TROUBLESHOOTING

When difficulty is encountered by trained personnel using Optalign®, the problems generally fall into one of several classes. These are, in order of decreasing probability, as follows:

- Machinery defects in the equipment being aligned
- Incorrect use or installation of brackets
- Defect or failure in the Optalign® system

Show Transparency 17 (Figure 18). Point out the required dimensions.

Show trainees how to perform a vertical alignment using a laser.

Have trainees practice performing a vertical alignment using a laser. This laboratory corresponds to Performance Task 5.

Assign reading of Section 8.0.0.

Ensure that you have everything required for teaching this session.

Review the types of problems encountered during laser alignments.

Discuss the character-
istics of each type of
machinery defect, and
point out remedies for
each.

Review the guidelines
for proper bracket in-
stallation.

8.1.0 Machinery Defects

Given that the Optalign® is operating correctly and the brackets are correct, the measurement repeatability problems are reflections of machinery conditions. Some common sources of repeatability are as follows:

- Soft feet
- Coupling strain
- Externally induced strains
- Loose bearings
- Other sources of repeatability

8.1.1 Soft Feet

The strains that soft feet create on the machinery frames, bearings, and rotors have a variety of negative effects on the shaft center line. If a machine is bolted tight with a soft foot, it will alter alignment accuracy severely.

8.1.2 Coupling Strain

Under misaligned conditions, couplings can transmit strains to the shaft ends. These forces bend the shaft very slightly. When the laser and prism are rotated, they ride around with the bend. If a center of rotation for the prism or laser were drawn, it would not pass through the bearing centers. Instead, it would follow the tangent of the shaft bend at the point at which the prism was mounted. Because the computer has no way of knowing that the shafts are under strain, it computes the misalignment between the tangent lines that the laser and prism rotated about. Even with small shaft strains, by the time the center line projection, or tangent, reaches the back feet, the error can be well over 100 mils.

If coupling strain is suspected, unbolt the coupling and take new readings. If the readings differ from those taken when the coupling was bolted up, use the unbolted values until the shafts are almost aligned; then bolt up the coupling and continue. The coupling helps eliminate backlash, which in turn improves accuracy.

8.1.3 Externally Induced Strains

The effects of external equipment, pipe, or braces straining either the MTBM or STAT are unpredictable and can be surprising. Eliminate all externally induced strains.

8.1.4 Loose Bearings

The effects of loose bearings on alignment readings can be partially eliminated by always rotating to each measuring position in the same direction. If you overshoot, back all the way up and come to the point from the original direction a second time.

8.1.5 Other Sources of Repeatability

Worn, eccentric, or damaged gears, thermal expansion or contraction caused by temperature changes between readings, a weak or loose baseplate, or a bearing out of roundness or otherwise damaged can also cause repeatability problems.

8.2.0 Incorrectly Installed Brackets

Incorrectly installed brackets are the most common source of repeatability problems. This comes from either lack of training, carelessness, or, in some cases, broken or defective brackets. This section covers guidelines to remember when installing or checking bracket installation.

The risers must be rigidly mounted on the rotor or shaft. Be certain that all looseness or wobble is removed from the brackets. Chain brackets should be installed firmly and snugly. Magnetic brackets should be clean and mounted on reasonably clean, flat surfaces. If the surface is badly damaged by hammer marks or wrench marks, the magnetic bracket can still be used, but take care to avoid bracket wobble.

The brackets must touch the rotor only at their designed surfaces. Chain brackets and sides should never touch any part of the rotor. All four points of the V should touch the shaft at one time without wobble, even before the chain is tight.

The brackets and everything they support must not touch or rub on any stationary part during rotation. Clip the cable to the riser near the base of the bracket. The cable between the clip and the laser must have some small amount of slack. Ensure that the chain and cable do not hang up on anything and watch that the knob clears the baseplate near the 6-o'clock position. Also ensure that risers do not rub the bearing housing bolts or other obstructions as they turn.

Ensure that risers and the laser and prism houses are free of grit. Accumulated grit, grease, dirt, and other buildup can cause the laser or prism to rock on the risers.

Install the brackets with the risers at the same angle. Eliminate backlash and keep the brackets synchronized at each reading point.

2.20 MILLWRIGHT ◆ LEVEL FIVE

Instructor's Notes:
_____
_____
_____

8.3.0 System Failure or Defect

Although on any given alignment the least likely thing to go wrong is the Optalign® system electronics, the inexperienced Optalign® user often suspects this source of trouble first. Like most semiconductor-based electronics, Optalign® for the most part either measures very accurately or it does not measure at all. Follow these steps to determine if your Optalign® is working correctly:

Step 1   Ensure that the computer and laser are properly connected.

Step 2   Turn on the computer.

Step 3   Compare the serial numbers of the computer and laser to ensure that they match. If no serial number is displayed on the computer, check the batteries. If they are good, call Ludeca for service.

Step 4   Press the / key (inch mode); then press the M key. The computer screen should display OFF.

Step 5   Ensure that red LED on the laser housing is blinking, indicating that the laser is powered. If the LED is not blinking, check the cable, paying special attention to the end connectors. Replace the cable with the spare if necessary. If the LED still does not blink, call Ludeca for service. The laser/detector is not user-serviceable and can be rendered unrepairable if it is incorrectly disassembled.

Step 6   Set up the laser and prism on a straight shaft, such as the riser storage tube, with the laser-to-prism distance less than 6 inches.

Step 7   Press the M key to zero the beam.

> **NOTE**
> If the beam zeros, the red LED on the laser housing will be blinking. Proceed to Step 12. If it does not zero, proceed to Step 8.

Step 8   Place the beamfinder over the detector portion of the laser/detector so that it is facing the incoming beam.

Step 9   Adjust the prism until the beamfinder shows the beam centered in the detector. If you cannot find the beam, using the beamfinder, call Ludeca for service.

Step 10   Remove the beamfinder. Both red and green lights on the laser housing should be blinking. If they are not, verify that the beam is centered in the detector, and call Ludeca.

Step 11   Turn the prism thumbwheel slowly to zero the numbers on the Optalign®. If numbers do not appear or the beam cannot be zeroed, call Ludeca.

Step 12   Press the H/V key when the display reads 0.0 0.0, ensuring that you do not touch, jiggle, or vibrate the laser-prism-shaft system.

Step 13   Record the displayed value, and watch the numbers for approximately 2 to 3 minutes. If the display varies by more than 0.2 mils, call Ludeca.

If you have not called Ludeca by this point, your Optalign® measures correctly. It is still possible that a given key does not work or some part of the display is burned out. These types of failures, although hindering or barring use of Optalign®, do not affect accuracy or stability.

An alternate and effective substitute test procedure is to place Optalign® into soft foot mode and watch it for stability. Do not move the equipment or loosen any bolts. Watch the Optalign® soft foot reading to ensure that it remains from 0.0 to 0.5 mils for 2 to 3 minutes.

8.3.1 Damp Keyboard or Screen

If the computer appears to respond as though you pressed a different key from the one actually pressed or the wrong number appears on the screen, it is often due to water in the keyboard or screen. Place the computer in a welding rod box, small cabinet with a light bulb, or some other similar device that holds ambient temperature below 140°F to dry it out. If the problem does not disappear after about 6 hours of drying, call Ludeca.

8.3.2 Numbers to OFF without END

On occasion, while taking measurements, the beam goes from the rough coordinate numbers to OFF without displaying END. This is caused by the beam reaching the edge of the prism. It is not reflecting back to the detector. A beamfinder can be used to prove that the condition exists. To correct the problem, adjust the brackets and start the measurements over.

Describe the procedures for determining if the Optalign® system is working properly.

Review Questions

1. For machinery installation, only the _____ of different machines are aligned.
   a. feet
   b. shaft surfaces
   c. bearings
   d. rotating shaft center lines

2. A condition in which tightening or loosening bolt(s) of a single foot distorts the machine frame is _____.
   a. wrong foot
   b. serial foot
   c. shaking feet
   d. soft foot

3. Bent foot can be corrected using _____.
   a. step shims
   b. a hammer
   c. extra grout
   d. welding

4. A condition that shows up as soft foot on the laser but in which few or no feeler gauges can be inserted is called _____.
   a. flat foot
   b. shot foot
   c. moist foot
   d. squishy foot

5. As a standard rule, no more than _____ shims should be used under one foot.
   a. two
   b. four
   c. six
   d. eight

6. External forces on a machine create a condition called _____ soft foot.
   a. indicated
   b. internal
   c. extruded
   d. induced

7. The expansion of materials caused by a rise in temperature is called thermal _____.
   a. sizing
   b. growth
   c. seizure
   d. current

8. Incorrect pipe weldment or pipe hangers, defective anchor bolts, and bad bearings or foundations cause _____.
   a. rotational noise
   b. vibration
   c. coupling stress
   d. bent shafts

9. An amplified and collimated beam of coherent light energy that travels perfectly straight is a _____.
   a. flashlight
   b. welding arc
   c. laser beam
   d. quartz halogen light

10. All lasers, because of the _____ beam energy, can damage the eye irreparably.
    a. red, pulsing
    b. hot, continuous
    c. collimated, intense
    d. accurate, quick

11. The Optalign® laser beam is a class _____ beam.
    a. I
    b. II
    c. III
    d. IV

12. Actual misalignment is virtually always a combination of _____ misalignment.
    a. inside and outside
    b. vibrational and rotational
    c. seasonal and frictional
    d. angularity and offset

13. When the laser prism is moved up or down, the return beam moves _____ the distance the prism moved.
    a. half
    b. three times
    c. twice
    d. one tenth

Instructor's Notes:

_____

_____

_____

14. Before laser alignment is attempted, the coupling offset should be less than _____.

    a. ³⁄₃₂ inch
    b. 0.010 inch
    c. 8 mm
    d. 15 degrees

15. Even with small shaft strains, by the time the center line projection reaches the back feet, the error can be _____.

    a. as little as 100th of an inch
    b. less than ³⁄₃₂ inch
    c. almost 50 mils
    d. more than 100 mils

MODULE 15502-09 ◆ LASER ALIGNMENT   2.23

Summarize the major concepts presented in the module.

Administer the Module Examination. Be sure to record the results of the Examination on Craft Training Report Form 200, and submit the results to the Training Program Sponsor.

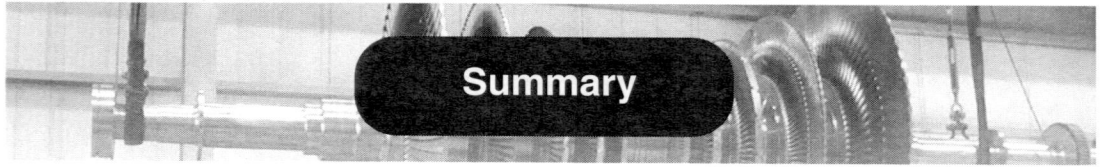

Administer the Performance Tests, and fill out Performance Profile Sheets for each trainee. If desired, trainee proficiency noted during laboratory sessions may be used to complete the Performance Test. Record the results on Craft Training Report Form 200, and submit the results to the Training Program Sponsor.

Summary

The goal of any alignment system is to align the rotating center lines of connected machines as closely as possible. The closer the rotational centers, the more energy is saved, wear is avoided, and trouble-free operation is achieved. Methods for achieving this close alignment vary, but the laser alignment coupled with the power of a computer can give results equal to those of the best reverse-indicator alignment and in far less time.

The versatility of laser alignment lends itself to other tasks, such as measuring shaft sag, thermal growth, or aligning long shafts, that indicator methods cannot reach. The laser system can only be as good as the procedures followed; therefore, like anything that makes a millwright's work easier, care and thoroughness are required for consistent results.

Notes

2.24 MILLWRIGHT ◆ LEVEL FIVE

Instructor's Notes:

_____

_____

_____

Trade Terms Introduced in This Module

Coherent: Composed of only one color or wavelength and all the waves in phase with each other.

Collimate: To make straight and parallel.

Laser: A device that amplifies and concentrates coherent light waves in an intense beam of parallel, nonscattering energy. The word *laser* is an acronym for Light Amplification by Stimulated Emission or Radiation.

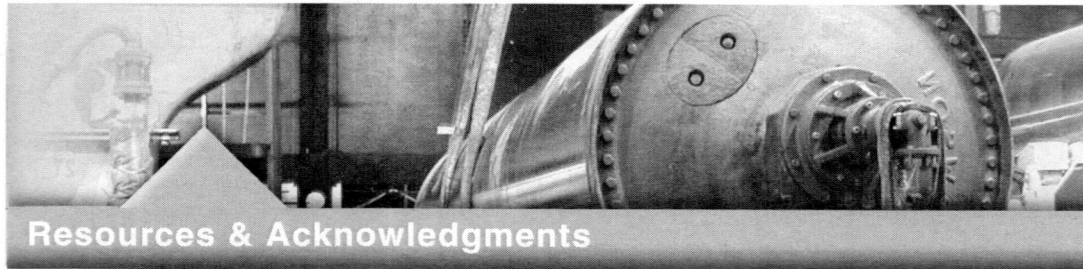

Resources & Acknowledgments

Additional Resources

This module is intended to be a thorough resource for task training. The following reference work is suggested for further study. This is optional material for continued education rather than for task training.

The Optalign Training Book. Galen Evans and Pedro Casanova, Miami, FL: Ludeca, Inc.

Figure Credits

Photo courtesy of Ludeca, Inc. – www.ludeca.com, 502F05 (photos), 502F11(left), 502F13, 502F16, 502F17

Ed LePage, 502F11 (right)

Instructor's Notes:

_____

_____

_____

The following are suggested activities or instructional methods to help you teach the material in this module.

General

When you call on someone to answer a question, the rest of the class relaxes or even tunes out because they expect that the question and answer will take place only between you and the trainee you called on. Instead, use this technique to involve more trainees in answering questions and to keep them on their toes.

1. Ask trainees to define a term or explain a concept.
2. After one trainee has answered, ask a trainee seated nearby if the answer is right. Then ask whether a trainee in the back of the room agrees.
3. Ask trainees to explain why they think an answer is right or wrong.
4. Use the session to clear up incorrect ideas and encourage trainees to learn from their mistakes.

Section 2.0.0 *Types of Soft Foot*

This Quick Quiz will familiarize trainees with laser diagnostics for types of soft foot. You will need photocopies of the quiz provided on the following page. Trainees will need pencils. If you allow trainees to use the Trainee Guide, decrease the amount of time you give them to complete the quiz, and remind them to bring their books to class.

1. Make a photocopy of the quiz for each trainee.
2. Give trainees between 5 and 10 minutes to complete the quiz.
3. Go over the answers to the quiz.
4. Ask trainees if they have questions.

Answers to Quick Quiz

1. c
2. a
3. d
4. b
5. e

Quick Quiz   ***Soft Foot***

Identify the types of soft foot that correspond to the soft foot symptoms. From the list below, write the corresponding letter in the space provided.

Symptoms

_____ 1. Shows as a soft foot measurement on the laser, but when measured with a feeler gauge, little or no gap is found.

_____ 2. Found as two diagonally opposed readings.

_____ 3. Occurs as two soft feet on the same side of the machine, and the feeler gauge indicates a gap that is parallel or nearly parallel.

_____ 4. Appears as three readings being the same and one reading being different.

_____ 5. One foot shows a visible gap before tightening the bolt, but no soft foot reading after bolts have been tightened.

Types of Soft Foot

a. Parallel air gap
b. Bent foot
c. Squishy foot
d. Induced soft foot
e. Gaps without soft foot

Section 6.0.0 *Offset and Angularity*

This Quick Quiz will familiarize trainees with the designations for offset and angularity directions. You will need photocopies of the quiz provided on the following page. Trainees will need pencils. If you allow trainees to use the Trainee Guide, decrease the amount of time you give them to complete the quiz, and remind them to bring their books to class.

1. Make a photocopy of the quiz for each trainee.
2. Give trainees between 5 and 10 minutes to complete the quiz.
3. Go over the answers to the quiz.
4. Ask trainees if they have questions.

Answers to Quick Quiz

1. b
2. e
3. h
4. a
5. f
6. g
7. c
8. d

Offset and Angularity

Identify the locations of coordinates that correspond to the readings of angularity and offset. In the space provided, write the corresponding letter selected from the list below.

Readings of Angularity and Offset

_____ 1. Offset is shown as +x–y

_____ 2. VA is shown as positive

_____ 3. HA is shown as negative

_____ 4. Offset is shown as +x+y

_____ 5. HA is shown as positive

_____ 6. VA is shown as negative

_____ 7. Offset is shown as –x–y

_____ 8. Offset is shown as –x+y

Locations of Coordinates

a. Offset between 12 o'clock and 3 o'clock

b. Offset between 3 o'clock and 6 o'clock

c. Offset between 6 o'clock and 9 o'clock

d. Offset between 9 o'clock and 12 o'clock

e. Gap is open at 12 o'clock

f. Gap is open at 3 o'clock

g. Gap is open at 6 o'clock

h. Gap is open at 9 o'clock

MODULE 15502-09 — ANSWERS TO REVIEW QUESTIONS

Answer		Section
1.	d	1.0.0
2.	d	2.0.0
3.	a	2.2.0
4.	d	2.3.0
5.	b	2.3.0
6.	d	2.4.0
7.	b	3.0.0
8.	c	4.1.0
9.	c	5.0.0
10.	c	5.1.0
11.	b	6.0.0
12.	d	6.1.0
13.	c	7.0.0
14.	a	7.1.1
15.	d	8.1.2

The NCCER makes every effort to keep these textbooks up-to-date and free of technical errors. We appreciate your help in this process. If you have an idea for improving this textbook, or if you find an error, a typographical mistake, or an inaccuracy in NCCER's Contren® textbooks, please write us, using this form or a photocopy. Be sure to include the exact module number, page number, a detailed description, and the correction, if applicable. Your input will be brought to the attention of the Technical Review Committee. Thank you for your assistance.

Instructors – If you found that additional materials were necessary in order to teach this module effectively, please let us know so that we may include them in the Equipment/Materials list in the Annotated Instructor's Guide.

Write: Product Development and Revision
National Center for Construction Education and Research
3600 NW 43rd St., Bldg. G, Gainesville, FL 32606

Fax: 352-334-0932

E-mail: curriculum@nccer.org

Craft	Module Name	
Copyright Date	Module Number	Page Number(s)

Description

_____

_____

_____

_____

(Optional) Correction

_____

_____

_____

(Optional) Your Name and Address

_____

_____

_____

Advanced Blueprint Reading

NCCER STANDARDIZED CRAFT TRAINING PROGRAM

The National Center for Construction Education and Research (NCCER) provides a standardized national program of accredited craft training. Key features of the program include instructor certification, competency-based training, and performance testing. The program provides trainees, instructors, and companies with a standard form of recognition through a National Craft Training Registry. The program is described in full in the *Guidelines for Accreditation*, published by NCCER. For more information on standardized craft training, contact the NCCER by writing us at 3600 NW 43rd St., Bldg. G, Gainesville, FL 32606; calling 352-334-0911; or emailing info@nccer.org. More information may be found at our website, www.nccer.org.

HOW TO USE THIS ANNOTATED INSTRUCTOR'S GUIDE

Each page presents two sections of information. The larger section displays each page exactly as it appears in the Trainee Module. The narrow column ties suggested trainee and instructor actions to each page and provides icons (detailed below) to call your attention to material, safety, audiovisual, or testing requirements. The bottom of each page includes space for your notes.

The **Audiovisual** icon indicates an appropriate time to show a transparency or other audiovisual aid.

The **Classroom** icon prompts you to define a term, stress a point, ask trainees to explain a concept, or give examples.

The **Demonstration** icon directs you to show trainees how to perform tasks.

The **Examination** icon tells you to administer the written module examination.

The **Homework** icon is placed where you may wish to assign reading for the next session, assign a project, or advise trainees to prepare for an examination.

The **Laboratory** icon is used when trainees are to practice performing tasks.

The **Materials** icon is a reminder for you to gather materials needed for classes, labs, and testing.

The **Performance Testing** icon tells you to administer a performance test or a portion thereof.

The **Safety** icon is used to emphasize safety issues. It is often keyed to *Caution* and *Warning!* statements in the Trainee Module.

The **Teaching Tip** icon indicates additional guidance is available, such as how to conduct an exercise, get the most educational value from a field trip, or encourage class participation. Teaching Tips may expand on a feature (*Think About It*, *Did You Know?*) or provide *Quick Quizzes* or similar exercises. You will be referred to the Teaching Tips section at the back of the module if there is additional material.

The **Combination** icon indicates that the laboratory listed corresponds with a performance task. If desired, you can note the proficiency of the trainees during the laboratory, and use it to satisfy performance testing requirements.

PREPARATION

Before teaching this module, you should review the Objectives, Performance Tasks, Materials and Equipment List, and Module Outline. Be sure to allow ample time to prepare your own training or lesson plan and gather all required materials and equipment.

MODULE OVERVIEW

This module builds on the skills developed in earlier training, providing the millwright with the information needed to determine the specific machine and parts required for a repair. Various facets of advanced blueprint reading are covered, including numbering systems, drawing hierarchy, machine drawing information, and drawing system usage and practices.

PREREQUISITES

Prior to training with this module, it is recommended that the trainee shall have successfully completed *Core Curriculum; Millwright Level One; Millwright Level Two; Millwright Level Three; Millwright Level Four;* and *Millwright Level Five,* Modules 15501-09 and 15502-09.

OBJECTIVES

Upon completion of this module, the trainee will be able to do the following:

1. Explain the use of a drawing numbering system.
2. Identify the types of drawings in a drawing package.
3. Read and interpret plant or foundation layout drawings.
4. Read and interpret assembly drawings.
5. Read and interpret detail drawings.
6. Identify and explain the parts of a machine drawing.
7. Locate individual components on a plant layout.
8. Locate an assembly drawing using a detail part.

PERFORMANCE TASKS

Under the supervision of the instructor, the trainee should be able to do the following:

1. Find detail drawings using assembly drawings.
2. Find assembly drawings using detail drawings.
3. Use a bill of materials to perform a materials takeoff.

MATERIALS AND EQUIPMENT LIST

Overhead projector and screen

Transparencies

Blank acetate sheets

Transparency pens

Whiteboard/chalkboard

Markers/chalk

Pencils and scratch paper

Set of drawings to show hierarchy

Samples of various drawing types

Sketches of parts with different types of dimensioning

Detail drawings

Assembly drawings

Bill of materials

Copies of the Quick Quizzes*

Module Examinations**

Performance Profile Sheets**

* Located at the back of this module.

** Located in the Test Booklet.

SAFETY CONSIDERATIONS

Ensure that the trainees are equipped with appropriate personal protective equipment and know how to use it properly.

ADDITIONAL RESOURCES

This module is intended to present thorough resources for task training. The following reference work is suggested for both instructors and motivated trainees interested in further study. This is optional material for continued education rather than for task training.

Geometrics II, The Application of Geometric Tolerancing Techniques. Lowell Foster. Reading, MA: Addison-Wesley Publishing Co., 1986.

TEACHING TIME FOR THIS MODULE

An outline for use in developing your lesson plan is presented below. Note that each Roman numeral in the outline equates to one session of instruction. Each session has a suggested time period of 2½ hours. This includes 10 minutes at the beginning of each session for administrative tasks and one 10-minute break during the session. Approximately 25 hours are suggested to cover *Advanced Blueprint Reading*. You will need to adjust the time required for hands-on activity and testing based on your class size and resources. Because laboratories often correspond to Performance Tasks, the proficiency of the trainees may be noted during these exercises for Performance Testing purposes.

Topic	Planned Time
Sessions I and II. Introduction; Numbering System; Drawing Hierarchy	
A. Introduction	_____
B. Numbering System	_____
C. Drawing Hierarchy	_____
D. Laboratory	_____
Have trainees practice identifying types of drawings from examples.	
Sessions III and IV. Drawing Information	
A. Lines	_____
B. Dimensions	_____
C. Notes and Symbols	_____
D. Scale	_____
E. Revisions	_____
F. Vendor Information	_____
G. Material Specifications	_____
H. Laboratory	_____
Have trainees practice reading various types of drawings.	
Sessions V–VIII. Drawing System Usage	
A. Finding Details	_____
1. Laboratory	_____
Trainees find detail drawings using assembly drawings. This laboratory corresponds to Performance Task 1.	
B. Finding Assembly Drawings	_____
1. Laboratory	_____
Have trainees find assembly drawings using detail drawings. This laboratory corresponds to Performance Task 2.	

Session IX. Materials Takeoff

    A. Bill of Materials   _____

       1. Laboratory   _____

          Have trainees use a bill of materials to perform a materials takeoff. This laboratory corresponds to Performance Task 3.

Session X. Review and Testing

    A. Trade Terms Quick Quiz   _____

    B. Review   _____

    C. Module Examination   _____

       1. Trainees must score 70% or higher to receive recognition from NCCER.

       2. Record the testing results on Craft Training Report Form 200, and submit the results to the Training Program Sponsor.

    D. Performance Testing   _____

       1. Trainees must perform each task to the satisfaction of the instructor to receive recognition from NCCER. If applicable, proficiency noted during laboratory exercises can be used to satisfy the Performance Testing requirements.

       2. Record the testing results on Craft Training Report Form 200, and submit the results to the Training Program Sponsor.

15503-09

Advanced
Blueprint Reading

Assign reading of Module 15503-09.

15503-09
Advanced Blueprint Reading

Topics to be presented in this module include:

Overview

DIAP		DIAPHRAGM
KNFE		KNIFE GATE
SLDE		SLIDE
PNCH		PINCH
RTRY		ROTARY
DLGE		DELUGE
DRYP		DRY PIPE
PACT		PREACTION
RELF		AIR RELIEF

Blueprint reading is a basic skill for millwrights. This module explains how to find a specific machine in a drawing set as well as information on parts needed for repairs. The different drawings in a drawing package are examined and explained. Trainees will learn how to follow the drawing hierarchy and interpret the information presented on these drawings.

Instructor's Notes:

_____

_____

_____

_____

Objectives

Upon completion of this module, you will be able to do the following:

1. Explain the use of a drawing numbering system.
2. Identify the types of drawings in a drawing package.
3. Read and interpret plant or foundation layout drawings.
4. Read and interpret assembly drawings.
5. Read and interpret detail drawings.
6. Identify and explain the parts of a machine drawing.
7. Locate individual components on a plant layout.
8. Locate an assembly drawing using a detail part.

Trade Terms

Balloon
Computer-aided drafting (CAD)
Computer-aided engineering (CAE)
Computer-aided machining (CAM)
Computer-integrated manufacturing (CIM)

Datum
Dimension
Drawing package
Machined parts
Purchased parts
Revision
Title block

Required Trainee Materials

1. Pencil and paper
2. Appropriate personal protective equipment

Prerequisites

Before you begin this module, it is recommended that you successfully complete *Core Curriculum; Millwright Level One; Millwright Level Two; Millwright Level Three; Millwright Level Four;* and *Millwright Level Five*, Modules 15501-09 and 15502-09.

This course map shows all of the modules in the fifth level of the Millwright curriculum. The suggested training order begins at the bottom and proceeds up. Skill levels increase as you advance on the course map. The local Training Program Sponsor may adjust the training order.

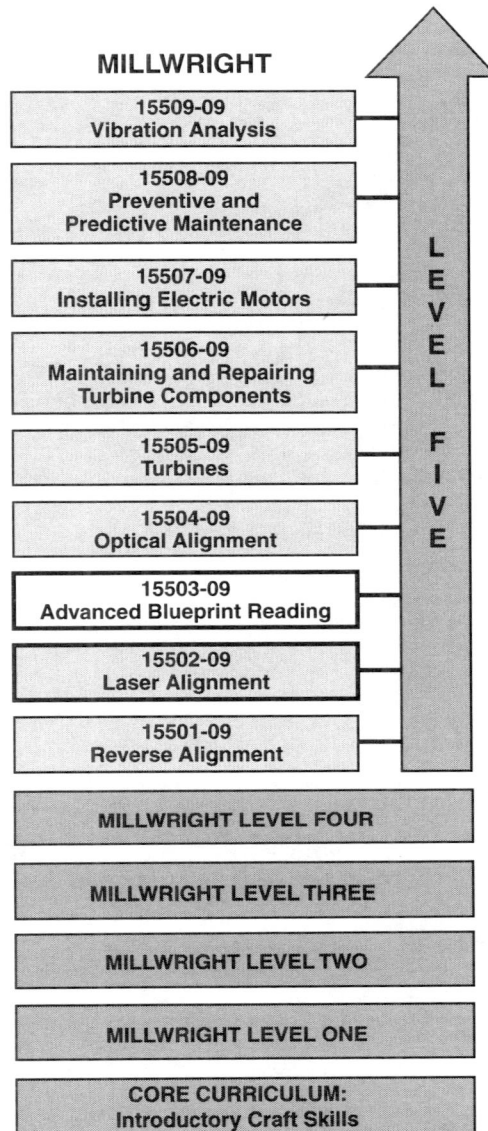

MILLWRIGHT

LEVEL FIVE

15509-09 Vibration Analysis
15508-09 Preventive and Predictive Maintenance
15507-09 Installing Electric Motors
15506-09 Maintaining and Repairing Turbine Components
15505-09 Turbines
15504-09 Optical Alignment
15503-09 Advanced Blueprint Reading
15502-09 Laser Alignment
15501-09 Reverse Alignment

MILLWRIGHT LEVEL FOUR

MILLWRIGHT LEVEL THREE

MILLWRIGHT LEVEL TWO

MILLWRIGHT LEVEL ONE

CORE CURRICULUM: Introductory Craft Skills

503CMAP.EPS

Ensure that you have everything required to teach the course. Check the Materials and Equipment List at the front of this module.

See the general Teaching Tip at the end of this module.

Explain that terms shown in bold are defined in the Glossary at the back of this module.

Show Transparency 1, Objectives, and Transparency 2, Performance Tasks. Review the goals of the module, and explain what will be expected of the trainees.

Review the modules covered in Level Five, and explain how this module fits in.

See the Teaching Tip for Sections 1.0.0–6.0.0 at the end of this module.

Explain why it is necessary for millwrights to understand engineering drawings and design drafting.

Discuss computer-aided drafting.

Show Transparency 3 (Figure 1). Describe how to read the number tree.

See the Teaching Tip for Sections 1.0.0–6.0.0 (Sources of Drawings) at the end of this module.

Explain how drawing hierarchy reflects an increasingly specific level of detail.

1.0.0 ◆ INTRODUCTION

The purpose of engineering drawings is to communicate information. If the drawing does not contain complete, detailed information, or if the person reading the drawing does not have the necessary skills to interpret it, the communication fails. Therefore, it is important for millwrights to understand all aspects of design drafting.

Computer-aided drafting (CAD) and all the variations of computer assistance have made hand-drawn blueprints virtually obsolete. The growing power of digital information and of the CAD interface with engineering, manufacturing, and integral process control demands standardization of this information exchange. Throughout this module, we refer to the information as drawings, but as a rule, most information is already on CAD, and some plants work entirely from computer workstations.

2.0.0 ◆ NUMBERING SYSTEM

A variety of numbering systems are used to number drawings. Each system is an almost infinite combination of numbers and letters, often specific to one company. Even within companies, the sequence and logic of the system can be confusing. The purpose of a numbering system is to make it easy to find any drawing and relate all of its parts to the whole machine.

There is no set rule on how a numbering system should be set up, but the intention is that the numbers and letters run in sequence and convey some relation of the detail, subassembly, and main assembly drawings so that anyone can find information that relates one to the other. In general, a numbering system starts at the main assembly, such as D-998-0001, and the subassemblies and details branch off from there (*Figure 1*).

The best method for interpreting a numbering system is to read through the number log where drawing numbers are assigned and logged in, whether it is in a notebook or on a computer. This should give an idea of the logical sequence and make it easier to find details that are branched from the assembly.

Another source for finding drawings related to each other is the bill of materials, which may be on the main assembly drawing, on a separate drawing, or may be a stand-alone document. When searching through a drawing system, always take notes and make your own path to trace where each item is located in the system.

3.0.0 ◆ DRAWING HIERARCHY

The level of detail and type of information is specialized on each type of drawing. The information on the layout is very general and gets increasingly more specific in the drawings that follow. For example, to replace a pillow block mount would require only the detail drawing of the mount and the bill of materials to find the pillow block part number.

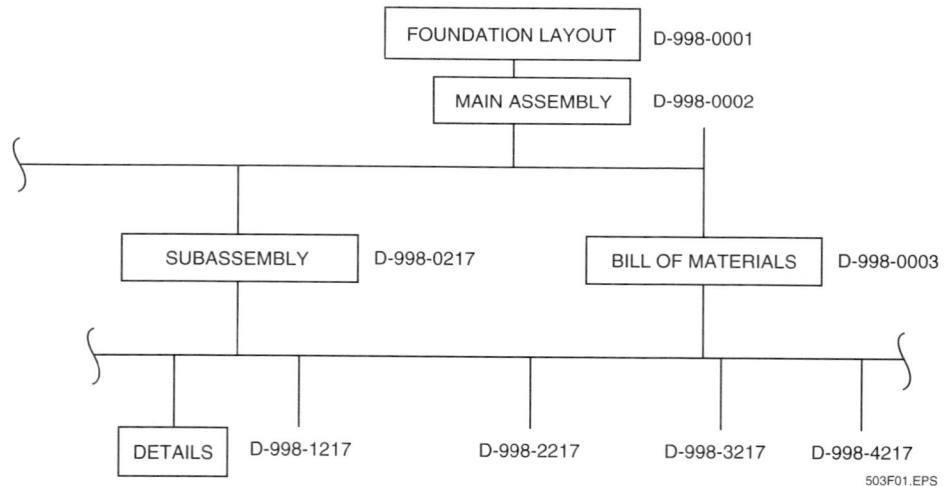

Figure 1 ◆ Drawing number tree.

Instructor's Notes:

_____

_____

_____

A new piece of equipment to be located near or connected to existing production lines would require the plant or foundation layout, the assembly drawings of the existing equipment, and details of areas that would interface with the new equipment. Each type of drawing is important to the drawing system and provides an accurate history of plant development as long as the drawings are kept up to date.

The organization of the **drawing package**, in descending order, is as follows:

- Plant or foundation layout drawing(s)
- Main assembly drawings
- Subassembly drawings
- Bill of materials
- Detail drawings

3.1.0 Plant or Foundation Layout Drawings

Plant or foundation layout drawings are the first details of the manufacturing area and are subject to many changes both in the construction stage, and later when equipment is replaced or re-designed. If the drawings are not kept up to date when **revisions** are made, they can provide misleading information and cause major problems.

A foundation layout (*Figure 2*) details the inside **dimensions**, structural column locations, and mounting pad locations for baseplates. These dimensions are necessary for machinery installation, running utility service drops, and all future changes in the equipment.

The foundation layout drawing shown shows overall dimensions, pad-to-pad dimensions, the equipment center line location, and structural column locations. Notes pertaining to the pads are in the bottom left; the scale in the **title block** shows 1":48", which means 1" = 48"; and the match line on the right-hand side means that Sheet 3 has more information that locates off the center line.

3.2.0 Assembly Drawings

An assembly drawing does not have enough dimensions to make the parts, but refers to, lists, and shows drawings that do. Assembly drawings provide the total picture of a machine or process, showing all the parts and the location of each. On large equipment that has many assemblies, there is usually a main assembly drawing that has the subassemblies **ballooned** out, which in turn have the details ballooned and referenced to the bill of materials. Individual assembly drawings may have to be spread out on several pages and have

the detail parts ballooned with letters or numbers that relate to the bill of materials. All machined and **purchased parts** are shown as they fit together, and other views illustrating adjustments, options, or process details can be added. The primary rule for assembly drawings is that they must be clear and concise. The hidden lines and some object lines may be omitted for clarity and noted. Views from any perspective or scale can be added to show details.

Subassembly drawings break down the assembly into more understandable segments. They usually have more detail or are in a larger scale. Individual processes or special items can be detailed here, and balloons can be used to reference the detail drawings on the bill of materials.

3.3.0 Bill of Materials

A bill of materials (*Figure 3*) is a list of all the parts, components, and assemblies that make up the whole machine or assembly. This list generally has a number or letter code for locating the part on the assembly drawing. The parts of a machine are generally divided into purchased parts and machined or modified parts. The purchased parts are referenced by part numbers, and the **machined parts** by assigned drawing numbers. Other information that may be included in the bill of materials are special processes, the type of material to be used, quantities, descriptions, and other reference information.

In practice, a millwright should be able to gather all the items in the bill of materials, assemble them, and finish with a complete machine or line of equipment. As with the detail and assembly drawings, changes and revisions must be kept up to date.

3.4.0 Detail Drawings

Detail drawings contain all the information needed to make an individual part. This includes from one to six views of the part, dimensions, section views, type of materials, tolerances, and special processes, such as heat treatment, surface plating, coating, or paint. The detail drawing may also include reference information such as where or what other parts it mates to and a revision history if it has been modified or changed.

The features on a detail drawing should be clear and dimensioned just enough to define what is needed in the final product. Dimensions that are repeated or not necessary must be noted as reference (ref) dimensions, meaning that they are not to be machined but offer some indirect or noncritical information.

Describe the hierarchy of a drawing package.

Review the information shown in a foundation layout.

Show Transparency 4 (Figure 2).

Discuss assembly drawings and how they relate to the bill of materials.

Show Transparency 5 (Figure 3).

Describe the information shown in detail drawings.

Provide a drawing set for trainees to examine. Explain the hierarchy and point out the different types of drawings in the package.

Have trainees practice identifying types of drawings from examples provided.

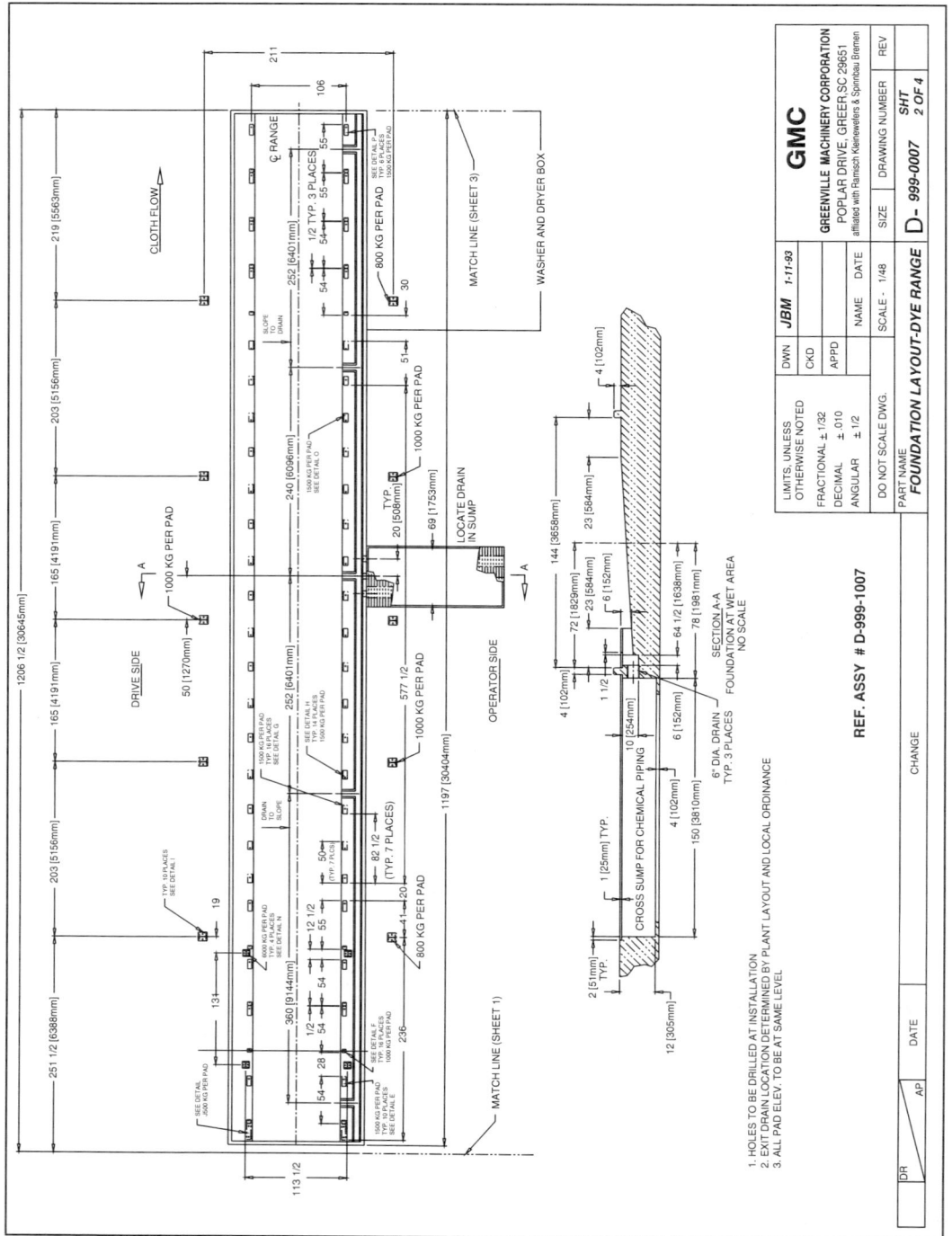

Figure 2 ◆ Foundation layout drawing.

Instructor's Notes:

GREENVILLE MACHINERY CORP. GREER, SOUTH CAROLINA USA	ASSEMBLY LIST			LMF	ORDER NO. D730022-81
MACHINE NAME 6" DIA. ROLL 14" COMPENSATOR ASSEMBLY (16" NIP ROLL ASSEMBLY)	ASSEMBLY NO. D-805-2115	ASSEMBLY LIST NO. 1		DATE 3/11/08	CUSTOMER SMITH INDUSTRIAL
PAINT ROYAL BLUE 5015 | VOLTAGE 415/3/50 | MATERIAL NOTED	FACE 90" | DRIVE SIDE R.H.	SHEET NO. 1 OF 3	REVISION	CHECKER JWM	RELEASE DATE 1/15/08

	REQ'D PER ORDER	REQ'D PER ORDER	PART NUMBER	FACE	REV. NO.	PROD. CODE	DESCRIPTION	DWG. ITEM NO.	REMARKS
1	4	1	C-785-0933-[01]		R1		SUPPORT, COMPENSATOR	1	AS SHOWN
	4	1					TUBE - 2 1/2" x 2 1/2" x 3/16" x 29"	2	H.R.S. BEV-1-E
	4	1					PLATE - 5/8" x 10 1/2" x 10 1/2"		H.R.S. TEMP.
2	4	1	C-785-0933-[02]		R1		SUPPORT, COMPENSATOR		OPP. HAND
	4	1					TUBE - 2 1/2" x 2 1/2" x 3/16" x 29"	1	H.R.S. BEV-1-E
	4	1					PLATE - 5/8" x 10 1/2" x 10 1/2"	2	H.R.S. TEMP.
3	4	1	B-721-0024-[01]	90"	R3		ROLL, COMPENSATOR		
	4	1					TUBE - 6" O.D x .120 WALL x [90 1/8"] LG.	1	T-304
	8	2					BAR - 1 3/4" x 9 1/4" LG.	2	T-304
	8	2					DISC - 5 7/8" O.D x 1 3/4" I.D x 3/8" THK.	3	T-304
	8	2					DISC - 5 7/8" O.D x 1 3/4" I.D x 3/8" THK.	5	H.R.S.
4	8	2				717	BEARING - FAFNIR VCJ T - 1 3/16"	4	2-BOLT FLANGE
5	4	1	B-796-0026	90"			SHAFT, COMPENSATOR		
	4	1					BAR - 1 7/16" DIA. x [121 5/8"] LG.	1	T-304
	8	2					BAR - 3/8" x 3/8" x 5 1/2" LG.	2	T-304 KEY
	8	2					BAR - 3/8" x 3/8" x 5 1/2" LG.	3	T-304 KEY
6	4	1	B-105-0003-1				ARM, COMPENSATOR (AS SHOWN)		
	4	1	B-105-0003-2				ARM, COMPENSATOR (OPP. HAND)		
	8	2				3809	ARM - CASTING # A64-45		CAST IRON

503F03.EPS

Figure 3 ◆ Bill of materials.

MODULE 15503-09 ◆ ADVANCED BLUEPRINT READING 3.5

Assign reading of Sections 4.0.0–4.7.0.

Ensure that you have everything required for teaching this session.

Stress the importance of the notes on detail drawings.

Explain that machine drawings are used to manufacture parts.

Point out the basic parts of a machine drawing.

Describe the use of different types of lines on a machine drawing to convey specific information.

Show Transparency 6 (Figure 4). Identify and interpret the types of lines.

Explain that dimensions on a detail drawing show the precise size of an object and the sizes and locations of its features.

Note that *ANSI Y14.M-1994* is the standard dimensioning system.

The processes and special notes on detail drawings are very specific, and, if ignored, could cause the equipment to work incorrectly. Notes specifying heat treatment, plating, coating, special grinding, or processes, such as wire EDM or radiographic weld tests, are spelled out or referenced to internal specifications.

4.0.0 ◆ DRAWING INFORMATION

Machine drawings contain all the information needed to manufacture a part exactly as it is designed regardless of what machine shop makes it or where in the world it is made. The standardization of drawing practices and CAD programs ensures that engineering drawings are a common language that provides consistent results. The basic parts of a machine drawing are the following:

- Lines
- Dimensions
- Notes and symbols
- Scale
- Revisions
- Vendor information
- Material specifications

4.1.0 Lines

The lines on a machine drawing describe the object to be made (*Figure 4*). The lines differ in their thickness, or weight, and in their construction so that different types of information can be detailed all around and on the part without becoming confusing. Each type of line has a specific use intended to convey precise, universal communication. The lines may be continuous, dashed, or a combination of the two. The standard lines used on a machine drawing are defined as follows:

- *Object lines* – Heavy, continuous lines that show the outline or overall shape of the object. Object lines show every edge of the object that is visible from a particular view.
- *Hidden lines* – Dashed lines that show surfaces or edges of the object that are hidden or not visible in that view.
- *Center lines* – Thin lines that consist of alternating dashes and longer lines that show the center of a symmetrical or round object or part of the object for reference, clarification, or dimensioning.
- *Dimension lines* – Thin, continuous lines that terminate with an arrowhead at the extension lines and show the dimensions and geometric tolerances of the object.

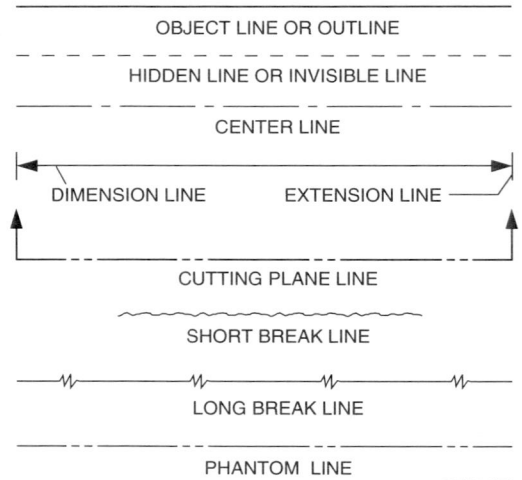

Figure 4 ◆ Types of lines used on machine drawings.

- *Extension lines, or leaders* – Thin, continuous lines that extend from the object at the exact locations between which dimension lines are drawn.
- *Cutting plane lines* – Heavy, dashed, or continuous lines that show where the object is cut to make the cutaway section.
- *Break lines* – Dashed or continuous lines that may be freehand or machine drawn to show cross sections or internal details.
- *Phantom lines* – Thin, dashed lines that denote alternate positions or details of an object that is not on that drawing.
- *Section lines* – Different patterns are used to denote various materials, such as aluminum, steel, or plastic. Section lines show a cutaway view of the inner structure of an object.

4.2.0 Dimensions

The precise size of an object is shown on a detail drawing by the use of accurate dimensions. Dimensions are also used to show the size and locations of holes, slots, threads, and other detailed features of the object.

There are several methods of dimensioning. The traditional method uses leaders coming off the object, with dimension lines connecting the leader to an arrowhead. A break in the dimension line allows the numerical value of the dimension to be inserted.

The standard dimensioning system used is *ANSI Y14.5M-1994*, which is commonly known as geometric tolerancing. The purpose of this system is to

Instructor's Notes:

_____

_____

_____

standardize engineering drawings and provide a target dimensioning system that makes precise, consistent results easier to obtain from a drawing.

With the advent of computer-controlled machining comes **datum** or coordinate dimensioning, which uses two outside, 90-degree intersecting surfaces from which to reference all the dimensions. Tabular dimensioning uses a target designation on each feature that is listed in a chart that contains the X-Y location of the feature and, if required, a description of the feature. *Figure 5* shows types of dimensioning.

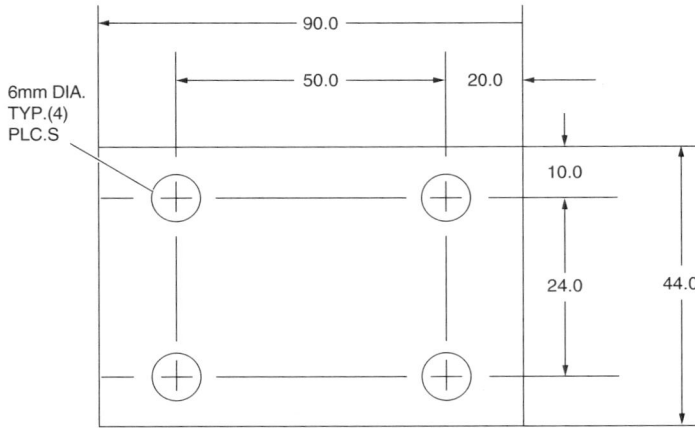

Describe geometric tolerancing, and explain why a standard dimensioning system is important.

Compare the traditional, coordinate, and tabular methods of dimensioning.

Show Transparency 7 (Figure 5).

Provide sketches of parts showing different types of dimensioning. Review the sketches with trainees and answer any questions.

TRADITIONAL DIMENSIONING

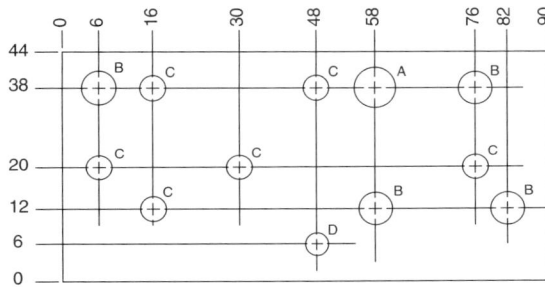

COORDINATE DIMENSIONING

HOLE SYMBOL	HOLE Ø
A	6
B	5
C	4
D	3

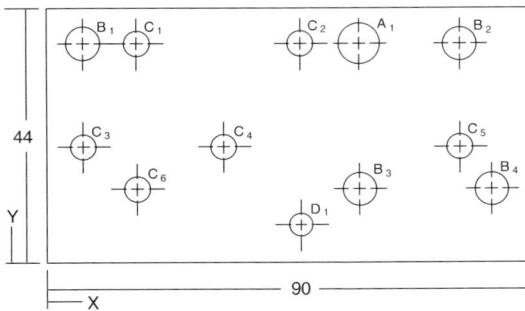

TABULAR DIMENSIONING

HOLE SYMBOL	HOLE Ø	LOCATION X	Y
A_1	6	58	38
B_1	5	6	38
B_2	5	76	38
B_3	5	58	12
B_4	5	82	12
C_1	4	16	38
C_2	4	48	38
C_3	4	6	20
C_4	4	30	20
C_5	4	76	20
C_6	4	16	12
D_1	3	48	6

503F05.EPS

Figure 5 ◆ Types of dimensioning.

MODULE 15503-09 ◆ ADVANCED BLUEPRINT READING   3.7

The most common dimensions and tolerances that are found on a detail drawing are the following:

- *Fractional dimensions* – Given as fractional parts of an inch, fractional dimensions are used on objects that do not require very close tolerances, such as furniture or utility items.
- *Decimal dimensions* – Expressed in thousandths of an inch, decimal dimensions are used predominately and are more practical to use. The conversion to metric, the use of *ANSI Y14.5M-1994*, and the almost complete dependency on computer interface has almost eliminated the use of fractions in machine design. Tolerances in thousandths or ten-thousandths of an inch are used on objects requiring greater precision.
- *Metric dimensions* – Both decimal dimensions and metric dimensions are used on many drawings so that they can be used outside of the United States. One of the dimensions is usually shown in parentheses. As more companies from other countries locate in the U.S. and world manufacturing is becoming a reality, more designs are generated with metric-dominant or metric-only dimensions.
- *Tolerances* – A tolerance is the total amount that a specific dimension may vary. These are standard limits specified in the title block or are specific high or low limits at individual dimensions. *Figure 6* shows examples of tolerances from machine drawings.

4.3.0 Notes and Symbols

Notes and symbols are used to provide information that is not conveyed by the views or dimensions on the drawing. Other than dimensions, this is the most important information on an

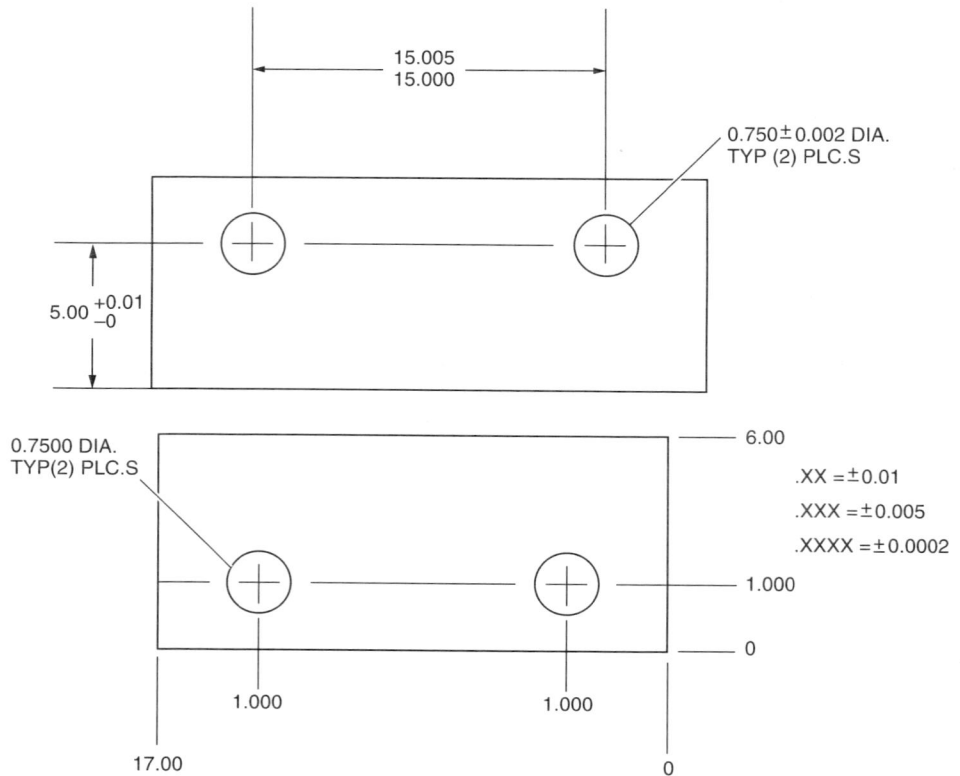

Figure 6 ◆ Tolerances.

503F06.EPS

Instructor's Notes:

engineering drawing. Three types of information that are conveyed by notes and symbols are the following:

- Finish marks
- Threads
- Geometric tolerancing

Finish marks (*Figure 7*) designate an average grade of surface roughness in microinches or micrometers. All machining processes cut the material, leaving grooves in the machined area. These grooves are measured and graded to provide a usable reference for the designer, machinist, and product-user to specify how parts fit or run together. The check symbol and surface grade are shown on the surface or on the dimension leader nearest the surface specified.

Detail drawings give the information required to fabricate or install a small part of system equipment. The details in *Figure 8* show methods to be used to install holders for temperature-sensing equipment in various contexts. Each detail gives welding and part preparation information, as well as thread specifications.

Geometric tolerancing is also defined by *ANSI Y14.5M-1994*. This standard of symbols and values standardizes the acceptable tolerance of fea-

tures for all aspects of machining and assembly. These symbols are designed to provide the maximum allowable tolerance in feature dimensions and still maintain a good working part. Surface features, such as flatness, parallelism, and perpendicularity, define the tolerance without being overstrict. Target tolerances for diameters define a circular zone in which mating parts can vary and still fit accurately. It is important to understand what the symbols and tolerances mean because they are not meant to replace numerical, plus/minus tolerances but enhance the definition of the part. *Figure 9* shows geometric symbols and terms.

4.4.0 Scale

Whenever possible, objects are drawn to their actual size. However, some objects are so small or so large that it is impossible to draw them to actual size. In either case, the object is drawn at some scale or exact proportion of its actual size. A drawing in which an object is represented as being enlarged, reduced, or actual size is known as a scale drawing. The ratio or proportion between the original and the scale drawing is known as the scale.

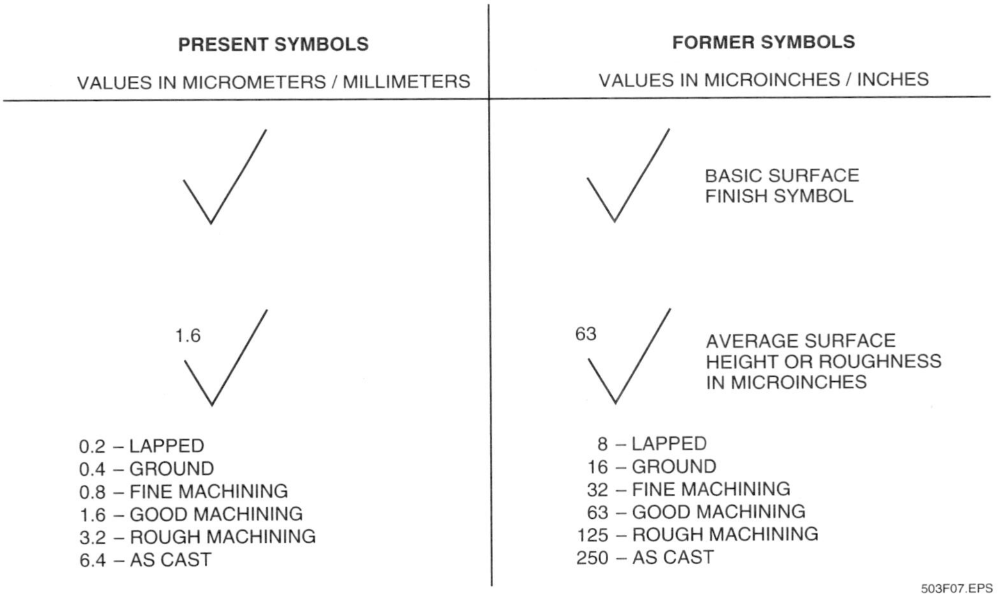

PRESENT SYMBOLS	FORMER SYMBOLS	
VALUES IN MICROMETERS / MILLIMETERS	VALUES IN MICROINCHES / INCHES	
✓	✓	BASIC SURFACE FINISH SYMBOL
1.6 ✓	63 ✓	AVERAGE SURFACE HEIGHT OR ROUGHNESS IN MICROINCHES
0.2 – LAPPED 0.4 – GROUND 0.8 – FINE MACHINING 1.6 – GOOD MACHINING 3.2 – ROUGH MACHINING 6.4 – AS CAST	8 – LAPPED 16 – GROUND 32 – FINE MACHINING 63 – GOOD MACHINING 125 – ROUGH MACHINING 250 – AS CAST	

503F07.EPS

Figure 7 ◆ Finish marks.

Explain how finish marks indicate a grade of surface roughness.

Show Transparency 9 (Figure 7).

Note that thread specifications are also indicated by notes and symbols.

Show Transparency 10 (Figure 8).

Explain how geometric tolerancing is shown using notes and symbols.

Stress that parts must be manufactured within acceptable tolerances in order to work properly.

Show Transparency 11 (Figure 9).

Explain what is meant by a scale drawing, and provide examples.

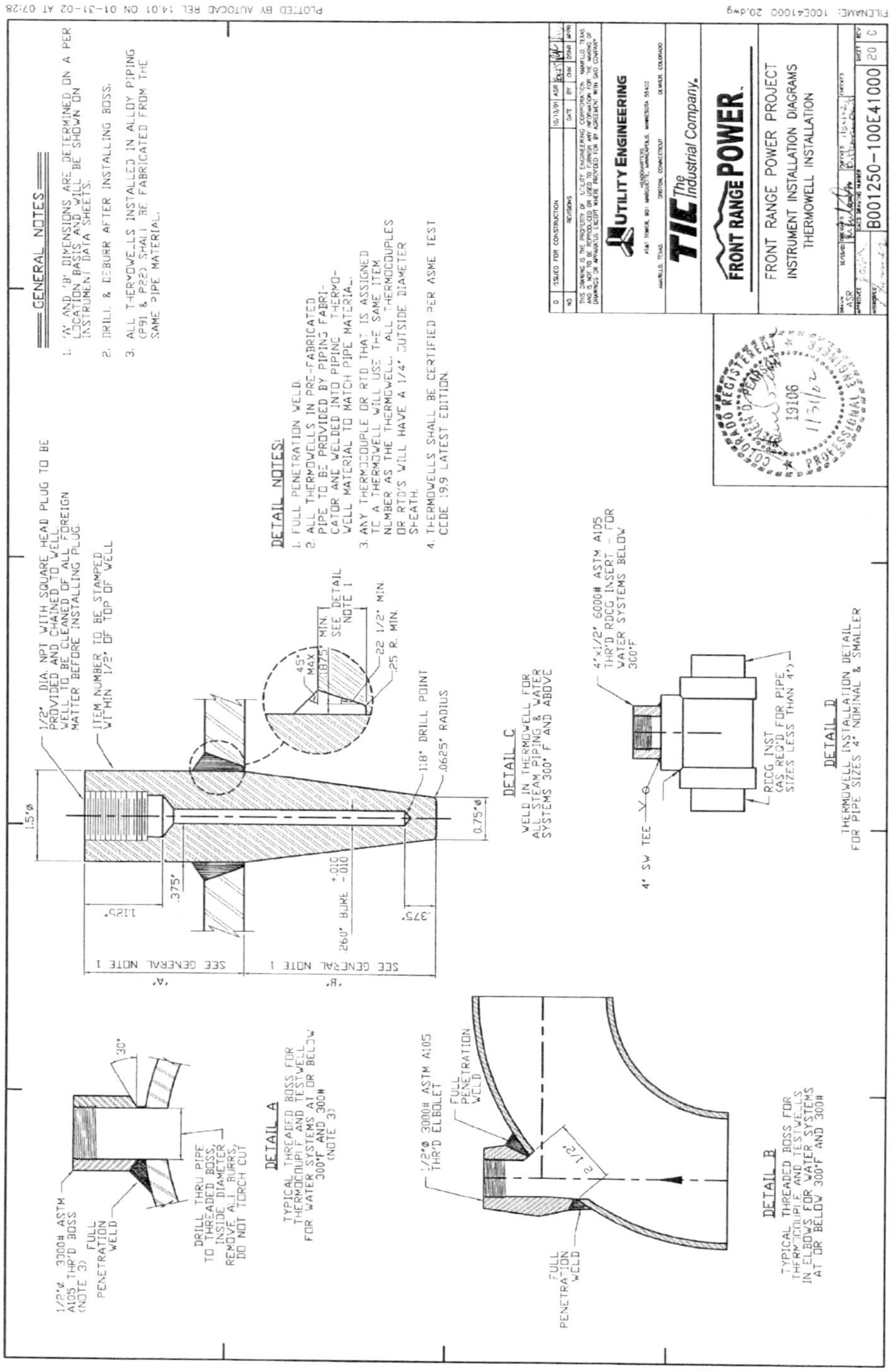

Figure 8 ◆ Detail drawings.

Instructor's Notes:

SYMBOL	CHARACTERISTIC	TYPES	KIND OF FEATURE
⏥	FLATNESS	FORM	INDIVIDUAL
—	STRAIGHTNESS	FORM	INDIVIDUAL
○	CIRCULARITY	FORM	INDIVIDUAL
⌭	CYLINDRICITY	FORM	INDIVIDUAL
⌒	PROFILE OF A LINE	PROFILE	INDIVIDUAL OR RELATED
⌓	PROFILE OF A SURFACE	PROFILE	INDIVIDUAL OR RELATED
⊥	PERPENDICULARITY (SQUARENESS)	ORIENTATION	RELATED
∠	ANGULARITY	ORIENTATION	RELATED
//	PARALLELISM	ORIENTATION	RELATED
↗	CIRCULAR RUNOUT	RUNOUT	RELATED
⌰	TOTAL RUNOUT	RUNOUT	RELATED
⌖	POSITION	LOCATION	RELATED
◎	CONCENTRICITY	LOCATION	RELATED
⌯	SYMMETRY (REPLACED BY POSITION IN 1982 STANDARD)	LOCATION	RELATED
Ⓜ	MAXIMUM MATERIAL CONDITION (MMC)		
Ⓢ	REGARDLESS OF FEATURE SIZE (RFS)		
Ⓛ	LEAST MATERIAL CONDITION (LMC)		
Ⓟ	PROJECTED TOLERANCE ZONE		
∅	DIAMETRICAL (CYLINDRICAL) TOLERANCE ZONE OR FEATURE		
[.605]	BASIC, OR EXACT, DIMENSION		
[-A-]	DATUM FEATURE SYMBOL		
[⌖|∅.005Ⓜ|A]	FEATURE CONTROL FRAME		
(A1)	DATUM TARGET		

503F09.EPS

Figure 9 ◆ Geometric symbols and terms.

MODULE 15503-09 ◆ ADVANCED BLUEPRINT READING 3.11

Emphasize the importance of keeping accurate revision records.

Describe how revisions are recorded on the title block.

Explain that revisions are often shown with a cloud shape around them.

Point out where important vendor information can be found on the drawings.

Provide examples that show where material specifications are located on drawings.

See the Teaching Tip for Section 4.0.0. at the end of this module.

Have trainees practice reading various types of drawings.

Assign reading of Sections 5.0.0–5.2.0.

Ensure that you have everything required for teaching this session.

Discuss how the drawing system can be searched to find specific information.

For example, if an object that is 12 inches actual size measures 6 inches on the drawing, the drawing is half the size of the object. This drawing is drawn to a scale of ½. This scale is noted on the drawing as Scale: 1" = 2" or 1:2. This means that 1 inch on the drawing equals 2 inches on the object.

If an object of 6 inches actual size measures 12 inches on the drawing, the drawing is said to be double scale, or 2 to 1. This scale is noted on the drawing as Scale: 2" = 1" or 2:1. This means that 2 inches on the drawing are equal to 1 inch on the object.

For example, if a manufacturing floor is 300 feet by 175 feet, but measures 37.5 inches by 21.875 inches on the drawing, the scale is 1 foot = ⅛ inch, or ⅛" = 1'.

[⅛ = .125; 300 × .125 = 37.5; 175 × .125 = 21.875]

No matter the scale of the drawing, the dimensions are always the actual size of the object. When it comes to keeping the dimensions accurate, the importance of accurate revision records cannot be overemphasized.

4.5.0 Revisions

If any person involved in the equipment or drawing package makes any change, revision, or modification to the equipment or drawing package without documenting it thoroughly, the whole system becomes unreliable and useless. If a part of a machine fails and someone redesigns or alters the part to correct the problem but fails to revise the drawings to detail what was done, it is very likely that the original mistake will be made again. This negates the effort and can be very costly. Working from sketches and marked-up prints is sometimes necessary to save time, but these changes have to be incorporated into the system as soon as possible.

Each drawing has in its title block a space for listing the revision history. This block should include the date, the initials of the person who made the revision, a letter or number that appears beside the area where the correction was made, and a brief description of what was done. If the revision is too complicated or there is no room on the drawing, there are two alternatives. The drawing can be obsoleted, redrawn, and referenced, or it can be extended with the same number and title showing "Sheet 2 of 2." It is most common for the revision to be shown with a cloud surrounding the changed portion.

4.6.0 Vendor Information

Vendor information consists of specifications, part numbers, and sizes of components that are purchased to be used in the equipment. This information is listed in the bill of materials and appears on assembly or detail drawings as needed. The only rule for this type of information is that it communicates what is necessary to facilitate the machining, assembly, and operation of the equipment. Vendor installation manuals, troubleshooting guides, or instructional courses may be listed as required resources. As manufacturing equipment becomes more complex and is driven by computers, the need for accurate and concise vendor information becomes critical.

4.7.0 Material Specifications

Material specifications are found in the title block, in a note, or on a separate sheet or set of sheets, and spell out the details of the types of materials to be used. In all classes of materials, there are standards and specifications that correlate to performance or machineabilty. These specified materials can have substitutes that perform the same, or the specification may require the use of one specific material only. For example, Delrin™ and Celcon™ are specific brand names for an acetal copolymer plastic. The products are very similar but differ in some ways. Therefore, the material specification on the blueprint may call for Delrin™ or equivalent if the need is unspecific, or it may denote Delrin™ if this is the only material acceptable.

5.0.0 ◆ DRAWING SYSTEM USAGE

Although the computer has changed the way all drawings are generated and stored, it has not eliminated the problems traditionally associated with engineering drawings. Updating information and changes is more critical now because of the immediate interface with other computer hierarchies such as **computer-aided engineering (CAE)**, **computer-integrated manufacturing (CIM)**, and **computer-aided machining (CAM)**. The purpose of the drawing system is to allow anyone to find information on any piece of equipment from any source within the system. The following approaches offer two ways to search the system for specific information. The first approach is to find details on vendor drawings of installed equipment, and the second approach is to locate construction details from the plant foundation drawings.

Instructor's Notes:
_____
_____
_____

5.1.0 Finding Details

To start from a layout or main assembly and track through the drawing numbers to a small detail is straightforward and simple if the drawing package has been kept up to date or never modified. In this case, this would involve finding the area the part is in, pinpointing the part, and, with the line item number, find the drawing number from the bill of materials. With the correct drawing number, the print can be located and copied or revised.

This becomes more difficult when the drawing package has evolved over a period of time with many revisions. The correct number may not even exist, or the part may have been redesigned without reference to what it was. In this case, extensive searching may be required to find the drawing, or it may be necessary to redraw it.

To demonstrate how to find details, assume that the following drawing package has been kept up to date. Follow the drawing package sequence to find the detail drawing. The drawing package consists of the following drawings:

- Foundation layout: D-999-0007
- Washer and dye range: D-999-1007
- Washboxes: D-805-2007
- Nip roll assembly: D-805-2115
- Bill of materials

Figures 10 through *14* show the drawings in the drawing package.

For this exercise, assume that there is a bad bearing on a section of a line that has stopped because the bearing failed. You must locate a new bearing quickly and replace the failed one, which is on the roll compensator on box 7 of the washer range.

Start with the plant layout drawing and look for the reference assembly number that corresponds to the washer range. You need to find both the detail drawing and the bill of materials. From the plant layout and your imaginary location in front of the bearing, find the next assembly, which is the washer and dye range, to find the sub-assembly. The subassembly is the washboxes, D-805-2007, which shows more detail and overall dimensions and balloons the detail drawing where the roll compensator is shown. On the detail assembly, D-805-2115, the flange bearing is ballooned as part number 4. Refer to the bill of materials under roll compensator to find the bearing manufacturer and part number (Fafnir VCJT-1 3⁄16). This information is used to find a spare flange bearing or to order one.

5.2.0 Finding Assembly Drawings

With construction foundation plans, there are different types of symbols and systems used to correlate the information. Construction foundation plans consist of many sheets of assembly, detail, and section detail drawings.

A plan view shows the foundation and column lines from above. These columns are in a grid whose lines relate to north-south and east-west. The north-south lines use letter callouts, and the east-west lines use numbers. The north wall is typically the main locator or datum line from which all other features are dimensioned or sequenced. In the elevation views and sections, the columns are designated with the letter and number from the plan view, such as H3a or F2a. The elevation is measured from zero feet sea level, so the ground level measurement will not be zero unless the site is at sea level, and the actual height of a column is added to the site elevation.

A section is commonly referred to as a cut, so the cut diagram is a detailed section view taken from the area called out. The cut symbol is a bisected circle with an arrow outside of it indicating the direction of view. It has numbers inside that designate the cut number and location or drawing where the cut is drawn. On the detail sheet, the circle is under the detail and designates the cut number and which assembly it is from.

Another feature used is a cloud. This is a detail that has a scalloped circle drawn around it that looks like a cloud. It designates a feature whose design is incomplete and not released for construction. A cloud should have a revision letter and a proposed release date on it.

Construction drawings also contain information that a variety of crafts or specialties use to make the entire process work. The details are split off from the foundation and elevation drawings for pipefitters, vendors, millwrights, and construction personnel to provide information specific to the craft. The material specifications for bolts, bases, and a variety of other details are called out on the detail drawings and referenced on the foundation drawings. The information is in the general notes or schedules. These are under column schedules, anchor bolt schedules, or materials criteria, which contains grade of bolts, weld specifications, client specifications, or bolt torque requirements.

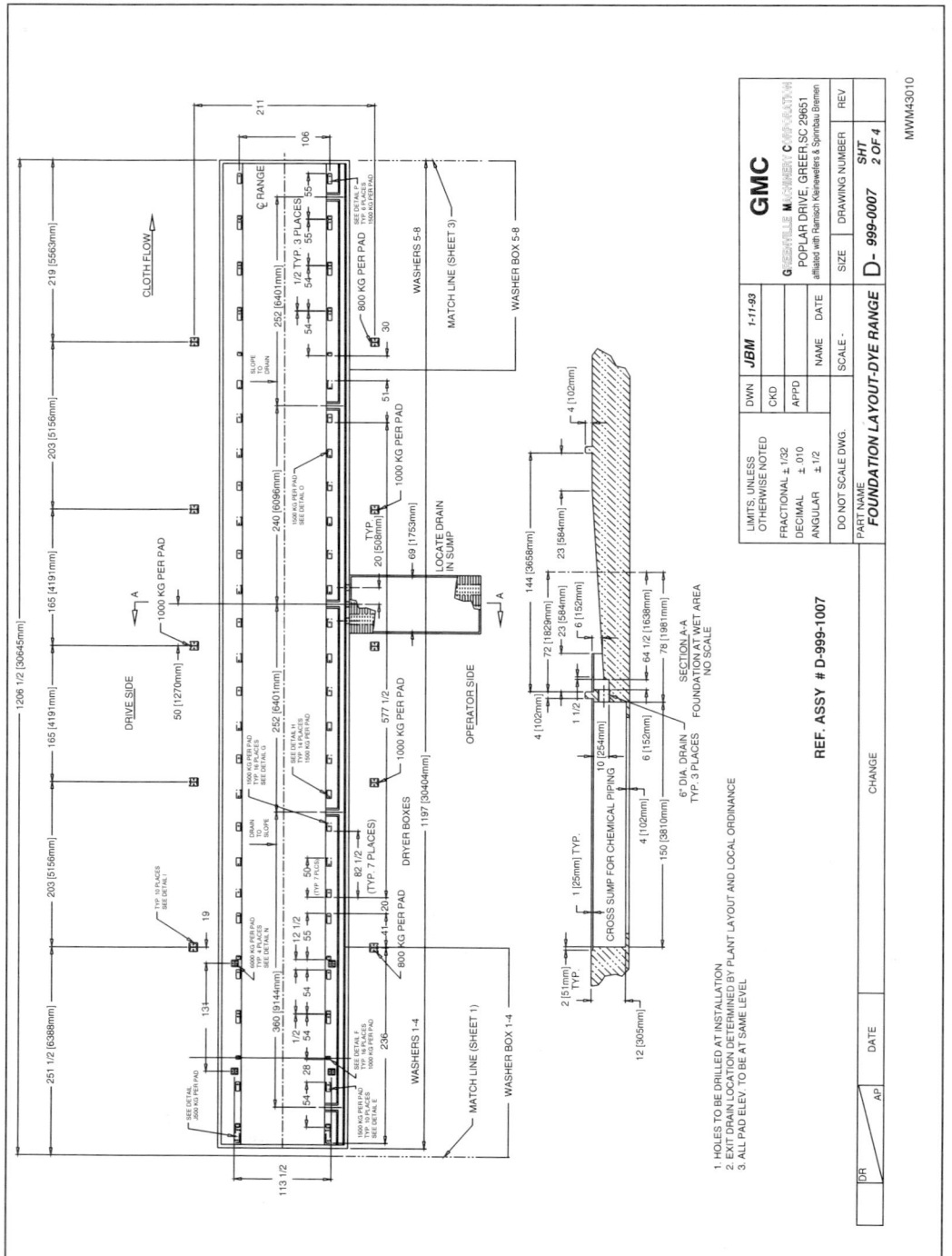

Figure 10 ◆ Foundation layout D-999-0007.

503F10.EPS

Instructor's Notes:

_____

_____

_____

_____

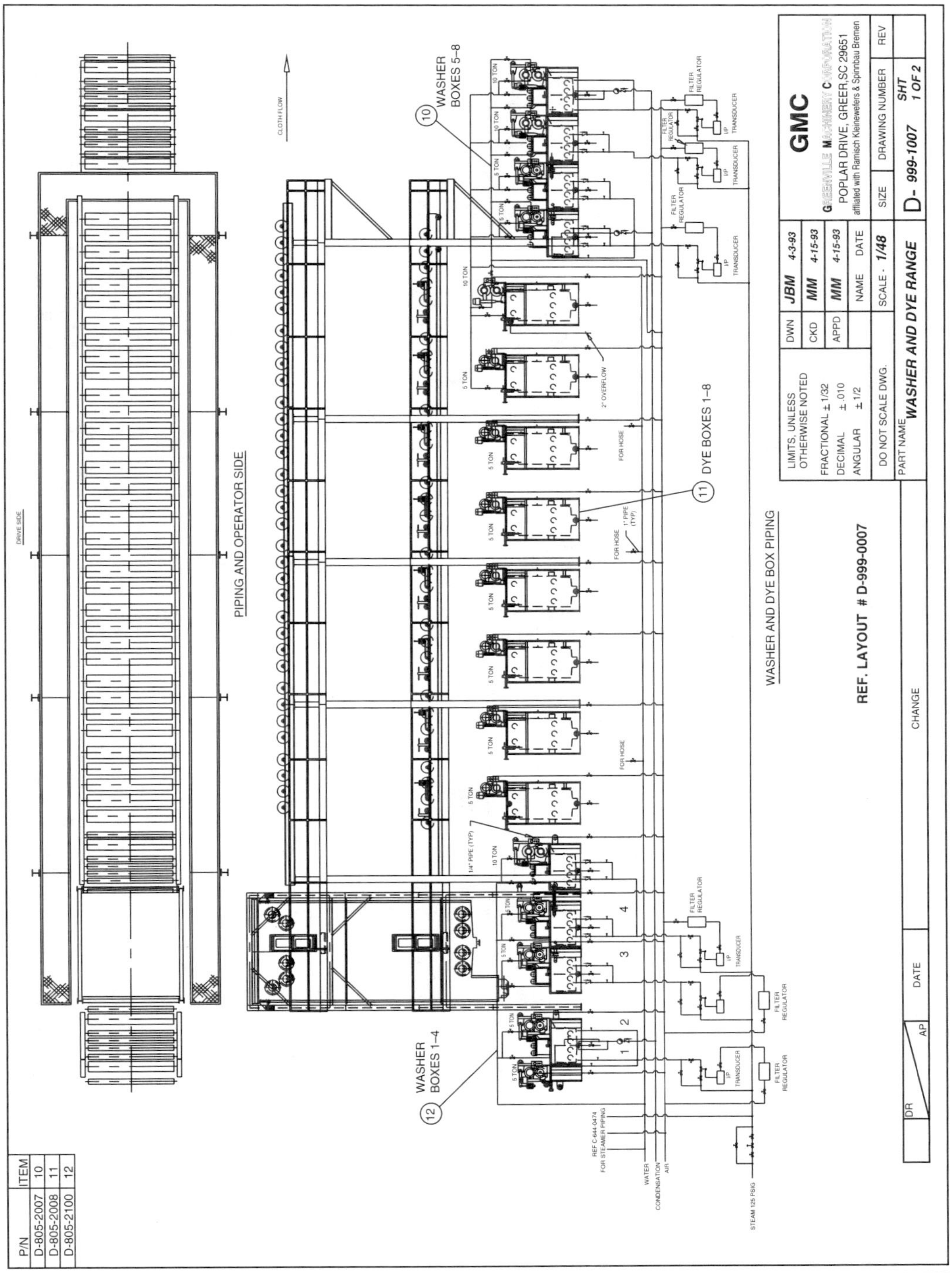

Figure 11 ◆ Washer and dye range D-999-1007.

PART NUMBER	ITEM
B-721-0023-01	1
B-105-0004-02	2
B-721-0023-03	3
B-105-0004-04	4
D-805-2111	5
D-805-2112	6
D-805-2113	7
D-805-2114	8
C-785-0132-09	9
C-785-0132-10	10
C-785-0933-01	11
C-785-0933-02	12
B-721-0025-01	13
B-721-0026-02	14
V-701-3021	15
T-174-0021	16

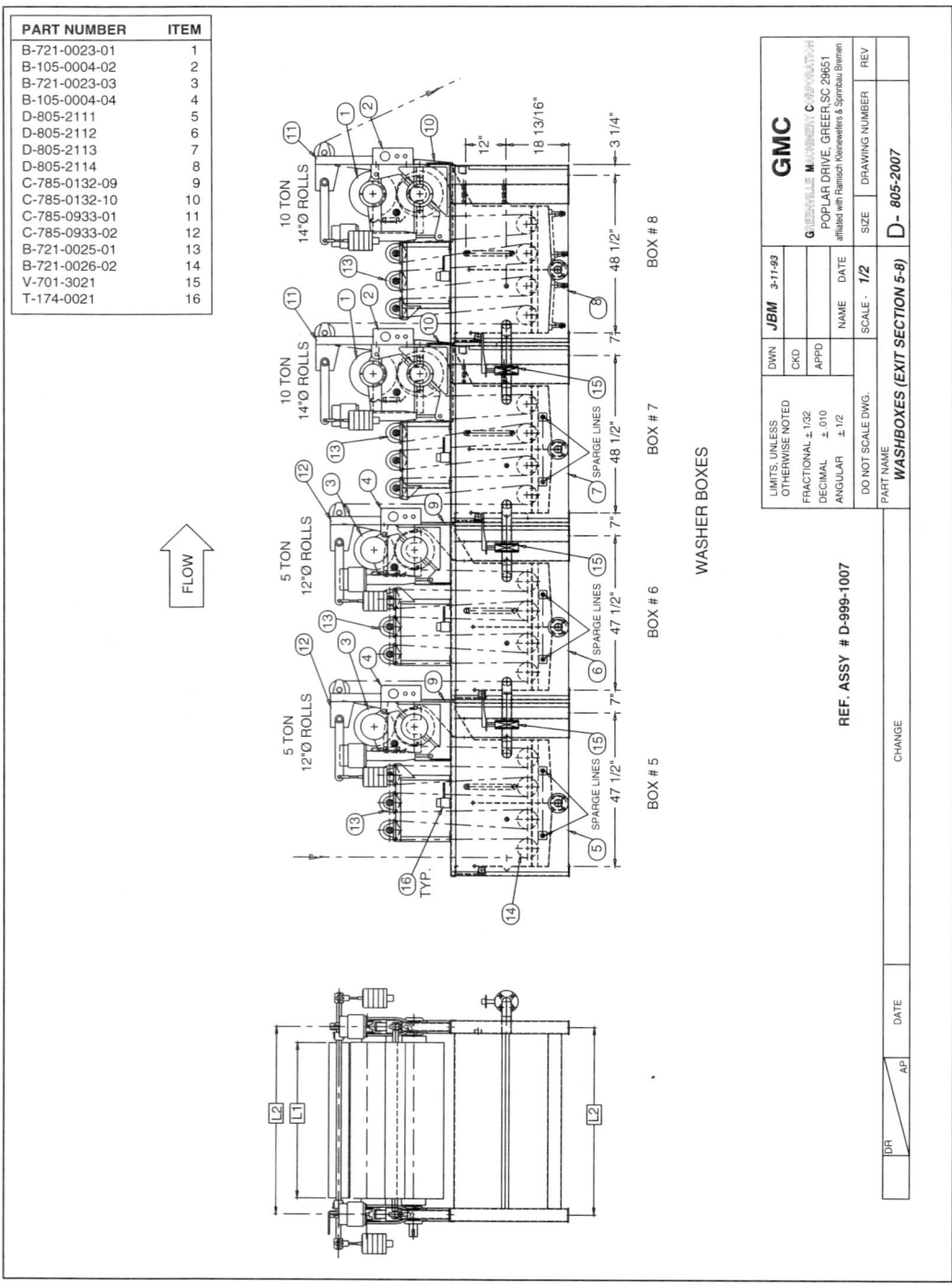

Figure 12 ◆ Washboxes D-805-2007.

503F12.EPS

3.16 MILLWRIGHT ◆ LEVEL FIVE

Instructor's Notes:

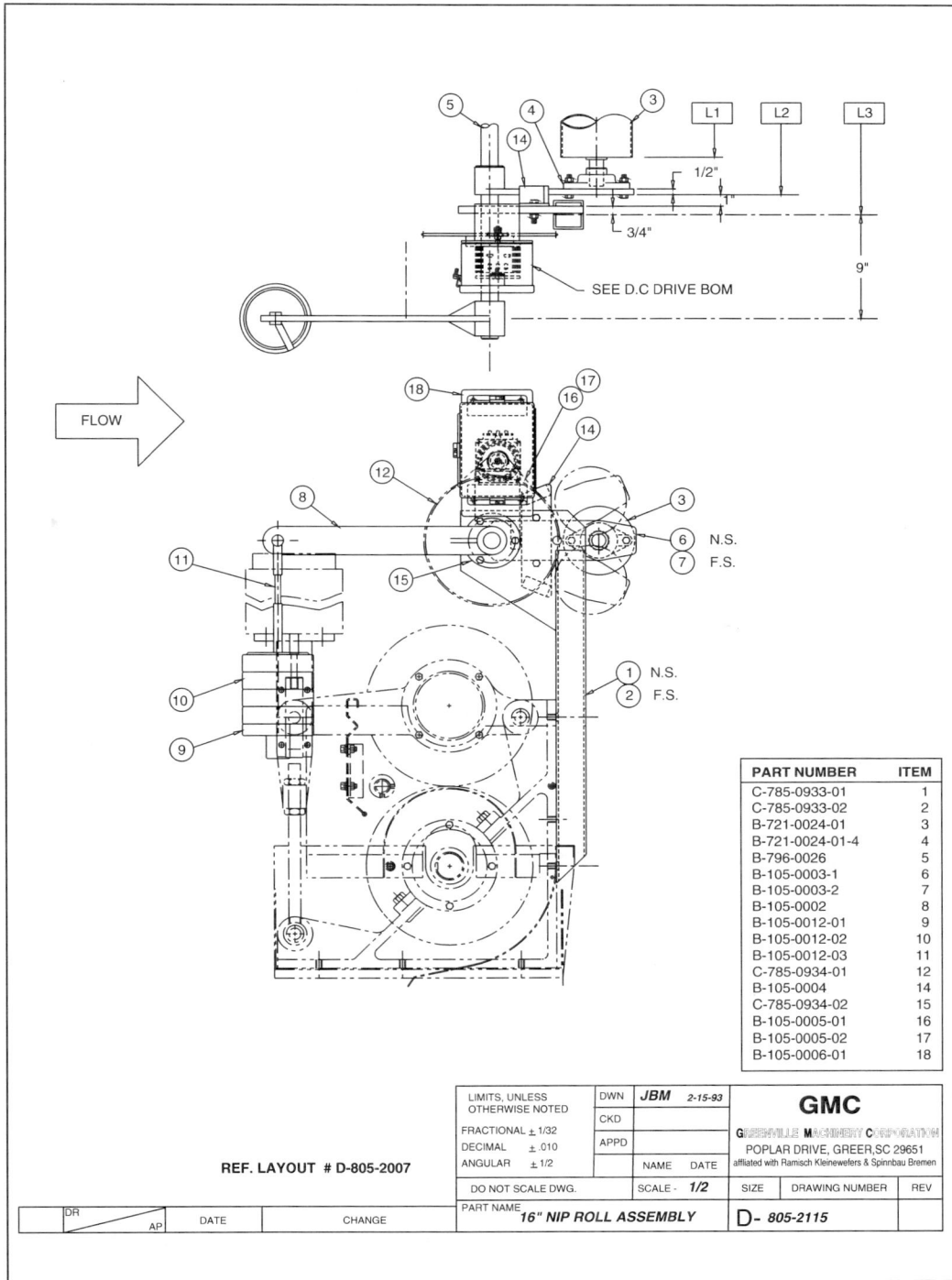

PART NUMBER	ITEM
C-785-0933-01	1
C-785-0933-02	2
B-721-0024-01	3
B-721-0024-01-4	4
B-796-0026	5
B-105-0003-1	6
B-105-0003-2	7
B-105-0002	8
B-105-0012-01	9
B-105-0012-02	10
B-105-0012-03	11
C-785-0934-01	12
B-105-0004	14
C-785-0934-02	15
B-105-0005-01	16
B-105-0005-02	17
B-105-0006-01	18

FLOW

SEE D.C DRIVE BOM

N.S.
F.S.

REF. LAYOUT # D-805-2007

LIMITS, UNLESS OTHERWISE NOTED	DWN	JBM	2-15-93	GMC
FRACTIONAL ± 1/32 DECIMAL ± .010 ANGULAR ± 1/2	CKD			GREENVILLE MACHINERY CORPORATION
	APPD			POPLAR DRIVE, GREER, SC 29651
		NAME	DATE	affliated with Ramisch Kleinewefers & Spinnbau Bremen

DO NOT SCALE DWG.   SCALE - 1/2   SIZE   DRAWING NUMBER   REV

PART NAME **16" NIP ROLL ASSEMBLY**   D- 805-2115

DR	AP	DATE	CHANGE

503F13.EPS

Figure 13 ◆ Nip roll assembly D-805-2115.

GREENVILLE MACHINERY CORP.
GREER, SOUTH CAROLINA USA

ASSEMBLY LIST

ORDER NO.	D730022-81
CUSTOMER	SMITH INDUSTRIAL
RELEASE DATE	1/15/08

ASSEMBLY LIST NO.	1	LMF
		DATE 3/11/08
		CHECKER JWM

MACHINE NAME	6" DIA. ROLL 14" COMPENSATOR ASSEMBLY (16" NIP ROLL ASSEMBLY)				
PAINT	ROYAL BLUE 5015	VOLTAGE 415/3/50			
		ASSEMBLY NO. D-805-2115	FACE 90"	DRIVE SIDE R.H.	SHEET NO. 1 OF 3
MATERIAL	NOTED				

ITEM	REQD PER ORDER	REQD PER ORDER	PART NUMBER	FACE	REV. NO.	PROD. CODE	DESCRIPTION	DWG. ITEM NO.	REMARKS
1	4	1	C-785-0933-[01]		R1		SUPPORT, COMPENSATOR	1	AS SHOWN
	4	1					TUBE - 2 1/2" x 2 1/2" x 3/16" x 29"	2	H.R.S. BEV-1-E
	4	1					PLATE - 5/8" x 10 1/2" x 10 1/2"		H.R.S. TEMP.
2	4	1	C-785-0933-[02]		R1		SUPPORT, COMPENSATOR	1	OPP. HAND
	4	1					TUBE - 2 1/2" x 2 1/2" x 3/16" x 29"	1	H.R.S. BEV-1-E
	4	1					PLATE - 5/8" x 10 1/2" x 10 1/2"	2	H.R.S. TEMP.
3	4	1	B-721-0024-[01]	90"	R3		ROLL, COMPENSATOR		
	4	1					TUBE - 6" O.D x .120 WALL x [90 1/8"] LG.	1	T-304
	8	2					BAR - 1 3/4" x 9 1/4" LG.	2	T-304
	8	2					DISC - 5 7/8" O.D x 1 3/4" I.D x 3/8" THK.	3	T-304
	8	2					DISC - 5 7/8" O.D x 1 3/4" I.D x 3/8" THK.	5	H.R.S.
4	8	2				717	BEARING - FAFNIR VCJ T - 1 3/16"	4	2-BOLT FLANGE
5	4	1	B-796-0026	90"			SHAFT, COMPENSATOR		
	4	1					BAR - 1 7/16" DIA. x [121 5/8"] LG.	1	T-304
	8	2					BAR - 3/8" x 3/8" x 5 1/2" LG.	2	T-304 KEY
	8	2					BAR - 3/8" x 3/8" x 5 1/2" LG.	3	T-304 KEY
6	4	1	B-105-0003-1				ARM, COMPENSATOR (AS SHOWN)		
	4	1	B-105-0003-2				ARM, COMPENSATOR (OPP. HAND)		
	8	2				3809	ARM - CASTING # A64-45		CAST IRON

Figure 14 ◆ Bill of materials.

503F14.EPS

Instructor's Notes:

_____

_____

_____

The relation of each drawing to the other should be apparent. A definite path of one drawing to the other is referenced from the assembly to the detail and back. The cut balloon or cut number shows the detail number and which drawing it is on. *Figure 15* shows cut numbers.

These construction and fabrication drawings provide specific information that is interwoven to make a precision final product that uses many different crafts and technologies. Understanding all aspects of the drawing system is important, and the ability to reference one drawing to the other should be practiced.

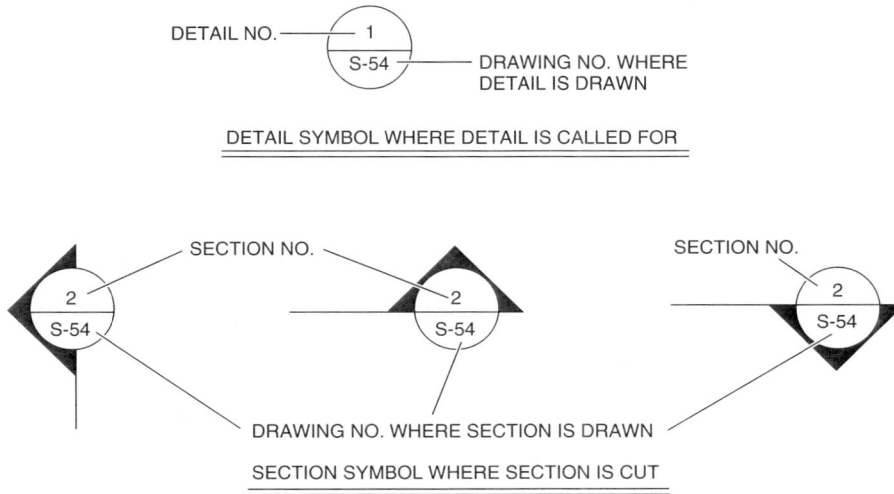

DETAIL NO. ——— 1
S-54 ——— DRAWING NO. WHERE
DETAIL IS DRAWN

DETAIL SYMBOL WHERE DETAIL IS CALLED FOR

SECTION NO.   SECTION NO.
2   2   2
S-54   S-54   S-54

DRAWING NO. WHERE SECTION IS DRAWN

SECTION SYMBOL WHERE SECTION IS CUT

503F15.EPS

Figure 15 ◆ Cut numbers.

Explain how to read cut numbers.

Show Transparency 17 (Figure 15).

Show trainees how to use a bill of materials to perform a materials takeoff.

Have trainees use a bill of materials to perform a materials takeoff. Note the proficiency of each trainee. This laboratory corresponds to Performance Task 3.

Review Questions

1. The purpose of engineering drawings is to _____.
   a. communicate
   b. discuss modifications
   c. serve as change orders
   d. provide employment

2. The purpose of a _____ system is to make it easy to find any drawing and relate all of its parts to the whole machine.
   a. lettering
   b. numbering
   c. paging
   d. machining

3. Foundation layout drawings, main assembly drawings, subassembly drawings, bills of material, and detail drawings together make up the drawing _____.
   a. list
   b. system
   c. package
   d. prints

4. Replacing a specific part would require the detail drawing of the part mounting and the _____.
   a. main assembly drawings
   b. foundation layout drawings
   c. subassembly drawings
   d. bill of materials

5. The drawing that details the inside dimensions, structural column locations, and mounting pad locations of a building is called the _____ drawing.
   a. detail
   b. preliminary layout
   c. assembly
   d. foundation layout

6. The drawings that provide the total picture of a machine or process and show the location of parts are called _____ drawings.
   a. detail
   b. layout
   c. assembly
   d. architectural

7. The list of all the parts, components, and assemblies that make up a whole machine or assembly is called the _____.
   a. assembly list
   b. bill of materials
   c. supply list
   d. detail list

8. Drawings that contain all the information needed to make an individual part are called _____ drawings.
   a. assembly
   b. subassembly
   c. layout
   d. detail

9. On a drawing, the heavy, continuous lines that show the outline or overall shape are called _____ lines.
   a. hidden
   b. section
   c. object
   d. center

10. The precise size of an object is shown on a detail drawing by the use of accurate _____.
    a. sketches
    b. dimensions
    c. notes
    d. views

11. The standard dimensioning system used is *ANSI Y14.5M-1994*, which is commonly known as _____.
    a. geometric tolerancing
    b. Cartesian coordinates
    c. standard targets
    d. computer entry

12. The dimensioning type that uses two outside, 90-degree intersecting surfaces from which to reference all dimensions is called _____ dimensioning.
    a. planar
    b. datum
    c. scalar
    d. quantum

Instructor's Notes:

_____

_____

_____

13. The three ways to numerically express dimensions are fractional, decimal, and _____.
    a. cryptic
    b. binary
    c. metric
    d. algebraic

14. Finish marks designate an average grade of surface _____ in microinches or micrometers.
    a. roughness
    b. roundness
    c. angularity
    d. tightness

15. Geometric tolerancing standards define surface features such as flatness, _____, and perpendicularity.
    a. cleanliness
    b. reflectivity
    c. parallelism
    d. pitch

16. The ratio or exact proportion between the actual part and the size of the drawing is called the _____.
    a. geometry
    b. cotangent
    c. specification
    d. scale

17. Any change or modification made to a part and documented on the drawing is called a _____.
    a. redesign
    b. revision
    c. revolution
    d. redraw

18. The accuracy of the drawing package is ensured by maintaining the _____ list in the title block.
    a. materials
    b. heat treatment
    c. detail
    d. revision

19. What a part is composed of is called out in the title block under the _____ specification.
    a. paint
    b. revision
    c. object
    d. material

20. The first place to look for a detail drawing number is the _____.
    a. bill of materials
    b. instruction manual
    c. title block
    d. foundation layout

Summarize the major concepts presented in the module.

Administer the Module Examination. Be sure to record the results of the Examination on Craft Training Report Form 200, and submit the results to the Training Program Sponsor.

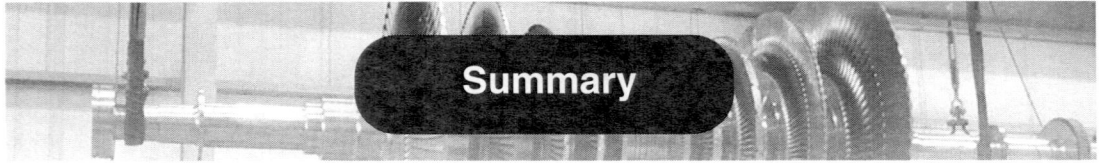

Administer the Performance Tests, and fill out Performance Profile Sheets for each trainee. If desired, trainee proficiency noted during laboratory sessions may be used to complete the Performance Test. Record the results on Craft Training Report Form 200, and submit the results to the Training Program Sponsor.

Summary

Although it is not necessary for millwrights to be able to design and generate CAD drawings, it is critical that they understand the information in a drawing system. The ability to find, read, revise, and help maintain the accuracy of the drawing system is important for many reasons. The future use of the drawings by millwrights, machinists, and engineers depends on how accurate and up-to-date the drawings for each machine are kept.

Each plant has its own system for incorporating modifications into the system and departments involved in document control. Accurate revisions in the drawings ensure the useful functioning of the drawing system. The relationship of the different drawings in a print package is important to understand, as is the ability to use them to locate all the related information.

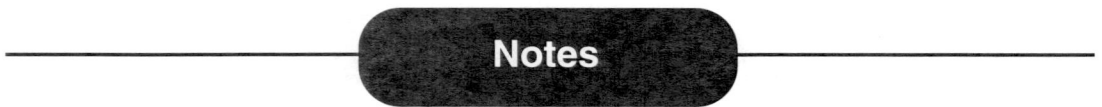

Notes

Instructor's Notes:

_____

_____

_____

Balloon: In machine drawings, a circle with letters or numbers in it that correlate to the parts in the bill of materials. It usually has a leader and arrowhead pointing to the part or subassembly.

Computer-aided drafting (CAD): Drawings that are used in all aspects of mechanical, civil, electronic, or architectural design.

Computer-aided engineering (CAE): A close interface with CAD that defines design and functionality by the use of finite element analysis, dynamic modeling, and mathematical simulation of all aspects in the use of the end product.

Computer-aided machining (CAM): CAD-generated detail drawings are fed directly into a CNC milling machine, punch press, lathe, or other computer-controlled manufacturing equipment.

Computer-integrated manufacturing (CIM): A total control system in a manufacturing environment that uses computers to link process and machine operations in an operating system.

Datum: A theoretically exact point, axis, or plane from which the location or geometric characteristics or features of a part are established and dimensioned.

Dimension: A numerical value indicated on a drawing along with lines, symbols, and notes to define the size or geometric characteristics of a part or feature.

Drawing package: A cohesive group of drawings comprised of assembly and detail drawings and a bill of materials that contain all the information needed to build or assemble a piece of equipment.

Machined parts: Parts and components that are designed and made for a specific function.

Purchased parts: Stock parts, components, or hardware that are purchased rather than made.

Revision: A change in a part of an engineering drawing that is noted on the drawing.

Title block: A section of an engineering drawing blocked off for pertinent information such as the title, drawing number, date, scale, material, draftsperson, and tolerances.

MODULE 15503-09 ◆ ADVANCED BLUEPRINT READING   3.23

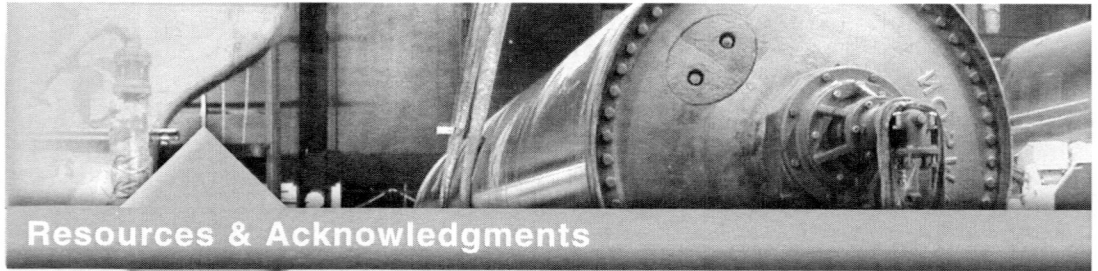

Resources & Acknowledgments

Additional Resources

This module is intended to be a thorough resource for task training. The following reference work is suggested for further study. This is optional material for continued education rather than for task training.

Geometrics II, The Application of Geometric Tolerancing Techniques. Lowell Foster. Reading, MA: Addison-Wesley Publishing Co., 1986.

Figure Credits

Utility Engineering Corp., a Zachry Group Company, 503F08

Instructor's Notes:

_____

_____

_____

_____

MODULE 15503-09 — TEACHING TIPS

The following are suggested activities or instructional methods to help you teach the material in this module.

General

When you call on someone to answer a question, the rest of the class relaxes or even tunes out because they expect that the question and answer will take place only between you and the trainee you called on. Instead, use this technique to involve more trainees in answering questions and to keep them on their toes.

1. Ask the trainees to define a term or explain a concept.
2. After one trainee has answered, ask a trainee seated nearby if the answer is right. Then ask whether a trainee in the back of the room agrees.
3. Ask the trainees to explain why they think an answer is right or wrong.
4. Use the session to clear up incorrect ideas, and encourage the trainees to learn from their mistakes.

Sections
1.0.0 – 5.0.0 *Trade Terms*

This Quick Quiz will familiarize trainees with trade terms commonly used in relation to construction drawings. You will need photocopies of the quiz provided on the following page. Trainees will need pencils. If you allow trainees to use the Trainee Guide, decrease the amount of time you give them to complete the quiz, and remind them to bring their books to class.

1. Make a photocopy of the quiz for each trainee.
2. Give trainees between 5 and 10 minutes to complete the quiz.
3. Go over the answers to the quiz.
4. Ask trainees if they have questions.

Answers to Quick Quiz

1. c
2. h
3. a
4. f
5. i
6. b
7. g
8. e

Quick Quiz   *Trade Terms*

For each description listed, identify the term that the text best describes. Write the corresponding letter in the blank provided.

_____ 1. A theoretically exact point, axis, or plane from which the location or geometrical characteristics or features of a part are established and dimensioned is a _____.

_____ 2. A change in part of an engineering drawing that is noted on the drawing is a _____.

_____ 3. A circle with letters or numbers in it that correlate to the parts in the bill of materials is a _____.

_____ 4. Parts and components that are designed and made for a specific function are called _____ parts.

_____ 5. The section of an engineering drawing that is blocked off for relevant information, such as the title, drawing number, date, scale, material, draftsperson, and tolerances is the _____.

_____ 6. A total control system in a manufacturing environment that uses computers to link process and machine operations in an operating system is called _____.

_____ 7. Components, hardware, or stock parts that are bought rather than made are called _____ parts.

_____ 8. A cohesive group of drawings that contains all of the information needed to build or assemble a piece of equipment, and that is made up of assembly and detail drawings and a bill of materials, is called a _____.

    a.   balloon
    b.   computer-integrated manufacturing
    c.   datum
    d.   dimension
    e.   drawing package
    f.   machined
    g.   purchased
    h.   revision
    i.   title block

Sections
1.0.0 – 5.0.0   *Sources of Drawings*

It is said that practice makes perfect. This is as true for blueprint reading as for any other skill. It will be of great benefit to the trainees if you can provide examples of various types of drawings. Time studying prints is time well spent. There are a number of possible sources for drawings:

1. Engineering departments of companies where your trainees work may be a good source of drawings for your class. Be sure to explain the purpose of your request.
2. Engineering companies often have old drawings that they are willing to give away.
3. Technical school CAD classes and university engineering schools may have drawings you could use to demonstrate drawing features.

Section 4.0.0   *Drawing Information*

This Quick Quiz will familiarize trainees with information found on drawings. You will need photocopies of the quiz provided on the following page. Trainees will need pencils. If you allow trainees to use the Trainee Guide, decrease the amount of time you give them to complete the quiz, and remind them to bring their books to class.

1. Make a photocopy of the quiz for each trainee.
2. Give trainees between 5 and 10 minutes to complete the quiz.
3. Go over the answers to the quiz.
4. Ask trainees if they have questions.

Answers to Quick Quiz

1. c
2. f
3. a
4. g
5. d
6. b
7. e

Quick Quiz   *Drawing Information*

For each description listed, identify the term that the text best describes. Write the corresponding letter in the blank provided.

_____ 1. Thin, dashed lines that denote alternative positions or details of an object that are not on the drawing are called _____ lines.

_____ 2. Dimensioning that uses a target designation on each feature that is listed on a chart containing the X-Y location of the feature and its description, if required, is called _____ dimensioning.

_____ 3. Dashed lines that show surfaces or edges of an object that are not visible in that view are called _____ lines.

_____ 4. The total amount that a specific dimension may vary is known as _____.

_____ 5. Documentation that contains the initials of the person making a correction, a letter or number that appears beside the area where the modification was made, and a brief description of the change is called the _____.

_____ 6. Heavy, continuous lines on a machine drawing that indicate the overall outline or shape are known as _____ lines.

_____ 7. Lines that show a cutaway view of the inner structure of an object are _____ lines.

a. hidden
b object
c. phantom
d. revision history
e. section
f. tabular
g. tolerance

MODULE 15503-09 — ANSWERS TO REVIEW QUESTIONS

Answer		Section
1.	a	1.0.0
2.	b	2.0.0
3.	c	3.0.0
4.	d	3.0.0
5.	d	3.1.0
6.	c	3.2.0
7.	b	3.3.0
8.	d	3.4.0
9.	c	4.1.0
10.	b	4.2.0
11.	a	4.2.0
12.	b	4.2.0
13.	c	4.2.0
14.	a	4.3.0
15.	c	4.3.0
16.	d	4.4.0
17.	b	4.5.0
18.	d	4.5.0
19.	d	4.7.0
20.	a	5.1.0

NCCER makes every effort to keep these textbooks up-to-date and free of technical errors. We appreciate your help in this process. If you have an idea for improving this textbook, or if you find an error, a typographical mistake, or an inaccuracy in NCCER's Contren® textbooks, please write us, using this form or a photocopy. Be sure to include the exact module number, page number, a detailed description, and the correction, if applicable. Your input will be brought to the attention of the Technical Review Committee. Thank you for your assistance.

Instructors – If you found that additional materials were necessary in order to teach this module effectively, please let us know so that we may include them in the Equipment/Materials list in the Annotated Instructor's Guide.

Write: Product Development and Revision
National Center for Construction Education and Research
3600 NW 43rd St, Bldg G, Gainesville, FL 32606

Fax: 352-334-0932

E-mail: curriculum@nccer.org

Craft _____ Module Name _____

Copyright Date _____ Module Number _____ Page Number(s) _____

Description _____
_____
_____
_____
_____

(Optional) Correction _____
_____
_____

(Optional) Your Name and Address _____
_____
_____

Optical Alignment

NCCER STANDARDIZED CRAFT TRAINING PROGRAM

The National Center for Construction Education and Research (NCCER) provides a standardized national program of accredited craft training. Key features of the program include instructor certification, competency-based training, and performance testing. The program provides trainees, instructors, and companies with a standard form of recognition through a National Craft Training Registry. The program is described in full in the *Guidelines for Accreditation*, published by NCCER. For more information on standardized craft training, contact the NCCER by writing us at 3600 NW 43rd St., Bldg. G, Gainesville, FL 32606; calling 352-334-0911; or emailing info@nccer.org. More information may be found at our website, www.nccer.org.

HOW TO USE THIS ANNOTATED INSTRUCTOR'S GUIDE

Each page presents two sections of information. The larger section displays each page exactly as it appears in the Trainee Module. The narrow column ties suggested trainee and instructor actions to each page and provides icons (detailed below) to call your attention to material, safety, audiovisual, or testing requirements. The bottom of each page includes space for your notes.

The **Audiovisual** icon indicates an appropriate time to show a transparency or other audiovisual aid.

The **Classroom** icon prompts you to define a term, stress a point, ask trainees to explain a concept, or give examples.

The **Demonstration** icon directs you to show trainees how to perform tasks.

The **Examination** icon tells you to administer the written module examination.

The **Homework** icon is placed where you may wish to assign reading for the next class, assign a project, or advise trainees to prepare for an examination.

The **Laboratory** icon is used when trainees are to practice performing tasks.

The **Materials** icon is a reminder for you to gather materials needed for classes, labs, and testing.

The **Performance Testing** icon tells you to administer a performance test or a portion thereof.

The **Safety** icon is used to emphasize safety issues. It is often keyed to *Caution* and *Warning!* statements in the Trainee Module.

The **Teaching Tip** icon indicates additional guidance is available, such as how to conduct an exercise, get the most educational value from a field trip, or encourage class participation. Teaching Tips may expand on a feature (*Think About It, Did You Know?*) or provide *Quick Quizzes* or similar exercises. You will be referred to the Teaching Tips section at the back of the module if there is additional material.

The **Combination** icon indicates that the laboratory listed corresponds with a performance task. If desired, you can note the proficiency of the trainees during the laboratory, and use it to satisfy performance testing requirements.

PREPARATION

Before teaching this module, you should review the Objectives, Performance Tasks, Materials and Equipment List, and Module Outline. Be sure to allow ample time to prepare your own training or lesson plan and gather all required materials and equipment.

MODULE OVERVIEW

This module covers optical alignment and the leveling instruments commonly used for accurately installing equipment. Basic procedures for setting up and using various types of leveling instruments are also introduced.

PREREQUISITES

Please refer to the Course Map in the Trainee Module. Prior to training with this module, it is recommended that the trainee shall have successfully completed *Core Curriculum*; *Millwright Level One*; *Millwright Level Two*; *Millwright Level Three*; *Millwright Level Four*; and *Millwright Level Five*, Modules 15501-09 through 15503-09.

OBJECTIVES

Upon completion of this module, the trainee will be able to do the following:

1. Explain how to use a theodolite, a precision tilting level, a total station, and an auto level.
2. Level equipment using optical alignment.

PERFORMANCE TASKS

Under the supervision of the instructor, the trainee should be able to do the following:

1. Check level using one of the following:

   - Theodolite
   - Precision tilting level
   - Total station
   - Auto level

MATERIALS AND EQUIPMENT LIST

Overhead projector and screen

Transparencies

Blank acetate sheets

Transparency pens

Whiteboard / chalkboard

Markers / chalk

Pencils and scratch paper

Appropriate personal protective equipment

Levels, including:
   Spirit level
   Coincidence level
   Plate level
   Circular level
   Laser level
   Precision tilting level
   Builder's level
   Automatic level

Optical tooling scales

Telescopic sights

Optical micrometer

Magnifying glasses

Tripods, mounting plates and rings

Double direct vernier

Theodolite

Theodolite with a digital display / optical plummet

Equipment needed for a plate bubble test, crosshair test, and optical plummet check

EDMI

Total station

Prism assembly

Wrenches

Copies of the Quick Quizzes*

Module Examinations**

Performance Profile Sheets**

* Located at the back of this module.

**Located in the Test Booklet.

SAFETY CONSIDERATIONS

Ensure that the trainees are equipped with appropriate personal protective equipment and know how to use it properly.

ADDITIONAL RESOURCES

This module is intended to present thorough resources for task training. The following reference works are suggested for both instructors and motivated trainees interested in further study. These are optional materials for continued education rather than for task training.

Brunson Instrument Company
www.brunson.us

Topcon Corporation
www.topcon.com

Trimble Navigation Limited
www.trimble.com

TEACHING TIME FOR THIS MODULE

An outline for use in developing your lesson plan is presented below. Note that each Roman numeral in the outline equates to one session of instruction. Each session has a suggested time period of 2½ hours. This includes 10 minutes at the beginning of each session for administrative tasks and one 10-minute break during the session. Approximately 25 hours are suggested to cover *Optical Alignment*. You will need to adjust the time required for hands-on activity and testing based on your class size and resources. Because laboratories often correspond to Performance Tasks, the proficiency of the trainees may be noted during these exercises for Performance Testing purposes.

Topic	Planned Time
Sessions I and II. Introduction; Establishing Line of Sight	
A. Introduction	_____
B. Establishing Line of Sight	_____
1. Collimation and Auto-Collimation	_____
2. Optical Instruments	_____
3. Builder's Level	_____
4. Tripods	_____
C. Laboratory	_____
Have trainees practice checking level using an auto level. This laboratory corresponds to Performance Task 1.	
Session III. Reading Theodolite Scales and Verniers; Initial Setup, Adjustment, and Checkout of a Transit/Theodolite	
A. Reading Theodolite Scales and Verniers	_____
1. Understanding Degrees, Minutes, and Seconds	_____
2. Reading Vernier Scales	_____
3. Reading Optical Scales and Digital Displays	_____
B. Initial Setup, Adjustment, and Checkout of a Transit/Theodolite	_____
1. Setting Up Using an Instrument with an Optical Plummet	_____
2. Checking Theodolite Calibration	_____

Sessions IV and V. Horizontal and Vertical Angle Measurements

A. Basic Horizontal and Vertical Angle Measurements _____

   1. Turning 90-Degree Angles _____

   2. Measuring Horizontal Angles _____

   3. Measuring Vertical Angles _____

   4. Mistakes Made When Making Angular Measurements _____

B. Laboratory _____

Have trainees practice checking level using a theodolite. This laboratory corresponds to Performance Task 1.

Session VI. Electronic Distance Measurement; Measuring Errors in Parts per Million; History of Total Stations; Prisms and Reflective Targets

A. Electronic Distance Measurement _____

   1. History _____

   2. Instruments _____

B. Measuring Errors in Parts per Million _____

C. History of Total Stations _____

D. Prisms and Reflective Targets _____

Session VII. Setup and Checkout of a Total Station

A. Total Station Controls _____

B. Initial Setup and Coarse Centering _____

C. Initializing the Total Station for Measurements _____

D. Laboratory _____

Have trainees practice checking level using a total station. This laboratory corresponds to Performance Task 1.

Session VIII. Alignment Instrument Field Checks; Trigonometric Leveling

A. Alignment Instrument Field Checks _____

   1. Geometry of Angle Measuring Instruments _____

   2. Instrument Field Checks _____

   3. Laser Beam Level Check _____

B. Trigonometric Leveling _____

Session IX. Checking Height; Using Optical Levels

A. Checking Height _____

   1. Coincident Lines of Sight _____

   2. Bucking-In _____

   3. Care of Optical Instruments _____

B. Using Optical Levels _____

C. Laboratory _____

Have trainees practice checking level using a precision tilting level. This laboratory corresponds to Performance Task 1.

Session X. Review and Testing

    A. Trade Terms Quick Quiz   _____

    B. Module Review   _____

    C. Module Examination   _____

       1. Trainees must score 70% or higher to receive recognition from NCCER.

       2. Record the testing results on Craft Training Report Form 200, and submit the results to the Training Program Sponsor.

    D. Performance Testing   _____

       1. Trainees must perform each task to the satisfaction of the instructor to receive recognition from NCCER. If applicable, proficiency noted during laboratory exercises can be used to satisfy the Performance Testing requirements.

       2. Record the testing results on Craft Training Report Form 200, and submit the results to the Training Program Sponsor.

Millwright Level Five

15504-09

Optical Alignment

Assign reading of Module 15504-09.

15504-09
Optical Alignment

Topics to be presented in this module include:

Overview

This module explains how to locate, align, and level equipment using telescopes, radio signals, or lasers to accurately determine the relative locations of points. The tools used for these tasks include the theodolite, the precision tilting level, the auto level, and the total station. Equipment made by different manufacturers will vary, but this module presents general principles that apply to the basic procedures for setting up and using each of these optical alignment tools.

Instructor's Notes:

_____

_____

_____

Objectives

When you have completed this module, you will be able to do the following:

1. Explain how to use a theodolite, a precision tilting level, a total station, and an auto level.
2. Level equipment using optical alignment.

Trade Terms

Aim	Objective lens
Alidade	Parallax
Azimuth	Peg test
Azimuth axis	Plate level
Buck-in	Position
Circular level	Reference line
Clamp	Reticle
Degree	Second
Diopter	Standards
Electro-optical	Station
instruments	Tangent screw
Elevation	Telescope axle
Elevation axis	Telescope direct
Focus	Telescope reversed
Horizontal	Telescopic sight
Infrared light	Vertical
Minute	Visible light

Required Trainee Materials

1. Pencil and paper
2. Appropriate personal protective equipment

Prerequisites

Before you begin this module, it is recommended that you successfully complete *Core Curriculum*; *Millwright Level One*; *Millwright Level Two; Millwright Level Three*; *Millwright Level Four*; and *Millwright Level Five*, Modules 15501-09 through 15503-09.

This course map shows all of the modules in the fifth level of the Millwright curriculum. The suggested training order begins at the bottom and proceeds up. Skill levels increase as you advance on the course map. The local Training Program Sponsor may adjust the training order.

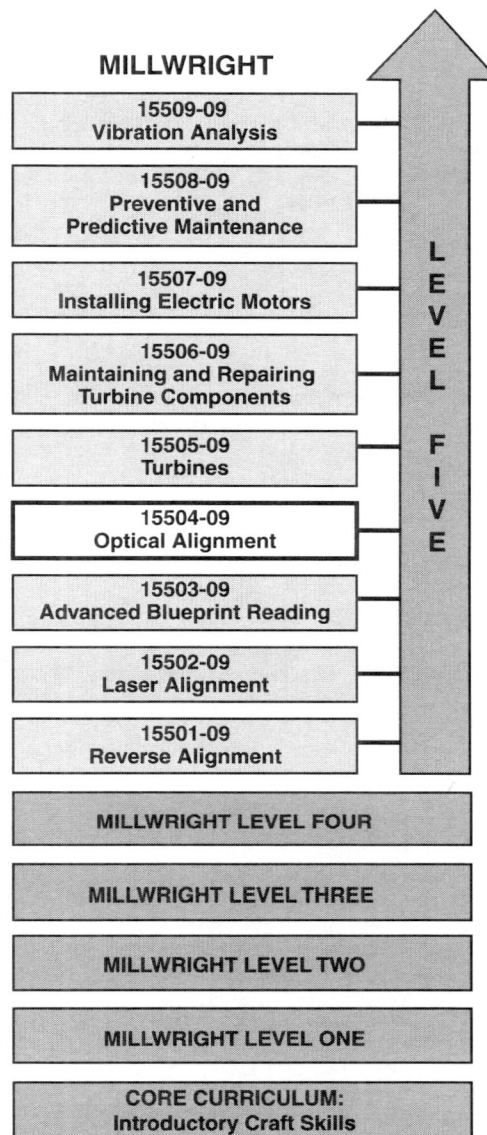

MILLWRIGHT

LEVEL FIVE

15509-09 Vibration Analysis
15508-09 Preventive and Predictive Maintenance
15507-09 Installing Electric Motors
15506-09 Maintaining and Repairing Turbine Components
15505-09 Turbines
15504-09 Optical Alignment
15503-09 Advanced Blueprint Reading
15502-09 Laser Alignment
15501-09 Reverse Alignment

MILLWRIGHT LEVEL FOUR

MILLWRIGHT LEVEL THREE

MILLWRIGHT LEVEL TWO

MILLWRIGHT LEVEL ONE

CORE CURRICULUM: Introductory Craft Skills

504CMAP.EPS

Ensure that you have everything required to teach the course. Check the Materials and Equipment List at the front of this module.

See the general Teaching Tip at the end of this module.

Explain that terms shown in bold are defined in the Glossary at the back of this module.

Show Transparency 1, Objectives, and Transparency 2, Performance Tasks. Review the goals of the module, and explain what will be expected of the trainees.

Review the modules covered in Level Five, and explain how this module fits in.

Discuss the two basic steps in the optical alignment of machinery and machine trains.

Explain the importance of keeping accurate records of alignment measurements.

Show Transparencies 3 and 4 (Figures 1 and 2). Describe the function and contents of alignment record forms.

Show Transparency 5 (Figure 3). Describe the components of a telescopic sight.

Provide examples of telescopic sights for trainees to examine.

Demonstrate and/or describe how to accurately focus a telescopic sight.

Define parallax and explain how it can be eliminated.

Describe the use of targets and the paired-line target design.

Show Transparency 6 (Figure 4).

Show Transparency 7 (Figure 5). Explain collimation and auto-collimation.

1.0.0 ◆ INTRODUCTION

Optical alignment of machinery and machine trains involves two basic steps. First, the optical equipment, either a theodolite or a level, must be set up and leveled. Then, the theodolite or total station and its reference target are used to establish **reference lines** of sight parallel to the machine. An optical level or automatic level is used to determine the levelness of a **horizontal** surface. The reference lines are used to align and mount machinery and equipment in the specified **position**.

An important part of any optical alignment procedure is documentation. The recordkeeping of each alignment measurement can be used to track changes over time and find problems, such as machine movements and foundation settling, that evolve slowly. Each plant needs to generate a form to keep alignment data consistent and to make the information accessible to everyone. *Figures 1* and *2* show samples of alignment records.

2.0.0 ◆ ESTABLISHING LINE OF SIGHT

Nearly every optical alignment procedure is based on the use of one or more lines of sight. These are established by **telescopic sights** (*Figure 3*). A telescopic sight consists of a tube with an objective lens near the front end, a **reticle** with cross lines or a similar pattern near the rear end, and an eyepiece mounted behind the reticle. A movable focusing lens is located between the **objective lens** and the reticle. Objects in front of the telescope are focused on the reticle by moving this lens. This forms an inverted image on the reticle pattern. The eyepiece erects and magnifies this image and the reticle pattern together so that the observer sees the cross lines apparently on the object at the point at which the line of sight strikes the object.

Certain precautions must be taken to accurately **focus** a telescopic sight. The eyepiece must be moved in or out by turning the eyepiece focusing ring nearest the eye to bring the reticle mark sharply into focus. This is not at an exact position, since the eye itself can change focus.

The focusing lens must be moved back and forth by turning the focusing knob to bring the object sighted into sharp focus. Because of the possibility of the eye changing focus, the image may not be on the reticle. This introduces **parallax** so that if the eye moves, the reticle pattern apparently moves with respect to the target sighted. Follow these steps to eliminate parallax:

Step 1 **Aim** the sight at a white object, and focus the eyepiece until the reticle pattern is sharp when the eye is relaxed.

Step 2 Aim the sight at the target, and focus the telescope.

Step 3 Move your eye slightly left and right or up and down.

> **NOTE**
> If there is apparent motion between the target and the reticle, eliminate it by changing the telescope focus slightly.

Step 4 Focus the eyepiece to obtain a sharp focus and continue to move the eyepiece and your eye alternately until the apparent movement is eliminated and both the object sighted and the reticle pattern are sharp. Write down the plus or minus reading on the **diopter** scale located around the focusing ring, and use this setting for future sights with this particular instrument.

The line of sight is aimed at a target. The special targets used are designed so that the error of pointing the line of sight at them is not more than 0.5 second of arc. There are many types of targets. The standard type is the paired-line target (*Figure 4*).

2.1.0 Collimation and Auto-Collimation

The principal focus of a lens is the plane at which parallel light rays that enter the lens meet behind the lens. Parallel rays can be created by a light source at an indefinite distance. When the focusing lens of a telescopic sight is placed so that parallel rays are focused on the reticle, the telescope is said to be set at infinite focus. Without a focusing lens, the same result can be attained by placing the reticle at the principal focus of the lens.

Since all optical effects are reversible, when the reticle is illuminated in the auto-collimator, the shadow of the cross lines is projected through the lens in parallel rays. Auto-collimation is a process of setting a mirror perpendicular to a telescopic line of sight. An eyepiece containing a semitransparent mirror and a light, called an auto-collimation eyepiece, is used on the telescope. When parallel rays meet on the reticle, the instrument is focused at infinity and it becomes a collimator (*Figure 5*).

When the light illuminates the reticle, the observer can see the reticle through the telescope. When the telescope is aimed at a mirror, the instrument serves as a telescopic sight and its image in the mirror serves as a collimator. When

4.2 MILLWRIGHT ◆ LEVEL FIVE

Instructor's Notes:

_____

_____

_____

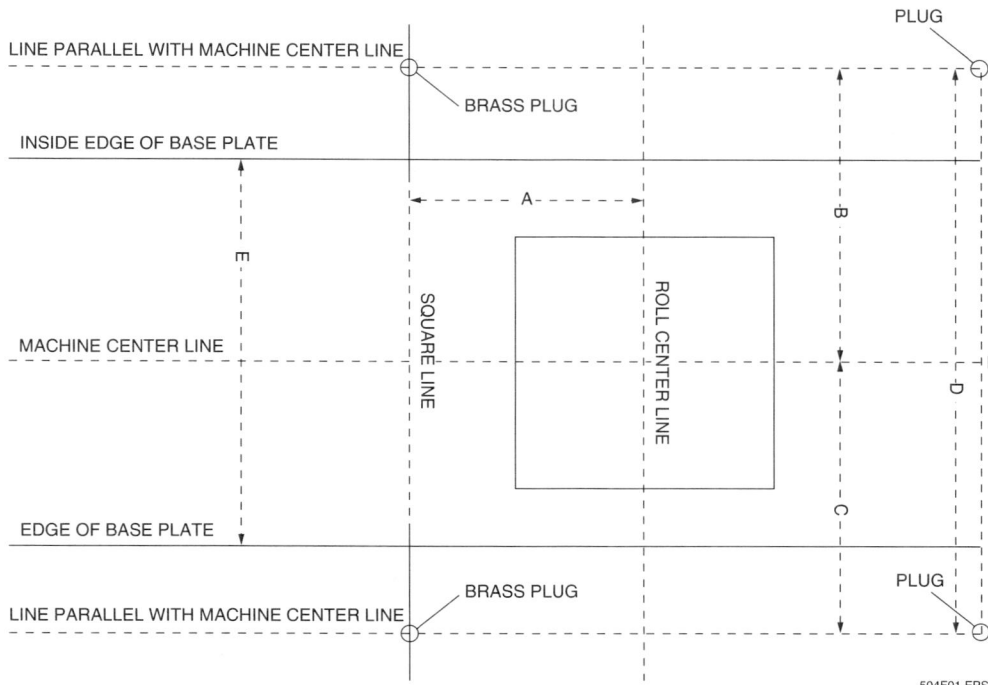

Figure 1 ◆ Alignment record, sample one.

Figure 2 ◆ Alignment record, sample two.

Discuss the accurate
formation of right
angles by auto-
collimation.

Show Transparency 8
(Figure 6).

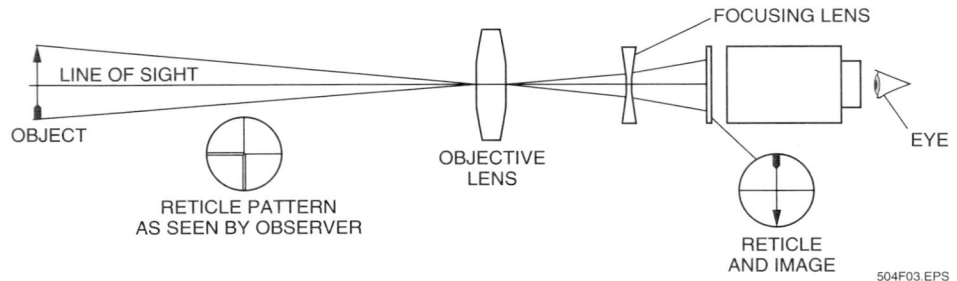

Figure 3 ◆ Telescopic sight.

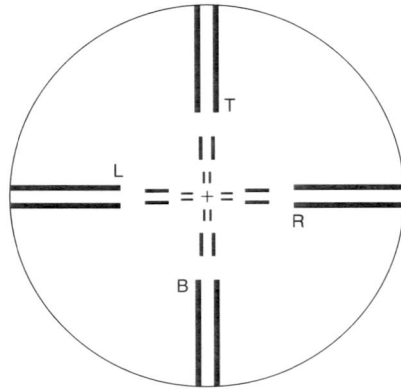

Figure 4 ◆ Paired-line target.

the observer focuses on the image of the cross lines as seen in the mirror, the instrument is at infinite focus and the observer can see the actual reticle and its reflected image. When the mirror is turned so that the reflection of the cross lines coincides with the actual cross lines, the mirror is perpendicular to the line of sight.

Accurate right angles are formed by auto-collimation, by using a pentaprism, or by using a jig transit optical auto-collimation eyepiece. In each method, a line of sight that sweeps a plane that is perpendicular to the original line of sight is established.

Figure 6 shows how a **telescope axle** mirror perpendicular to the **elevation axis** of a jig transit can be used for this purpose by auto-collimation or auto-reflection.

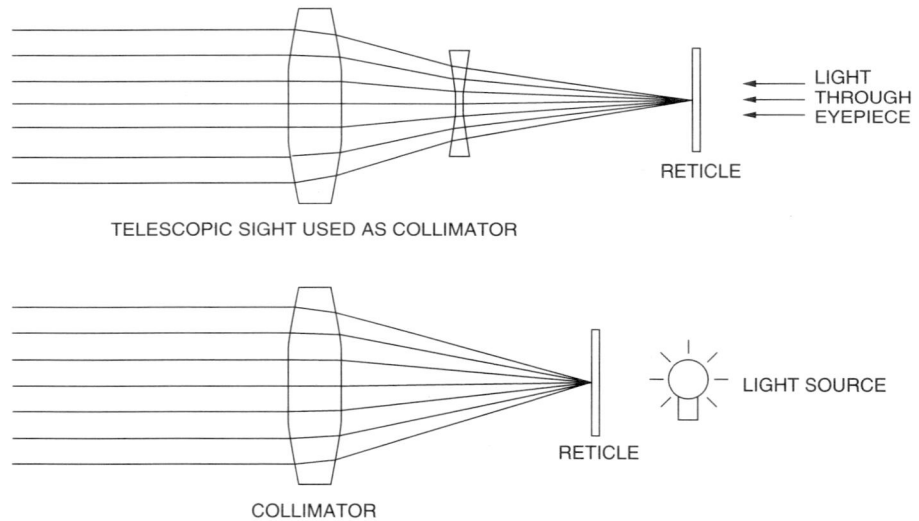

Figure 5 ◆ Collimation of light.

4.4   MILLWRIGHT   ◆   LEVEL FIVE

Instructor's Notes:
_____
_____
_____

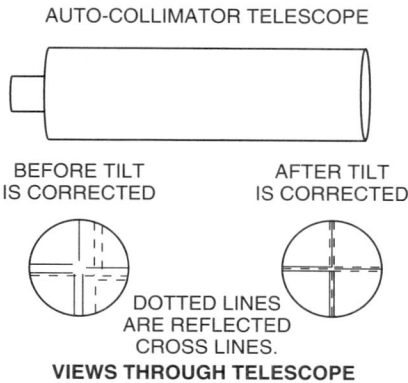

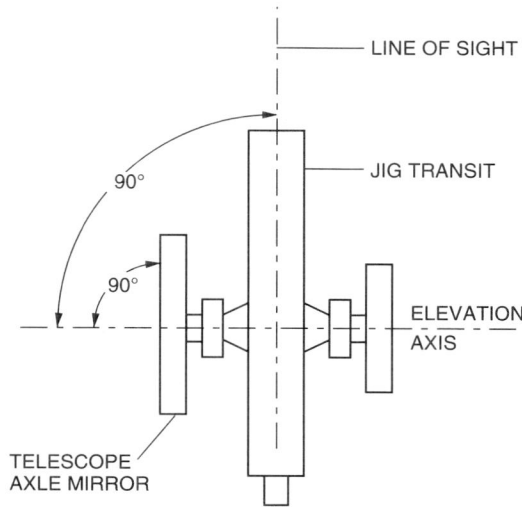

AUTO-COLLIMATOR TELESCOPE

BEFORE TILT
IS CORRECTED

AFTER TILT
IS CORRECTED

DOTTED LINES
ARE REFLECTED
CROSS LINES.

VIEWS THROUGH TELESCOPE

LINE OF SIGHT

JIG TRANSIT

90°

90°

ELEVATION
AXIS

TELESCOPE
AXLE MIRROR

504F06.EPS

Figure 6 ◆ Right angle by auto-collimation.

2.2.0 Optical Instruments

Optical equipment has many accessories used for specific alignment or leveling tasks. Since the precision and accuracy of the equipment must be maintained, special care and handling of the instruments are required.

Spirit levels provide a reference direction. Since the direction of gravity is determined so easily and accurately by spirit levels, gravity is usually used as a reference direction. In optical alignment, **vertical** means in the direction of gravity and horizontal means perpendicular to the direction of gravity. Spirit levels are used on many instruments to show the direction of gravity. Since the liquid in the vial increases in volume with increased temperature, the bubble changes length over a considerable range. Somewhat arbitrary scales are placed near or on the vials to show how far each end of the bubble has moved from a zero position. The vial is considered level when each end of the bubble is the same distance from its zero.

Coincidence levels (*Figure 7*) can be centered more accurately than spirit levels. When the two ends of a coincidence level are made to coincide, the bubble is centered. The two ends appear to move up or down together as the bubble changes length so that the device is undisturbed by changes in temperature.

Circular levels are used for approximate leveling. Their upper, inside surfaces are hemispherical.

Optical tooling scales (*Figure 8*) are used to measure short distances from lines of sight. They are graduated in units of 0.100 inch. At each 0.100 inch is a set of paired-line targets designed so that the error is not more than 0.5 second of arc. The scales are read with an optical micrometer attached in front of, or made part of, the telescopic sight used to observe them. Optical tooling scales must be held absolutely vertical when they are read with an optical instrument. If the scale is not vertical in relation to gravity, an error will be introduced into the reading. The person holding the scale must take care to level the tooling scale with respect to vertical by using an accurate circular level or two vertical levels.

Linear measurements too long to be measured with optical tooling scales are measured with optical tooling tapes or optical tooling bars.

Optical micrometers displace a line of sight parallel to the line of sight. The optical micrometer is mounted onto the front end of the telescope, with the micrometer drum on top or underneath to measure left or right or turned 90 degrees to measure up or down. The extent of the movement is read to 0.001 inch. Some optical micrometers have vernier scales that read to 0.0001 inch. *Figure 9* shows the principle of an optical micrometer.

Explain how spirit levels are used to provide a reference direction.

Describe the function and use of coincidence levels.

Explain how optical tooling scales are used to measure short distances from lines of sight.

Discuss the use of optical micrometers.

Show Transparencies 9-11 (Figures 7-9).

Provide examples of spirit levels, coincidence levels, optical tooling scales, and optical micrometers for trainees to examine.

Describe the components and operation of a builder's level. Refer to Figure 10.

Point out that a builder's level is used to establish elevations and measure horizontal angles.

Discuss the two types of leveling head systems used in builder's levels.

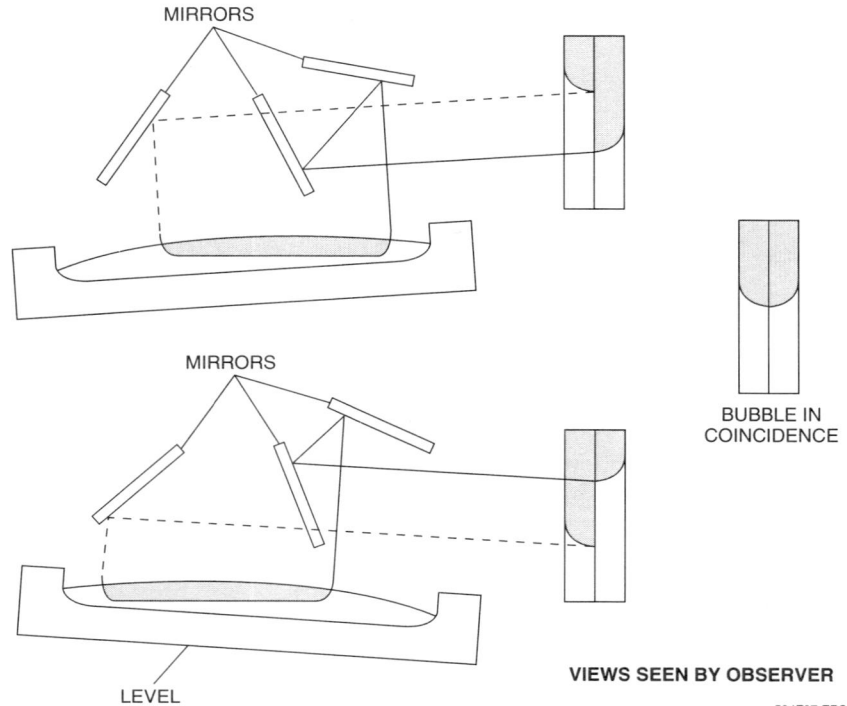

Figure 7 ◆ Coincidence level.

Figure 8 ◆ Scale pattern.

2.3.0 Builder's Level

The builder's level (*Figure 10*) is an instrument used to establish **elevations**. A builder's or optical level cannot be used to measure vertical angles. It consists of a telescope, a bubble spirit level (leveling vial) mounted parallel with the telescope, and a leveling head mounted on a circular base with a horizontal circle scale graduated in degrees. The telescope can be rotated 360 degrees for measuring horizontal angles and can be tilted slightly for sighting purposes. Builder's levels are mounted on a tripod when in use.

Depending on the model, builder's levels are made with telescope powers ranging from 12 power (12X) to 32 power (32X), with 20 power (20X) being the most common. The power of a telescope determines how much closer an object will appear when viewed through the telescope. Two types of leveling head systems are used in builder's levels: a four-screw system and a three-screw system. The advantage of the three-screw system is that it allows the instrument to be leveled more quickly. Four-screw systems are more common in older models. Coverage of the builder's level in this module focuses on its use in determining elevations.

Instructor's Notes:

_____

_____

_____

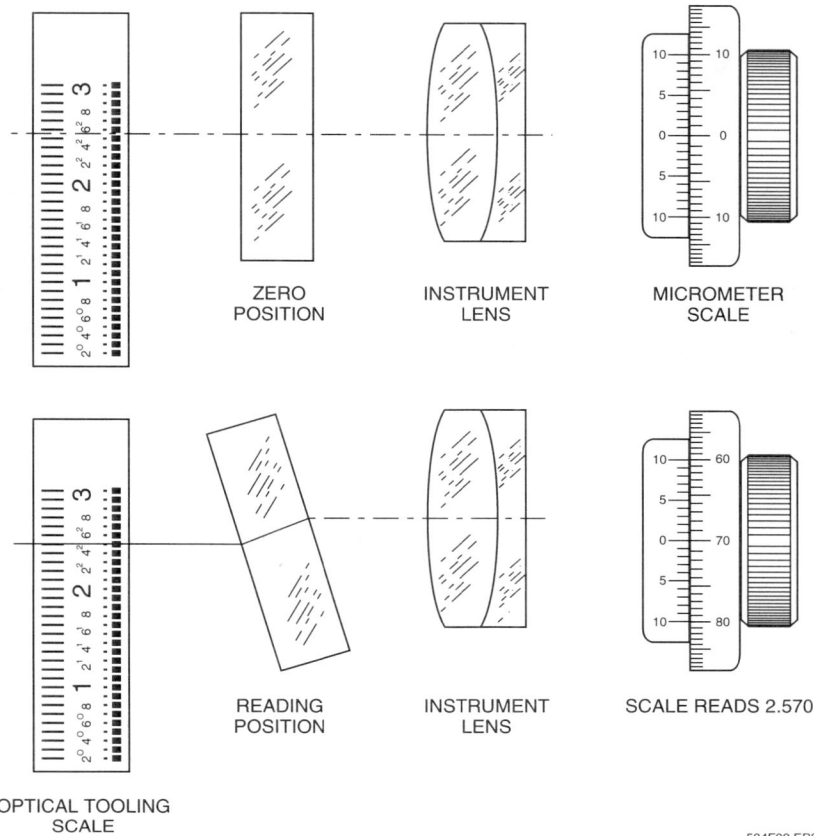

Figure 9 ◆ Optical micrometer.

ZERO POSITION

INSTRUMENT LENS

MICROMETER SCALE

READING POSITION

INSTRUMENT LENS

SCALE READS 2.570

OPTICAL TOOLING SCALE

504F09.EPS

Automatic levels (*Figure 11*) are used to perform the same measurements and operations as described for the builder's level. These instruments have a built-in compensator mechanism that works to automatically maintain a true level line of sight. Compensator instruments still have to be leveled within the range of the compensator by three screws located on the base. An automatic leveling instrument must be kept upright and should never be carried over your shoulder. It contains a prism that can be damaged if the level is carelessly handled.

2.4.0 Tripods

Because most alignment activities are performed in the field or out on the job site, it is not always easy to get sighting and leveling devices properly aligned to a given reference point. The environ-ments in which these reference points exist may damage the equipment if it is placed onto the ground or floor. Placing the equipment on the ground or floor may also make it very difficult for the instrument person to see through the sighting device; therefore, a tripod is used.

2.4.1 Tripod Basics

Many years ago, devices called tripods were developed so that the reference point could be off-set vertically and the instruments could be better positioned and more easily used. Most tripods are made of wood, aluminum, or fiberglass, and, as the name implies, they have three supporting legs. Each leg is adjustable so that the tripod can be set up on almost any surface. *Figure 12* shows a typical tripod that can be used with most alignment instruments.

Describe how automatic levels operate, and explain why they must be kept upright and handled very carefully. Refer to Figure 11.

Provide examples of a builder's level and automatic level for trainees to examine.

Show trainees how to establish elevations and measure horizontal angles using a builder's level and automatic level.

Explain the purpose and use of tripods.

Show Transparency 12 (Figure 12).

Figure 10 ◆ Builder's level.

504F10.EPS

504F11.EPS

Figure 11 ◆ Automatic level.

2.4.2 Tripod Mounting Plate and Leveling

The mounting plate on top of a tripod is typically flat and may or may not have small levels attached to it. Some mounting plates or heads have mounting rings on top of the plates for securing the instruments to the plates. Some of these mounting rings have threads inside them, and others have mounting threads outside the ring. Some form of hole in the center of the mounting plate allows for a plumb bob to be suspended from the instrument mounted on the tripod. *Figure 13* shows tripod mounting plates with instrument mounting rings for external or internal threading.

If the mounting plate of a tripod does not have levels attached to it already, a small construction

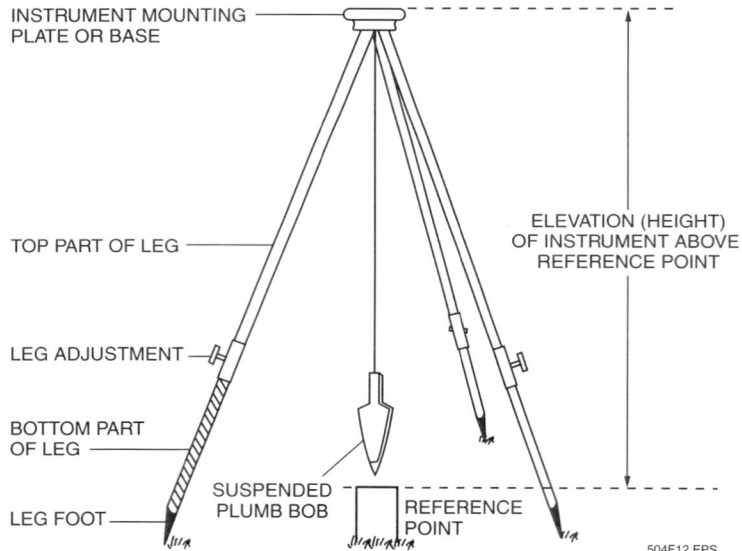

Figure 12 ◆ Tripod.

504F12.EPS

Instructor's Notes:

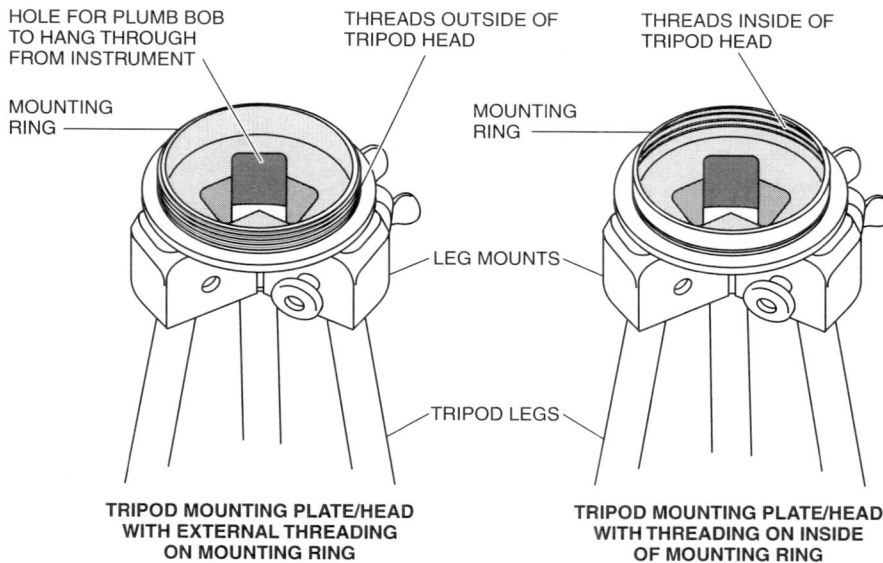

Figure 13 ◆ Variations of tripod mounting plates and rings.

504F13.EPS

level can be laid across the top of the plate to check the level of the plate. The legs of a tripod must be adjusted so that the mounting plate is as level as possible in all horizontal directions. Ensuring that the tripod is properly leveled and securely set is critical for good readings from the instruments attached to the tripod. Making sure that the tripod legs are securely set also helps ensure that the tri-pod does not get easily knocked over and possibly damage some very expensive instrument attached to it.

2.4.3 Tripod Maintenance

When the tripod is no longer needed, it needs to be removed from the work location, cleaned, lubricated as needed, inspected for damage, and properly stored if it is undamaged. Alignment instruments are only as good as the foundation on which the instruments are mounted. Taking care of the tripod on which the instruments sit is very important. Guidelines for the use and proper care and handling of tripods are as follows:

- When setting up the tripod, position the tripod legs properly. The legs should have about a 3-foot spread, positioned so that the top of the tri-pod head is horizontal.
- If the tripod's legs are adjustable, make sure that the leg levers are securely tightened.

- If setting up on dirt, make sure that the tripod points are well into the ground. Apply your full weight to each leg to prevent settlement.
- When setting up on a smooth floor or paved surface, secure the points of the legs by attach-ing chains between the legs or putting a brick or similar object in front of each leg.
- Attach the instrument to the tripod securely. Do not overtighten the attaching hardware.
- Frequently lubricate the joints and adjustable legs of the tripod using an appropriate lubri-cant.
- When not in use, protect the head of the tripod from damage.
- When transporting a tripod in a vehicle, never pile other materials on top of the tripod. Make sure to protect it from damage that can be caused by shifting equipment or materials.
- Keep the tripod clean and dry. When not in use, store the tripod in its protective case.

2.4.4 Tripod Setup

The first step in preparing to set up a tripod to be used with an instrument is determining which type of instrument is to be mounted on the tripod. Another consideration is ensuring that there is enough working space around the base of the tri-pod. The person using the instrument mounted

MODULE 15504-09 ◆ OPTICAL ALIGNMENT   4.9

on the tripod will need room to walk around the tripod legs.

Determine the following before selecting a tripod for the job:

- The kind of instrument to be mounted onto the tripod
- The environment in which the tripod is to be erected
- The location of the reference point over which the tripod is to be mounted

Perform the following steps to set up a tripod:

Step 1 Find a tripod with a mounting plate suitable for mounting the chosen instrument.

Step 2 Find a plumb bob with enough string to use with the selected tripod.

Step 3 Verify the following on the selected tripod:
- The tripod is undamaged and usable.
- The leg adjustments will both loosen and tighten properly.
- The adjustable legs can be adjusted and secured to hold the weight of the tripod and the instrument.
- The mounting plate appears to be usable for the instrument chosen.

3.0.0 ◆ READING THEODOLITE SCALES AND VERNIERS

The values for the angles measured by transits, theodolites, and other instruments are indicated in several different ways, depending on the type of instrument, its make, and the model. Common ways used by instruments to indicate the angular values are vernier scales, optical scales, and digital displays.

3.1.0 Understanding Degrees, Minutes, and Seconds

Before discussing how to read angular values using the scales and verniers associated with the different types of instruments, it is important to review the following relationships relative to angular measurements. A circle is divided into 360 parts with each part called a **degree**; therefore, one degree = $\frac{1}{360}$ of a circle. The degree is the unit of angular measurement commonly used for alignment and site layout work. The degrees are further divided into **minutes** and **seconds**, where one degree is equal to 60 minutes (60') and one minute is equal to 60 seconds (60"). This arrange-

ment, called the sexagesimal system, allows angles to be expressed in terms of degrees, minutes, and seconds. The sexagesimal system is used in the United States and most of the rest of the world to express angular measurements. However, some European countries express angular measurements using the centesimal system, in which a circle is divided into 400 parts called gons or grads. Also, in some military science work, angular measurements are based on a system that divides a circle into 6,400 parts, called mils.

3.2.0 Reading Vernier Scales

Horizontal and vertical angle vernier scales used on non-electronic instruments to indicate the values of angular measurements are the hardest to interpret and read accurately. Even though these instruments are being replaced by newer instruments that are easier to use, many are still in use. For this reason, you need to understand how to read their vernier scales. A magnifying glass may be required to read these scales.

3.2.1 Reading Horizontal Angles

To read horizontal angles, the transit's horizontal circle and vernier scales are used. Depending on the transit being used, the horizontal circle scale can be graduated in different ways. Most circle scales are labeled from 0 to 360 degrees and are graduated to the nearest 15, 20, or 30 minutes of arc. On some transits, the 360-degree circle scale is divided into four equal 90-degree quadrants, with each quadrant graduated from 0 to 90 degrees. The transit's horizontal vernier scale is used to obtain more precise readings. Vernier scales are graduated in minutes and typically can be read to either 20 or 30 seconds of arc, depending on the design. *Figure 14* shows the relationship between degrees, minutes, and seconds represented by the graduations on a simplified transit vernier.

One widely used style of vernier scale is the double direct vernier (*Figure 15*). The graduated circle scale used with this vernier is marked with two sets of numbers, one increasing in a clockwise (CW) direction from 0 to 360 degrees and the other increasing from 0 to 360 degrees in a counterclockwise (CCW) direction. This allows the vernier to be read both in the CW and CCW directions, depending on whether the eye end of the telescope is turned to the left or right, respectively. On some transits, the horizontal circle may be read by two verniers (verniers A and B) which are mounted 180 degrees apart on the transit **alidade**.

Instructor's Notes:

_____

_____

_____

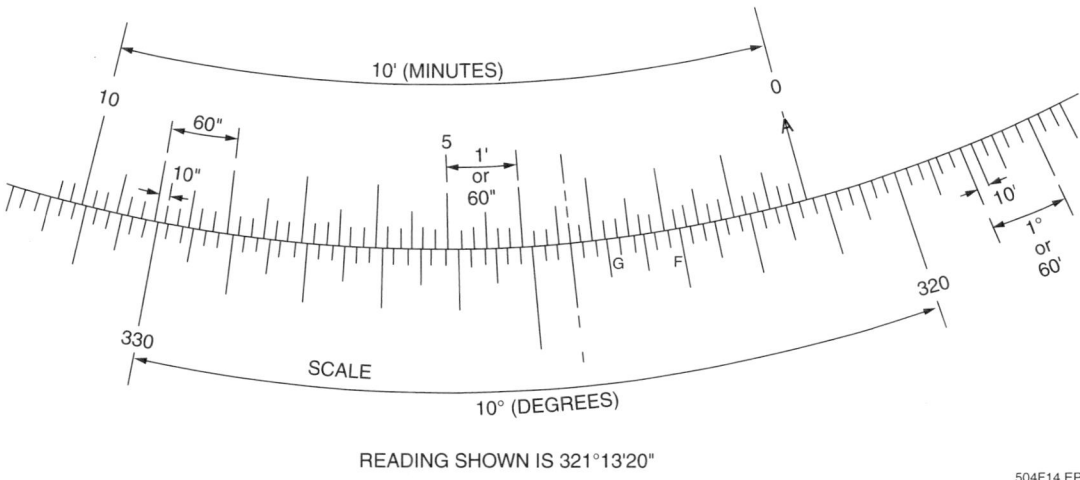

10' (MINUTES)

10° (DEGREES)

SCALE

READING SHOWN IS 321°13'20"

504F14.EPS

Figure 14 ◆ Graduation on a typical transit vernier.

The method for reading horizontal angles using a double direct vernier is described here. Reading other types of verniers is basically the same. To determine the degrees, minutes and seconds for an angular measurement, the readings from the main circle scales and vernier scales are added together. *Figure 15* shows an example of how to read a double direct vernier.

As shown in *Figure 15A*, the index mark on the vernier shows that a CW angle of 57°30' has been passed on the circle scale when read in the CW direction. When read in the CCW direction, the vernier index shows that an angle of 302 degrees has been passed.

To obtain a more precise measurement, the vernier degree scale reading is added to the circle degree reading. The vernier is read by finding a graduation on it that coincides (matches) with any division on the circle scale. On a double vernier there are two such matching lines, one for the CW angle and the other for the opposite CCW angle. Note that the graduations on a vernier are very fine and closely spaced together. For this reason, a magnifying glass may be needed in order to read the vernier accurately on older style transits.

The vernier shown in *Figure 15B* is graduated in 30 minutes capable of reading to one minute. As shown in *Figure 15B*, the vernier graduation that matches the circle graduation for a CW angle is 7'. Therefore, the CW angle being read is 57°37' (57°30' read on the circle scale +7' read on the vernier scale). The vernier graduation that matches the circle graduation for a CCW angle is 23'. Therefore, the CCW angle is 302°23' (302 degrees + 23').

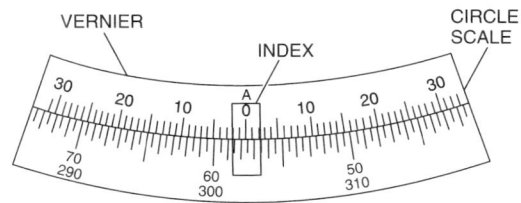

CLOCKWISE = 57°30'
COUNTERCLOCKWISE = 302°00'

(A) READING CIRCLE SCALE

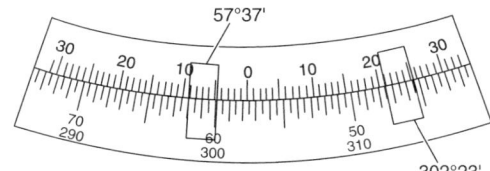

CLOCKWISE = 57°37'
COUNTERCLOCKWISE = 302°23'

(B) READING VERNIER SCALE

57°37'
302°23'
―――――――――
359°60' = 360°

(C) CHECK

504F15.EPS

Figure 15 ◆ Reading a double direct vernier.

Show Transparency 14 (Figure 14). Discuss the relationship between degrees, minutes, and seconds as shown on a transit vernier.

Describe the distinctive characteristics of the double direct vernier scale and discuss its advantages.

Show Transparencies 15 and 16 (Figures 15 and 16). Explain how to read the double direct vernier scales shown.

Provide a double direct vernier for trainees to examine.

It should be pointed out that the sum of the CW and CCW angles should always equal 360 degrees, otherwise an error has been made. To check the example, 57°37' + 302°23' = 359°60' = 360 degrees. *Figure 16* shows two more examples of double direct vernier styles.

3.2.2 Reading Vertical Angles

To measure vertical angles, theodolites have either a vertical circle scale or a vertical arc scale. Both types of scales move with the tilting motion of the telescope. Like horizontal circle scales, vertical circle scales are graduated from 0 to 360 degrees. Vertical arc scales are graduated from 0 to 45 degrees in two directions. A vertical vernier scale that is used in conjunction with the vertical circle/arc scale is attached to the frame and does not move when the telescope is tilted. Vertical angles measured with a theodolite are referenced to the horizon. When the telescope spirit level is centered, the vernier index on the vertical circle/arc should read 0°00'. Following this, the telescope is tilted up or down to the desired vertical

measurement angle and the angle is read using the vertical circle/arc and vernier scales. The manner in which the scales are read is basically the same as described previously for reading horizontal angles. Some mistakes commonly made while reading verniers include the following:

- Reading the wrong direction from zero
- Not using a magnifying glass for those models that require one
- Misinterpreting the scale graduation values
- Omitting 10, 15, 20, or 30 minutes on the circle scale when the index is beyond these marks
- Failing to read directly on the line (parallax)

3.3.0 Reading Optical Scales and Digital Displays

When compared to instruments with vernier-type scales, instruments with optical scales or electronic instruments with a digital display are simpler to use. For an electronic transit or theodolite with a digital display, the value of the measured

VERNIER READINGS

CW	342°35'	
CCW	17°25'	
CHECK	359°60' = 360°	

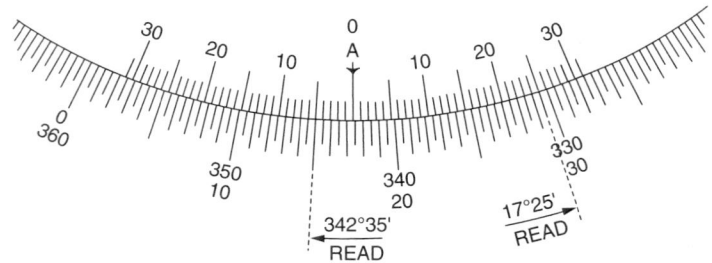

GRADUATED 30 MINUTES READING TO ONE MINUTE

VERNIER READINGS

CW	351°34'40"	
CCW	8°25'20"	
CHECK	359°59'60" = 360°	

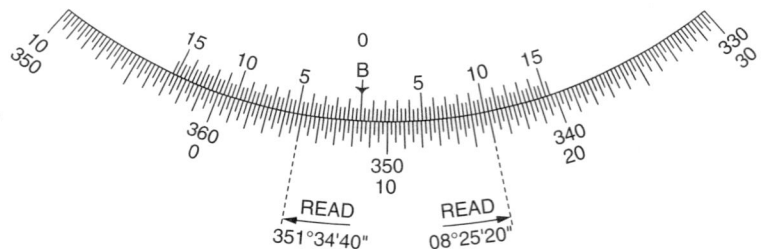

GRADUATED 15 MINUTES READING TO 20 SECONDS

504F16.EPS

Figure 16 ◆ Double direct vernier readings.

Instructor's Notes:

_____

_____

_____

horizontal or vertical angle is read directly in the same way you would read the display on a pocket calculator. *Figure 17* shows an example of an electronic transit/theodolite digital display.

Optical theodolites have specially designed scales that are viewed by looking through an optical microscope on the instrument. Depending on the design of the theodolite, there are different kinds of scales: scale reading, repeating, or directional. These scales are relatively easy to use and should be read as described in the operator's manual for the instrument being used. A brief explanation and examples of three common types of optical readouts are given here.

3.3.1 Reading a Scale Reading Theodolite

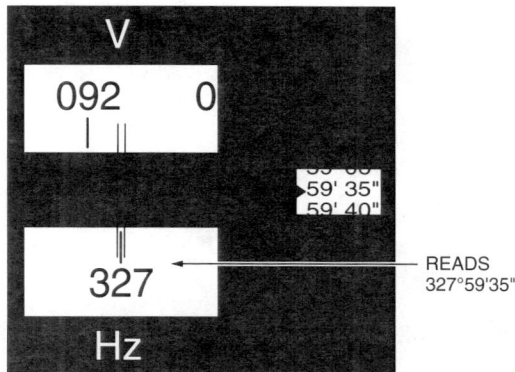

With a scale type of theodolite display, the vertical and horizontal degree circles are read directly with an optical microscope. As shown in *Figure 18A*, the scale is read where it is intersected by the degree reading from the circle.

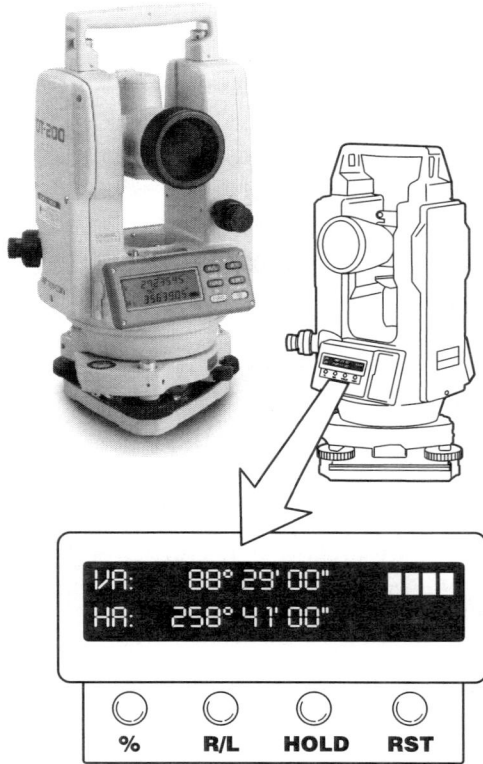

Figure 17 ◆ Electronic transit/theodolite display.

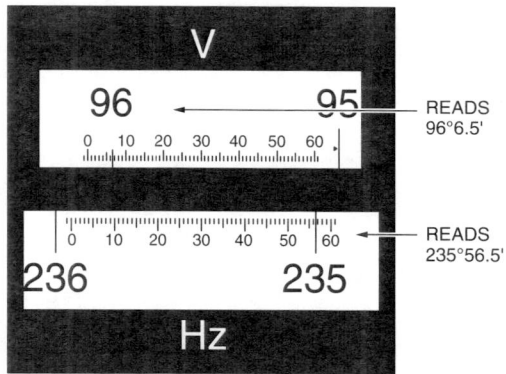

(A) SCALE READING THEODOLITE

(B) REPEATING THEODOLITE

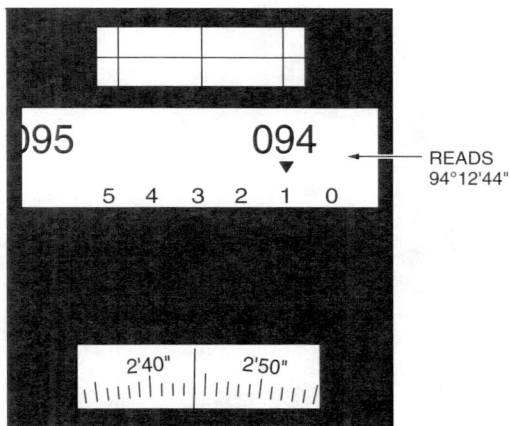

(C) DIRECTIONAL THEODOLITE

504F18.EPS

Figure 18 ◆ Common optical theodolite scales.

Show Transparency 17 (Figure 17). Point out that for an electronic transit or theodolite with a digital display, the value of the measured horizontal or vertical angle is read directly.

List the three different types of scales used on optical theodolites.

Show Transparency 18 (Figure 18). Explain how to read the scales on scale reading theodolites, repeating theodolites, and directional theodolites.

Provide a theodolite with a digital display for trainees to examine.

3.3.2 Reading a Repeating Theodolite

A repeating theodolite uses a glass degree circle and micrometer to read the turned angle. The micrometer is first used to align the degree index marks. Then the reading from the degree window and micrometer window are added together to obtain the angle, as shown in *Figure 18B*.

3.3.3 Reading a Directional Theodolite

The directional theodolite does not have a lower motion **clamp** and **tangent screw**. The horizontal circle remains fixed during a series of readings. The telescope is sighted on each of the measurement points and the directions are read. Horizontal angles are determined by calculating the differences between the directions. The reading of the angle is similar to that of the repeating theodolite. A micrometer is used to align the degree index marks. Then the reading from the degree window and micrometer window are added together to obtain the angle, as shown in *Figure 18C*.

3.3.4 Using an Optical Micrometer

Follow these steps to use the optical micrometer (*Figure 19*):

Step 1 Set the micrometer scale to zero.

Step 2 Aim the line of sight at the reference target.

Step 3 Position an optical tooling scale horizontally, with its zero end against the object whose distance from the reference line is required, and note where the line of sight strikes the scale.

Figure 19 ◆ Optical micrometer.

Step 4 Turn the elevation axis knob so that the line of sight is moved until it coincides with the smaller value of the two 0.100 inch marks on the optical tooling scale.

Step 5 Record the inches and tenths to this mark, and add the number of thousandths of an inch read from the micrometer scale.

> **NOTE**
> Units of 0.0001 inch can be read from the vernier scale.

4.0.0 ◆ INITIAL SETUP, ADJUSTMENT, AND CHECKOUT OF A TRANSIT/THEODOLITE

The initial setup of a transit and theodolite are basically the same. This section outlines the procedures for setting up a transit or theodolite directly over an established (fixed) point. This point will be the vertex for a subsequent horizontal angle measurement. Procedures are given both for setting up an older-style instrument over a point using a plumb bob and for setting up newer-style instruments over a point using a built-in optical plummet.

The initial setup of other site layout instruments such as builder's levels, automatic levels, and total stations is accomplished in basically the same way as described here for setting up a transit/theodolite.

> **NOTE**
> Always keep one hand on the instrument while setting it up.

4.1.0 Setting Up Over a Point Using an Instrument with an Optical Plummet

As explained earlier, newer transits/theodolites are equipped with a device called an optical plummet. This device allows the instrument to be optically lined up over a reference point by looking through an eyepiece and aligning the crosshairs over the point. An optical plummet consists of a set of lenses and mirrors which enable the user of the instrument to look into a viewing port on the side of the instrument (*Figure 20*). The optics and mirrors are located in the lower part of the instrument so when the base of the instrument is perfectly level, the crosshairs of the optical plummet fall on a point exactly under the center of the instrument.

504F19.EPS

Instructor's Notes:

_____

_____

_____

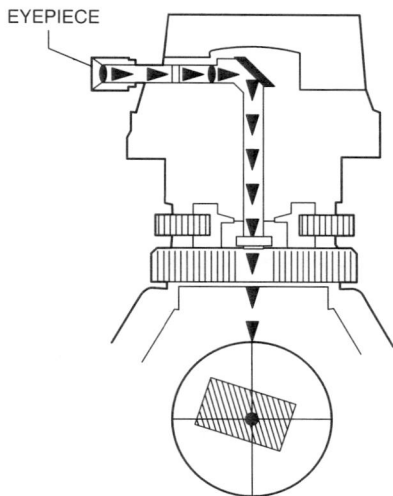

Figure 20 ◆ Optical plummet.

504F20.EPS

4.2.0 Checking Theodolite Calibration

Theodolites should be field tested for correct calibration and adjustment if you are using an instrument for the very first time or if you suspect the instrument is out of adjustment. To help you understand the principles upon which calibration of an instrument is based, a brief overview of the geometry of angle-measuring instruments is given here. Some tests commonly performed to check the calibration of an instrument are also briefly described.

4.2.1 Geometry of Angle Measuring Instruments

Transits, theodolites, and all other instruments used to make angular measurements consist essentially of an optical line of sight, which is perpendicular (at a right angle or 90 degrees) to and supported on a horizontal axis. As shown in *Figure 21*, the line of sight of the instrument is perpendicular to the horizontal axis, and the horizontal axis of the instrument is perpendicular to a vertical axis, about which it can rotate. The line of sight is perpendicular to the horizontal axis when the telescope level bubble is centered and when the vertical circle / arc is set at 90 degrees / 180 degrees / 270 degrees, or 0 degrees for vernier transits. Spirit levels mounted on the base of the instrument alidade are used to make the vertical axis coincide with the direction of gravity. These geometric relationships must be maintained in the instrument;

otherwise, the instrument will be out of calibration and any angles measured or laid out with it will be incorrect.

4.2.2 Plate Bubble Test

When in perfect adjustment, the plate bubbles on an instrument, once centered, should remain centered for all positions of the horizontal plate, unless the instrument settles or is otherwise disturbed. If either or both bubbles do not do this, adjustment is required. To determine whether the instrument requires adjustment, level the instrument carefully. Then, with one plate bubble centered over a pair of leveling screws, rotate the instrument through 180 degrees in the horizontal plane. The bubble should remain centered. If not, the instrument should be sent to a repair facility for adjustment.

4.2.3 Crosshair Tests

Crosshair tests are performed to make sure that the vertical and horizontal crosshairs are plumb and level, respectively. Both tests are easy to perform. The object of the horizontal crosshair test is to make sure that the instrument's horizontal crosshair is in a plane that is perpendicular to the vertical axis of the instrument. With a properly adjusted instrument, you should be able to place any part of the horizontal crosshair on the object or point being viewed with the telescope and still

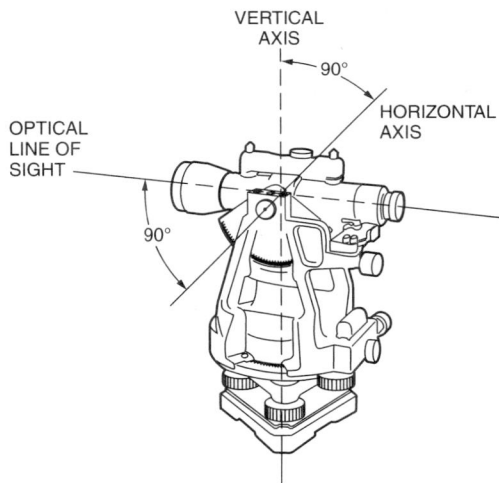

Figure 21 ◆ Geometry of an angle measurement instrument.

504F21.EPS

MODULE 15504-09 ◆ OPTICAL ALIGNMENT   4.15

Show Transparency 21 (Figure 22).

Demonstrate and/or describe the vertical crosshair test.

Explain how to check the alignment of the optical plummet's axis.

Show Transparency 22 (Figure 23).

Show trainees how to perform the optical plummet check.

Briefly describe the other calibration tests normally performed by an authorized person.

get an accurate reading. First, carefully level the instrument, then sight the horizontal crosshair on a distant target or other well-defined point (*Figure 22*). Once the crosshair is placed on the point, turn the instrument's horizontal tangent screw so that the instrument slowly rotates about its vertical axis. The horizontal crosshair should stay fixed on the point as the instrument is rotated. If any part of the crosshair moves above or below the reference point, the instrument needs adjustment and should be sent to a repair facility.

The object of the vertical crosshair test is to ensure that the instrument's vertical crosshair is in a plane that is perpendicular to the horizontal axis of the instrument. With a properly adjusted and leveled instrument, sight the vertical crosshair on a plumb bob string at rest. For a properly adjusted instrument, the vertical crosshair should coincide with the string. If any part of the crosshair does not completely coincide with the string (tilts to the left or right), the instrument needs adjustment and should be returned to a repair facility.

If either crosshair is out of adjustment, the instrument can still be used until the actual adjustment is made. This is done by using only that part of the crosshair which is nearest to the intersection point of both crosshairs.

4.2.4 Optical Plummet Check

The alignment of the optical plummet's axis relative to the vertical axis of a transit/theodolite should be checked periodically for accuracy. This is because an optical plummet can get out of adjustment, causing the instrument to be set over erroneous points.

The alignment of an optical plummet can be checked by placing its crosshairs over a reference point on the ground with the instrument at 0 degrees (*Figure 23*). Following this, the instrument

is rotated to 180 degrees, and the position of the crosshairs is checked again. If the crosshairs are superimposed on the reference point at both positions, the plummet is aligned. If not, the plummet should be adjusted in accordance with the instrument manufacturer's instructions. Typically, this is done by turning the adjustment screws on the optical plummet so that its crosshairs are positioned over a point (Point A) midway between the original points sighted for the 0- and 180-degree positions. This procedure should be repeated as necessary so that the crosshairs remain superimposed on the reference point when the instrument is at the 0 and 180-degree positions.

4.2.5 Other Tests

Other calibration checks that can be performed on an instrument are listed below. These checks are normally performed by an authorized person in accordance with the instructions given in the instrument manufacturer's operator's manual, or described in several readily available reference books pertaining to surveying.

- *Line of sight test* – Determines if the instrument's line of sight is perpendicular to the horizontal axis.

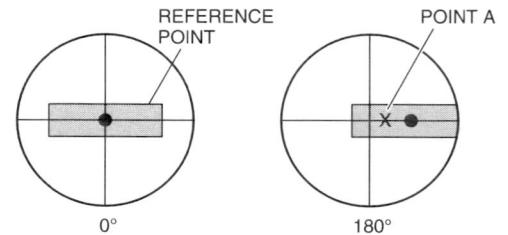

Figure 23 ◆ Optical plummet check.

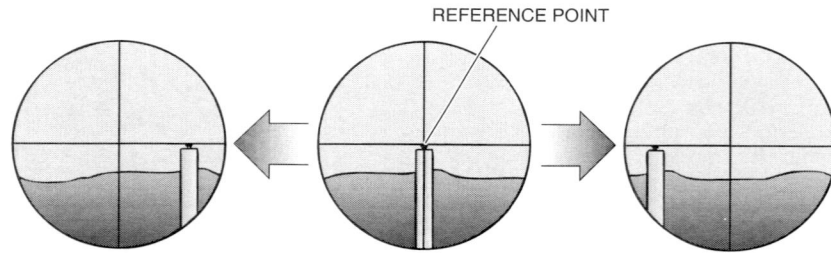

IF THE HORIZONTAL CROSSHAIR MOVES OFF THE
REFERENCE POINT, THE RETICLE NEEDS ADJUSTMENT

504F22.EPS

Figure 22 ◆ Horizontal crosshair test.

4.16 MILLWRIGHT ◆ LEVEL FIVE

Instructor's Notes:

_____

_____

_____

- *Horizontal axis test* – Determines if the instrument's horizontal axis is perpendicular to its vertical axis.
- *Telescope bubble test* – Determines if the axis of the telescope's level is parallel to its line of sight.
- *Vertical circle vernier test* – Determines if the vertical circle reads zero when the instrument is properly leveled and the telescope bubble is centered.

5.0.0 ◆ BASIC HORIZONTAL AND VERTICAL ANGLE MEASUREMENTS

This section describes basic procedures for using instruments to establish and/or measure horizontal and vertical angles. For the purpose of explanation, the procedures given here describe the use of a standard engineer's transit equipped with two independent horizontal motions (upper and lower) and vernier scales. Depending on the type of instrument actually being used, the procedures will vary somewhat. However, with the exception of the specific instrument operating procedures, the general methods described here apply to measurements made with most instruments.

5.1.0 Turning 90-Degree Angles

Turning 90-degree angles with a transit or other instrument is commonly performed when laying out the lines for a machine foundation. For decades, lines have been laid out using a transit and steel tape in a manner that traces the actual shape of the foundation. *Figure 24* shows an example of this method of layout. The procedure for zeroing the transit and turning the 90-degree angles required to perform the layout task is given here. Line AB in the figure represents the lot line and point E one corner of the foundation. Point C is the point where the line forming one side of the foundation intersects the wall.

Step 1 Set up and level the transit directly over a bench mark at point C. Sight the telescope so that the vertical crosshair is sighted on a rod or target that is on line with point B.

Step 2 Loosen both the upper and lower horizontal motion clamps, then turn the horizontal circle until the 0-degree graduation is nearly aligned with the vernier index. Tighten the upper motion clamp, then use the upper tangent screw to bring the horizontal circle 0-degree graduation into

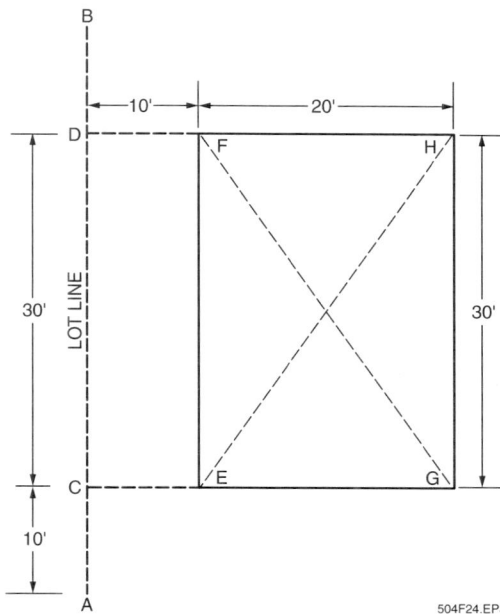

Figure 24 ◆ Site layout of machine foundation lines.

exact alignment with the vernier index. Once the vernier is set to zero, do not touch or loosen the upper motion clamp, since this can change the vernier setting.

Step 3 With the lower clamp loosened, rotate the telescope 90 degrees clockwise (to the right) as indicated by the horizontal circle so that the vertical crosshair is sighted on line CEG. Tighten the lower motion clamp, then use the lower tangent screw to bring the horizontal circle 90-degree graduation into exact alignment with the vernier index. Measure and lay out the distances from point C to points E and G along this line with a tape. Mark points E and G.

Step 4 Set up and level the transit directly over the mark at point G and sight the telescope so that the vertical crosshair is exactly on point E.

Step 5 Zero the transit horizontal circle and vernier as previously described in Step 2.

Step 6 With the lower clamp loosened, rotate the telescope 90 degrees clockwise as indicated by the horizontal circle so that the vertical crosshair is sighted on line GH.

Assign reading of Section 5.0.0.

Ensure that you have everything required for teaching this session.

Introduce the basic procedures for measuring horizontal and vertical angles.

Show Transparency 23 (Figure 24). Using this diagram, review the steps for turning 90-degree angles.

Tighten the lower motion clamp, then use the lower tangent screw to bring the horizontal circle 90-degree graduation into exact alignment with the vernier index. Measure and lay out the distance from point G to point H along this line with a tape. Mark point H.

Step 7 Set up and level the transit directly over the stake at point H and sight the telescope so that the vertical crosshair is exactly on point G.

Step 8 Zero the transit horizontal circle and vernier as previously described in Step 2.

Step 9 With the lower clamp loosened, rotate the telescope 90 degrees clockwise as indicated by the horizontal circle so that the vertical crosshair is sighted on line HF. Tighten the lower motion clamp, then use the lower tangent screw to bring the horizontal circle 90-degree graduation into exact alignment with the vernier index. Measure and lay out the distance from point H to point F along this line with a tape. Mark point F. This completes the layout. If the layout has been done correctly, the distances for lines CE and DF will be equal. Check the layout for square by measuring the length of diagonal lines EH and FG. These lines should be equal if the layout has been done correctly.

5.2.0 Measuring Horizontal Angles

To measure an unknown horizontal angle between two lines, proceed as follows. For the purpose of an example, assume that the instrument is located at point A (*Figure 25*), and it is desired to measure the angle between lines AB and AD.

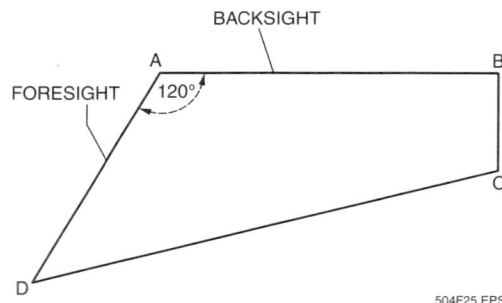

Figure 25 ◆ Measuring an angle.

Step 1 Set up and level the transit over the established reference point (point A).

Step 2 Loosen both the upper and lower horizontal motion clamps, then turn the horizontal circle until the 0-degree graduation is nearly aligned with the vernier index. Tighten the upper motion clamp, then use the upper tangent screw to bring the horizontal circle 0-degree graduation into exact alignment with the vernier index. Once the vernier is set to zero, do not touch or loosen the upper motion clamp or tangent screw because this can change the vernier setting.

Step 3 With the lower clamp loosened, rotate the telescope so that it is pointed at a target located at point B. Sight the telescope vertical crosshair so that it intersects the target, then tighten the lower motion clamp. Use the lower tangent screw to bring the vertical crosshair to an exact setting on the target. The line of sight for the telescope is now on one line or leg of the angle to be measured, and the instrument vernier scale is set to zero. This position, called the backsight, is the starting or reference point for the measurement of the angle.

Step 4 With the lower motion clamp still tightened, loosen the upper motion clamp and rotate the telescope so that it is pointed at a target located at point D. Sight the telescope vertical crosshair so that it is placed on the target; then tighten the upper motion clamp. Use the upper tangent screw to bring the vertical crosshair to an exact setting on the target. This position, called the foresight, is the ending point or line for the measurement of the angle.

Since the horizontal circle is clamped by the lower motion clamp, the zero still points toward the initial point. Because the upper motion clamp was loosened while the telescope was being rotated to the foresight position, the vernier on the upper plate moved over the circle as the transit was turned, thus indicating the value for the angle turned.

Step 5 Read and record the value of the angle turned using the circle and vernier scales. Be sure to read both the circle scale and the vernier in the direction that the vernier passed over the scale. For our example, the angle measured is 120°00'.

Instructor's Notes:

Step 6   Reverse or plunge the telescope (turn it upside down) by loosening the vertical motion clamp, then rotating it 180 degrees around its horizontal axis.

Step 7   Loosen the lower motion clamp and rotate the telescope back to the initial backsight position (point B). Sight the telescope's vertical crosshair so that it intersects the target and tighten the lower motion clamp. Use the lower tangent screw to bring the vertical crosshair to an exact setting on the target. Note that the value of the first angle turned (120 degrees) is still fixed on the vernier scale.

Step 8   Loosen the upper motion clamp, then rotate the telescope back to the foresight position (point D). Sight the telescope vertical crosshair so that it intersects the target, then tighten the upper motion clamp. Use the upper tangent screw to bring the vertical crosshair to an exact setting on the target.

Step 9   Read and record the value of the angle turned using the circle and vernier scales. The difference now is that the vernier indicates an angular value that is twice that of the first reading. For our example, the first angle reading measured 120°00', so the second reading should measure 240°00'. The values of the double angle and two times the value measured for the single angle should agree within tolerances. Also, this doubled angular value must be divided by two to obtain the average value for the angle measured.

5.2.1 Measuring Angles by Repetition

Measurement of angles by repetition is used when increased accuracy is required. It is used when it is desired to gain accuracy beyond the least count of the instrument being used. The least count is the finest reading that can be made directly on a vernier of a transit or micrometer of a theodolite.

Measurement of an angle by repetition is identical to the procedure described above for measurement by doubling except there are from four to eight repetitions made instead of only two. When recording the values measured for each angle, normal practice is to record only the first and last readings. Following this, the value for the accumulated (summed) angular measurements is divided by the number of repetitions to derive the average value for the angle. For example, assume that after six repetitions (three direct and three reversed) the summed angular value is 240°00'. The average value for the angle is then equal to 40°00' (240°00' ÷ 6). Note that it is often necessary to add 360 degrees, or multiples of 360 degrees, to the final instrument reading in order to account for the number of complete 360-degree revolutions the telescope has been turned horizontally while making the repeated measurements. For example, a 60°00' angle measured eight times causes the instrument to be turned through 480°00'; however, the instrument's scale would read only 120°00'. Therefore, to calculate the average value for the angle being measured, it is necessary to add 360° to 120°00' before dividing by 8 (480°00' ÷ 8 = 60°00').

5.2.2 Closing the Horizon

A technique called closing the horizon can be used to check the accuracy of angular measurements. Closing the horizon means that the unused angle is measured to complete the circle (*Figure 26*). When the horizon is closed and all angles at the **station** are added together, the sum should be exactly 360 degrees. Normally there will be some small error. Should the error be large (more than 30 seconds), a mistake has been made in the measurements, and they should be redone.

5.3.0 Measuring Vertical Angles

Vertical angles are measured in a similar way to horizontal angles. Vertical angles are measured with reference to the horizon when using a transit. A measurement from the horizon to a high point is a positive (+) vertical angle, and from the horizon to a low point is a negative (–) vertical angle. Obviously, it is important to record whether a vertical angle is positive or negative.

A theodolite must be carefully set up and leveled to measure a vertical angle. The telescope bubble should be centered and the vernier on the vertical vernier scale should read 0°00'. The procedure for measuring a vertical angle requires that the horizontal and vertical motion clamps be loosened. The telescope is rotated and vertically positioned so

Describe how to perform measurement of angles by repetition.

Define what is meant by the least count.

Review the technique called closing the horizon. Explain how the unused angle is measured to complete the circle.

Show Transparency 25 (Figure 26).

Describe how vertical angles are measured.

Define positive and negative vertical angles.

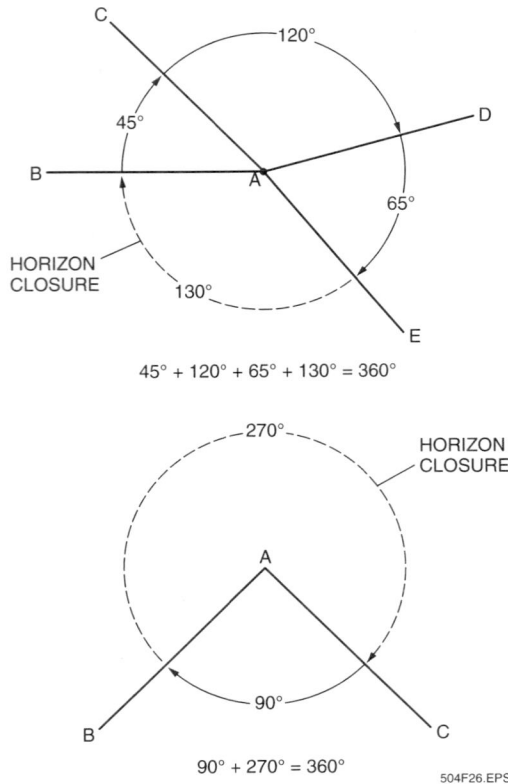

$$45° + 120° + 65° + 130° = 360°$$

$$90° + 270° = 360°$$

504F26.EPS

Figure 26 ◆ Simplified examples of horizon closures.

that the horizontal crosshair rests approximately on the point to which the vertical angle is to be measured. With the vertical motion clamp or clamps tightened, the vertical tangent screw is adjusted to set the horizontal crosshair exactly on the point. Following this, the value of the angle is read from the vertical circle and vernier. Note that if the transit has a full vertical circle and the telescope can be plunged, a more accurate reading can be obtained by measuring the vertical angle twice, once with the **telescope direct** (upright position) and once **telescope reversed** (inverted position), then averaging the two readings.

With some instrument models, the vertical angular reading with the telescope level is 90 degrees instead of 0 degrees. If this is the case, when measuring a positive vertical angle it is necessary to subtract the angle reading on the instrument from 90 degrees to obtain the actual angle. For example, if the instrument angular reading is 72°30', the actual angle being measured is 17°30' (90°00' – 72°30').

When measuring a negative vertical angle, it is necessary to subtract 90 degrees from the instrument reading to obtain the actual angle. For example, if the instrument angular reading is 118°30', the actual angle being measured is 28°30' (118°30' – 90°00').

5.4.0 Common Mistakes Made When Making Angular Measurements

Some common mistakes made when making angular measurements include the following:

- Poor setup and leveling of instruments
- Misreading the instrument scale indications
- Transposing and/or recording the wrong angle values in field notes
- Sighting on the wrong targets, marks, or lines when measuring horizontal or vertical angles
- Using the wrong instrument tangent screw
- Failure to center the telescope bubble before measuring a vertical angle
- Failure to take into account the algebraic signs for the values of vertical angles measured with a transit

6.0.0 ◆ ELECTRONIC DISTANCE MEASUREMENT

Today, electronic distance measurement (EDM) is a widely used technology that provides a fast and extremely accurate method for making long-distance measurements, including measurements over obstacles such as lakes, ravines, and roadways. EDM involves the use of instruments called electronic distance measurement instruments (EDMIs).

6.1.0 The History of Electronic Distance Measurement

The development of today's EDMIs was made possible through technology initially developed during and after World War II for radar systems. Radars are electronic systems that transmit beams of radio-frequency (RF) energy into space with the expectation that the energy will strike a target, such as an airplane in its path (*Figure 27*). When the RF energy strikes a target, the energy bounces off the target and is reflected back toward the radar transmitter. The radars are also equipped with a sensitive RF receiver that is tuned to receive the returning energy reflected off the targets. By measuring the amount of time it took for the energy to go out, bounce off the target, and return to the radar's receiver, the radar's receiver circuits compute the distance of the target from the radar.

Instructor's Notes:

_____

_____

_____

_____

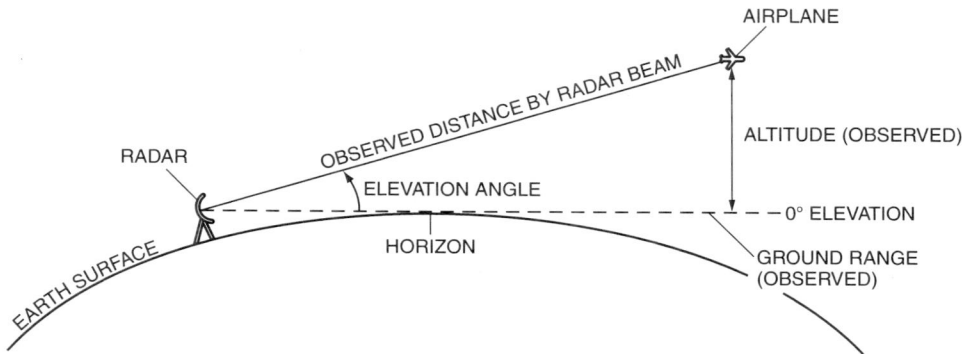

Figure 27 ◆ Radar system concept.

When the radar transmits the RF energy out into space, hits a target, receives the returned signal, and determines the distance from the radar to the target, the radar essentially creates the hypotenuse of an elevation right triangle. Since the airplanes being tracked by the radars are in the air and the radars are on the ground, the radar transmitters have to be pointed up above the horizon to detect the airplanes. In this situation, the horizon from the radar is the same as the 0-degree level line of a transit. When the radar's transmitter is pointed toward the airplane, the vertical movement of the transmitter above the horizon creates an elevation angle that is the angle theta (θ) of the right triangle.

Since radar systems can determine the hypotenuse and angle θ for each situation, trigonometric functions are used to compute the altitude of the airplane, which is the opposite side of the elevation right triangle. By also computing the adjacent side of the right triangle, the radar system can determine how far away the airplane is from the radar, measured along the ground. The adjacent side is the ground range.

The distance-measuring processes developed for radars led to the development of the first electronic distance meter by AGA Geodimeter (now part of Trimble Navigation) in 1947. If an instrument could transmit a signal into space and then time how long it took for the signal to return, then the instrument's internal processor circuits could compute the observed distance between the instrument and the object. As the distance-measuring process evolved, circuits were added to the instruments to automatically calculate the opposite and adjacent sides of the right triangle associated with elevation angles.

6.2.0 Electronic Distance Measurement Instruments

Today, there are two types of EDMIs commonly used for construction surveying and alignment work: **infrared light** and **visible light**. These types of instruments are called **electro-optical instruments**. The difference between the two types is in the wavelength of the distance measurement signal transmitted by the device. The EDM system consists of a transmitter/receiver housed in the EDMI (*Figure 28*) and one or more remotely located prism reflectors. Note that the EDMI shown in *Figure 28* is an older style instrument designed for use either as a stand-alone distance measurement instrument or in combination with an electronic theodolite. When both distance and angle measurements are required, it is mounted on top of the **standards** of an electronic theodolite to provide a total station capability. This arrangement is called a modular total station. Use of modular total stations was a common practice during the 1970s and 1980s. However, this type of EDMI remained in production by some manufacturers well into the 1990s. Modern total stations combine the EDMI and theodolite into a single instrument called a self-contained total station. Some manufacturers have also combined the EDM capability into certain models of theodolites, allowing them to be used to measure not only angles, but dimensions and distances as well.

Regardless of whether the EDMI is part of a modular or self-contained total station, the principle of operation is the same. When in use, the EDMI is set up on an appropriate tripod at one end of the line to be measured, and the reflector prism is placed at the other end (*Figure 29*).

Show Transparency 26 (Figure 27).

Compare infrared ight and visible light EDMIs.

Describe the elements of an EDM system. Refer to Figure 28.

Compare modular total stations and self-contained total stations.

Show Transparency 27 (Figure 29). Describe the setup and operation of a basic EDM system.

If possible, provide an EDMI for trainees to examine.

Demonstrate and/or describe how to use an EDMI to perform basic distance measuring.

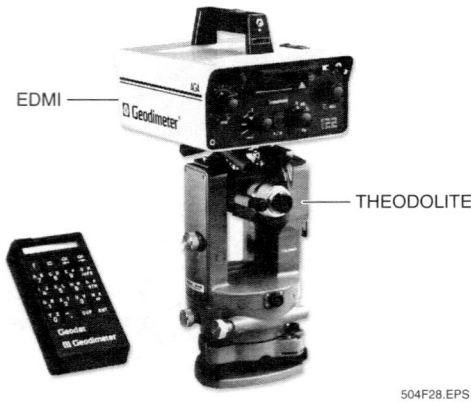

Figure 28 ◆ Older style modular total station.

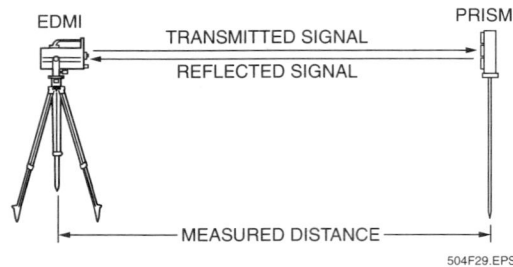

Figure 29 ◆ Basic EDM system.

The characteristics of reflective prisms are described in more detail later in this module. After the instrument and prism(s) are properly set up, a slope distance measurement is initiated by pressing the measure button and waiting a few seconds for the results to appear in the EDMI display. During a distance measurement, the EDMI directional signal is aimed at the reflector. Depending on the instrument's design, the EDMI transmits either a modulated, visible low-power light signal or an invisible infrared light signal. When the signal strikes the reflector's prism(s), it is reflected back to the EDMI in the direction from which it came. However, a specific time interval has elapsed, and a phase shift has been imparted to this reflected signal relative to the phase of the transmitted signal. Within the EDMI, the time and phase relationships between the transmitted and received signals are compared by a built-in computer (microprocessor). The resulting time and phase differences in the signals are processed electronically, and the distance value is shown on the instrument's LCD or LED display. Note that shorter distances can be measured by some laser-type EDMIs without the need for a reflector. In this case, the laser beam is reflected directly off the object back to the instrument for measurement.

Obviously, measuring distances using EDMIs is much easier than measuring distances by taping or chaining, but EDMIs do have some limitations. They need a clear line of sight in the space between the transmitter and reflector. Airborne substances, such as dust, snow, fog, or rain, can sometimes divert or block the transmitted or received signal, causing inaccurate readings. Other natural conditions, such as changes in temperature, humidity, and atmospheric pressure, can also cause errors in the readings produced by an EDMI. However, the effects of these changes are small and can be easily corrected. Some EDMIs automatically correct for changes in such variables, but some older, less sophisticated models require the operator to dial in corrections to the instrument for changing conditions.

7.0.0 ◆ UNDERSTANDING PARTS PER MILLION IN DISTANCE MEASUREMENTS

Before getting into the basics of total station setup and operation, it is necessary to briefly discuss the measurement of errors expressed in parts per million (ppm). The use of total stations and their software requires an understanding of ppm to use them properly. When using total stations, corrections involving ppm are made to compensate for errors induced in distance measurements caused by changes in the atmosphere. This is because the speed of light in air, and thus the distance measured, changes with the density of the air. Measuring the temperature and pressure allows an atmospheric correction to be calculated. This correction factor is expressed in terms of ppm.

One part per million is simply a fraction of $\frac{1}{1,000,000}$. Therefore, an error expressed as 5 ppm is $\frac{5}{1,000,000}$, an error of 10 ppm is $\frac{10}{1,000,000}$, and so on. It is easy to calculate the error related to a specific distance when the error is expressed as ppm. Simply take the distance being measured, divide by 1,000,000, and multiply by the ppm value. For example, given an error of +5 ppm and a measured distance of 1,000 feet, the error is 0.005' (1,000 ÷ 1,000,000 × 5 = 0.005'). For a measured distance of 5,000', the error is 0.025' (5,000 ÷ 1,000,000 × 5 = 0.025').

Surveyors often like to express error as a fraction always having one (1) as the numerator. For example, the error of 5 ppm ($\frac{5}{1,000,000}$) can also be

Instructor's Notes:

_____

_____

_____

_____

expressed as $\frac{1}{200,000}$; an error of 10 ppm ($\frac{10}{1,000,000}$) as $\frac{1}{100,000}$. These two fractions are equal. They are just expressed differently. Any error fraction with one in the numerator can easily be converted to its equivalent ppm value by dividing the denominator into 1,000,000. For example, the equivalent ppm value for an error fraction of $\frac{1}{200,000}$ is 5 ppm (1,000,000 ÷ 200,000 = 5).

8.0.0 ◆ THE HISTORY OF TOTAL STATIONS

As described earlier, modern total stations (*Figure 30*) are electronic instruments that combine the functions of a theodolite, an EDMI, and an internal computer (data collector) into a single instrument that can be used to make both distance and horizontal and vertical angular measurements. The telescope component of the total station serves two functions: first as an optical scope for sighting purposes, and second as a sending and receiving unit for the EDM signal.

Total stations, originally called electronic tachometers, were first introduced in Europe and the United States in the 1970s. The name total station, now widely used when referring to this type of instrument, was first used by the Hewlett-Packard Company when naming their Model 3810A instrument. The early total stations were manual total stations in which it was necessary to read the horizontal and vertical angles by eye, similar to reading angles when using a conventional optical theodolite. The only value that could be read electronically from a digital display was the slope distance. Later developments in technology led to the production of total stations that were semiautomatic instruments. With these instruments, the user had to read horizontal angles from a horizontal circle manually, but the vertical angle readings were sensed electronically and the values shown digitally. The slope distance was measured electronically and displayed. In addition to the slope distance, these instruments had an added capability that enabled them to calculate the horizontal and vertical distance components.

Modern total stations are battery-powered, automatic (smart) instruments. They measure the slope distance from the instrument to the reflector, along with the vertical and horizontal angles. The unit's microprocessor then computes the horizontal and vertical components of the slope distance. Using the computed components of the slope distance and the **azimuth** of the line, the microprocessor determines the north-south and east-west components of the line and the coordinates of the new point. These coordinates are then stored in memory. These instruments typically use infrared as the carrier signal for distance measurements.

The distance accuracy of a total station normally depends on the price and quality of the instrument. Total stations used for construction site layout work can generally measure distances ranging between 2,000' (600 meters) and 3,600' (1,100 meters) when using a single prism. Manufacturers state the accuracy of their instruments in terms of a constant and scalar instrument error. For example, the distance accuracy may be stated as ± (5 mm +5 ppm). The constant error part (5 mm) does not change regardless of whether a long or short distance is being measured. The scalar error part (+5 ppm) is proportional to the distance being measured. This means that the standard deviation of a single measurement with this instrument is a combination of 5 mm (0.016') and 5 ppm, which varies depending on the distance being measured. For example, in a measurement of 1,000', the error from the scalar part is 0.005'. The constant and scalar parts are added to determine the error for a single measurement. For our example of 1,000', the combined error is ±0.021' (0.016' + 0.005').

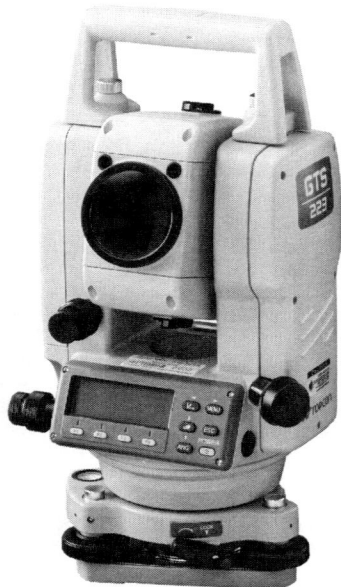

504F30.EPS

Figure 30 ◆ Modern total station.

Explain how to convert an error fraction to its equivalent ppm value.

Review the history of the development of total stations.

Discuss the function and operation of modern total stations. Refer to Figure 30.

Provide a total station for trainees to examine, if available.

Show trainees how to make distance and horizontal/vertical measurements using a total station.

Note that manufacturers state the accuracy of their total stations by means of constant and scalar instrument error. Provide examples for the trainees.

> **NOTE**
>
> To convert the constant error value given in milli-
> meters into an equivalent foot value, multiply the
> constant error value by 0.00328. For example, a
> constant error value of 5 mm is equal to 0.016'
> ($5 \times 0.00328 = 0.016$). To determine the scalar
> error for a specific distance being measured,
> divide the distance by 1,000,000, then multiply by
> the given ppm value. For example, when measur-
> ing a distance of 1,000' with a given error of
> +5 ppm, the error is 0.005' ($1,000 \div 1,000,000 \times 5 = 0.005$).

Modern total stations have a built-in memory capable of storing data for thousands of layout points and/or data records. The amount of memory in a unit is normally determined by the price of the instrument, with more expensive units having larger memory capabilities. Most total stations can be interfaced with electronic field books and external data collectors to store data for thousands of points or to lay out previously calculated information.

Under software control, two-way data flow between the total station and a remote location is possible via a standard (RS-232) communication interface. This capability enables the data collected and stored in the instrument to be recalled and downloaded to a local or remote printer for hard copy, or to a computer for calculations. It also enables data that has been processed at a remote location to be sent to the instrument and downloaded for field use.

Some of the latest technology improvements for total stations are as follows:

- Units that allow high-accuracy distance measurements to be made with or without the use of prisms or reflective sheet targets.
- Wireless control of a total station that allows one person to control the total station operation from a remote handheld controller at the target point itself, or from any other desired location.
- Real-time positioning via total stations that are integrated into a global positioning system.

9.0.0 ◆ PRISMS AND REFLECTIVE TARGETS

One or more prisms are used with total stations to reflect the transmitted distance measurement signal back to the instrument. A single prism is typically a glass cube corner prism that will reflect light rays back exactly in the same direction as they are received. When higher measurement accuracy is required, the prism should be mounted on a tripod via a tribrach. Typically, the tribrach is equipped with an optical plummet used to center the prism directly over the point of measurement. If lower levels of measurement accuracy are acceptable, the prism can be attached to an adjustable prism pole. The prism pole is then held vertically as indicated by a bull's eye level on the point of measurement. The pole should be held as motionless as possible. Regardless of the mounting method, the height of the prism is normally set to match the line of sight of the total station.

Some prism assemblies are stationary, while others have the capability to be tilted up and down (*Figure 31*) so that they can be positioned perpendicular to the light signal being transmitted from a total station located at a much higher or lower position. The length of the distance to be measured normally determines the number of prisms used. The longer the distance, the more prisms required. For example, the maximum range for one manufacturer's total station is stated as 3,000 meters (9,900'), 4,000 meters (13,200'), or 5,000 meters (16,400') when using one, three, or nine prisms, respectively. The distances specified are achieved with conditions of a slight haze with visibility of about 12.5 miles and moderate sunlight with light heat shimmer. The quality and cleanliness of the prisms being used also greatly affects the range and accuracy of the distance measurement.

Prism manufacturers specify a prism offset constant for each type of prism. This constant relates to the location of the prism in relation to the center of the prism pole. Offsets of 0 mm, –18 mm, –30 mm and –40 mm are common. The prism offset is

504F31.EPS

Figure 31 ◆ Prism.

Instructor's Notes:
_____
_____
_____
_____

done to help reduce any angular error caused when the prism is not pointed perfectly at the instrument. When using a total station, the correct prism constant for the prism in use must be entered into the total station software to achieve accurate measurement results. Failure to do so can result in an automatic error in every measurement. If you do not know the correct prism offset constant for the prism you are using, refer to the prism manufacturer's specifications.

Reflective target sheets can also be used instead of a prism when measuring distance with a total station. Reflective target sheets are self-adhesive sheets that come in a variety of sizes ranging from about 5 × 5 mm up to 90 × 90 mm. These sheets are designed to stick to almost any type of dry surface. Reflective sheets can also be mounted in a variety of different holders that allow them to be rotated, mounted on steel surfaces via a magnetic base, or mounted in female bolt holes. The size of the reflective sheet used determines the maximum range and accuracy of the distance measurements that can be made.

10.0.0 ◆ SETUP AND CHECKOUT OF A TOTAL STATION

There are numerous manufacturers and different models of total stations available. The controls and capabilities can differ widely with each make and model of total station. Before operating a particular make or model of instrument for the first time, you should thoroughly study the operator's/user's instruction manual (*Figure 32*) to familiarize yourself with the instrument, its controls, and software menus. Many manufacturers also have pocket-sized guides that summarize important instructions for operating their instruments. This section gives some general guidelines for setting up and initializing a typical total station. Because of the vast differences in the controls, displays, and software of total station instruments, you must operate the instrument you are using as directed in the manufacturer's instructions for that instrument.

10.1.0 Total Station Controls

Figure 33 shows the names and functions for the controls and indicators of a typical total station. Excluding the display unit and related keys, the functions of most other controls for a total station are the same as those used with transits and theodolites. These controls include the telescope focusing knob, leveling screws, horizontal and vertical motion clamps and related tangent screws, the optical plummet, and the plate and circular levels.

504F32.EPS

Figure 32 ◆ Total station instruction manual.

The tribrach fixing lever shown in *Figure 33* is used to attach the instrument to its tribrach base. The serial RS-232C connector shown is used to connect the instrument to a computer or data collector. This feature enables the total station to send measured data to the computer or the computer to send a set of preset data to the total station.

The display unit (*Figure 34*) consists of a liquid crystal diode (LCD) display and related operator keys. A display typically has the capability of displaying about four lines of text or characters at about 20 characters per line. Normally, the top three lines are used to display the values for measured data, while the bottom line is used to display programmed options controlled by soft key functions.

The number of keys on a display unit and their functions vary widely among manufacturers and models. All total stations have two categories of operating keys: hard keys and soft keys. Hard keys are used to turn the instrument on and off, select modes of operation, make menu selections, enter alpha-numeric characters, and similar operations. Soft keys have functions that change depending on the selected mode of instrument operation. Soft keys are typically labeled F1, F2, and so on. The display unit shown in *Figure 34* has six hard keys with the functions indicated and four soft keys, designated F1 through F4. As mentioned earlier, the program-driven functions for soft keys F1 through F4 are shown on the bottom line of the display. The specific function associated with each key is indicated by the displayed message. In addition to task-specific hard keys,

Describe how reflective target sheets can be used instead of a prism when measuring distance with a total station.

Assign reading of Sections 10.0.0–10.3.0.

Ensure that you have everything required for teaching this session.

Emphasize the importance of following the manufacturer's instructions for a specific total station.

Discuss the controls and indicators on total stations.

Describe the function of the tribrach fixing lever and serial connector.

Show Transparency 28 (Figure 33).

Show Transparency 29 (Figure 34). Describe the functions of the display keys on a typical total station.

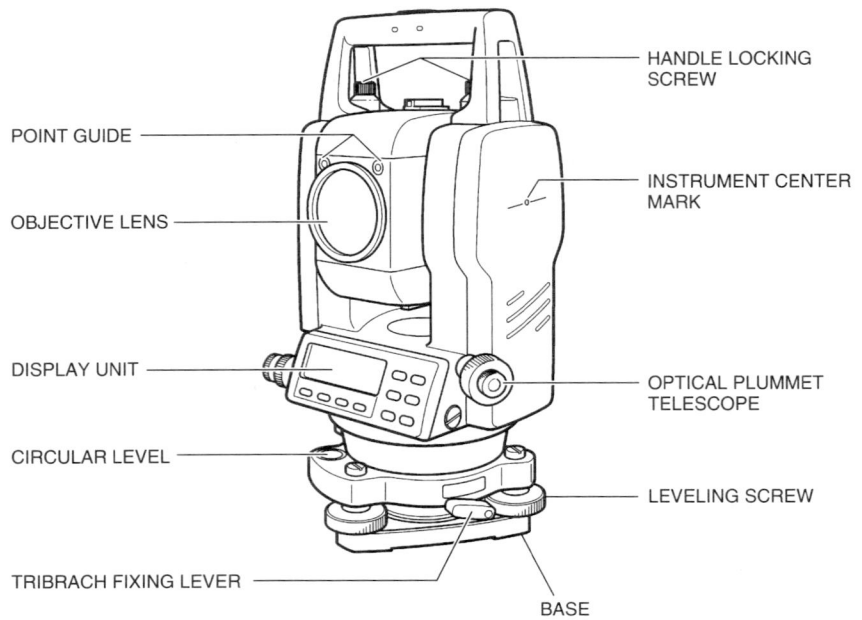

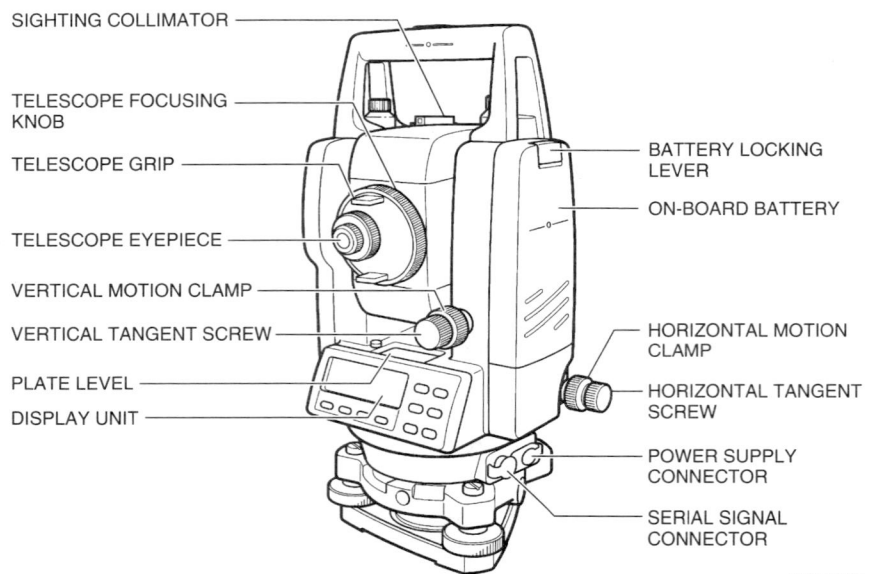

Figure 33 ◆ Total station controls and indicators.

504F33.EPS

Instructor's Notes:

_____

_____

_____

_____

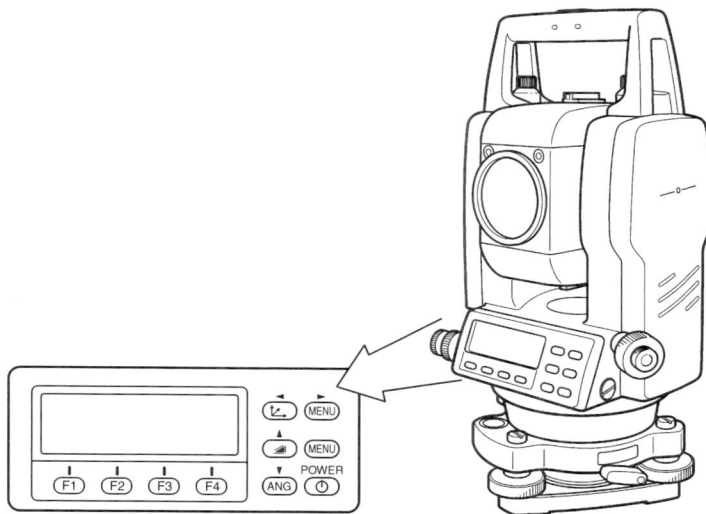

Keys	Name of Key	Function
⌐↳	Coordinate meas. key	Coordinate measurement mode
◢	Distance meas. key	Distance measurement mode
ANG	Angle meas. key	Angle measurement mode
MENU	Menu key	Switches between menu mode and normal mode. Also used to set application measurements and adjust in menu mode.
ESC	Escape key	• Returns to the measurement mode or previous layer mode from the mode set. • Used to switch to DATA COLLECTION mode or LAYOUT mode directly from normal mode.
⏻	Power source key	ON/OFF power source
F1~F4	Soft key (function key)	Responds to the message displayed.

504F34.EPS

Figure 34 ◆ Total station display unit.

Describe the abbreviations and modes on a total station display unit. Note that these can vary with manufacturer and model.

Show Transparency 30 (Table 1).

Show Transparency 31 (Figure 35).

Show trainees the controls and indicators on a total station, if available. Demonstrate the display unit's functions and capabilities.

some instruments also have a full alpha-numeric hard key set used to enter alpha-numeric characters and message strings directly from the instrument keyboard. Others, like the one shown, enter alpha-numeric characters and text strings via soft keys and menu selections.

Abbreviations used to identify measured parameter values and selectable functions on the instrument display vary with manufacturer and model of the instrument. *Table 1* shows an example of the abbreviations used for an instrument made by one manufacturer to identify the mea-

sured parameters and their units. *Figure 35* shows example angle, distance, and coordinate measurement mode screens that show the use of some of these abbreviations. Note that the bottom line of the display in *Figure 35* shows some typical abbreviations used with the soft keys. Abbreviations used with soft keys are numerous and vary with the mode of instrument operation selected. For this reason, you should always refer to the operator's instructions for the instrument being used to identify the exact meaning of each abbreviation.

Table 1 Example Abbreviations Used to Identify Measured Parameters

Display	Parameter	Display	Parameter
V	Vertical angle	HR	Horizontal angle right
N	N coordinate	HL	Horizontal angle left
E	E coordinate	HD	Horizontal distance
Z	Z coordinate	VD	Relative elevation
*	EDM working	SD	Slope distance
M	Meter	Ft	Feet
		Fi	Feet and inches

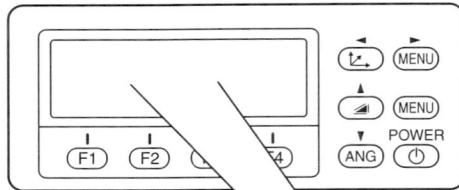

Angle measurement mode

```
V  :  90° 10' 20"
HR:  120° 30' 40"

OSET  HOLD  HSET P1↓
  TILT   REP    V%   P2↓
     H-BZ  R/L  CMPS  P3↓
        ⋮      ⋮     ⋮      ⋮
      [F1]   [F2]   [F3]   [F4]
```

Soft keys

Distance measurement mode

```
HR : 120° 30' 40"
HD*: [ r ]      < < m
VD :              m
MEAS MODE S/A   P1↓
   OFSET S.O. m/f/i    P2↓
```

Coordinates measurement mode

```
N  :      123.456 m
E  :       34.567 m
E  :       78.912 m
MEAS MODE  S/A    P1↓
   R.HT INSHT OCC      P2↓
      OFSET --- m/f/i      P3↓
```

504F35.EPS

Figure 35 ◆ Total station display.

Instructor's Notes:

10.2.0 Total Station Initial Setup and Coarse Centering

The basic leveling and angle-positioning controls of a total station are similar to those of an electronic theodolite. A total station mounted on a tripod must still be initially leveled before the built-in auto-leveling (compensator) device activates when the instrument is turned on. General instructions for the initial setup and coarse centering of a total station mounted on a tripod over an established point are given here.

> **NOTE**
>
> Before attempting to set up and initialize a total station, make sure that the battery for the instrument you intend to use is fully charged, or has sufficient charge so that you can complete the measurement task. Have a second fully charged battery available should it be needed. Depending on the instrument, battery charge life can vary widely. Typically, a fully charged battery will last about 5 to 10 hours when making a combination of distance and angle measurements. If making mostly distance measurements, the battery discharges at a much faster rate.

Step 1 Open up the tripod and extend its legs. Adjust the length of the legs so that the total station telescope will be at eye level. Set one leg (anchor leg) of the tripod firmly on the ground. Holding the other two legs and keeping the anchor leg in the same position, move the tripod until is centered over the ground setup point. Anchor the two remaining legs. The tripod should be exactly over the setup point with the tripod head plate as level as possible.

Step 2 Carefully remove the instrument from its case and attach it to the tripod head plate. The instrument leveling screws should be in mid-position.

Step 3 Coarse level the instrument using the circular level by adjusting the length of the two tripod legs (not the anchor leg) to bring the bubble to the center of the circular level.

Step 4 Level the instrument by using the **plate level**. Loosen the horizontal motion clamp and rotate the instrument so that the plate level is parallel with two of the

leveling screws (*Figure 36A*). Level the instrument by simultaneously turning the two leveling screws at the same rate in opposite directions as needed to center the plate level bubble. When rotating the two leveling screws, the plate level bubble will always follow the direction of your left-hand thumb. Turning both screws in (counterclockwise) moves the bubble to the right. Turning both screws out (clockwise) moves the bubble to the left. The left thumb rule also applies when the left hand is used to adjust the remaining single leveling screw.

Step 5 Turn the instrument 90 degrees horizontally so that the plate level is perpendicular to the first two leveling screws (*Figure 36B*). Level the instrument again using the remaining leveling screw.

Step 6 Repeat Steps 4 and 5 for each 90 degrees of instrument rotation, and check that the plate level bubble is centered at all four points.

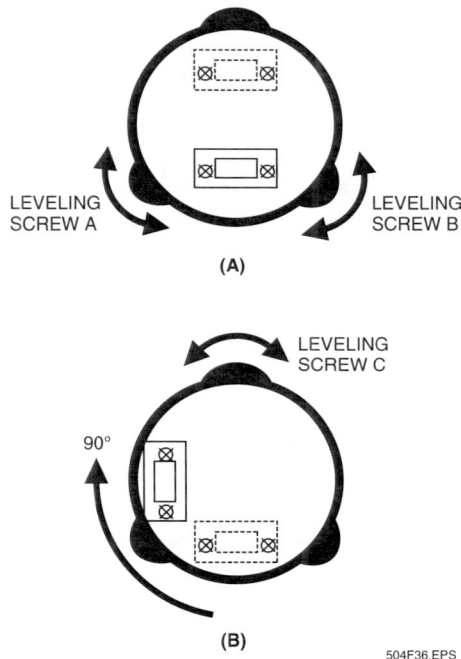

Figure 36 ◆ Leveling instrument with plate level.

Review the steps for the initial setup and coarse centering of a total station mounted on a tripod over an established point.

Note that sufficiently charged batteries/backup batteries should be on hand before setting up and initializing a total station.

Show Transparency 32 (Figure 36). Describe the procedure for leveling a total station using the plate level.

Point out that if the total station is out of its automatic compensation range, the manual leveling process will have to be repeated.

Warn trainees against sighting the total station telescope on the sun or other strong light source.

Note that parallax causes layout errors. Explain how parallax can be corrected.

Discuss the procedures required to achieve accurate measurements.

Stress the need to allow the total station to adjust to the ambient temperature.

Describe the general procedure for initializing a total station.

Step 7   Adjust the eyepiece of the optical plummet. Loosen the tripod screw, then slide (do not rotate) the instrument as necessary so that the optical plummet crosshairs are centered directly on the setup point.

Step 8   Repeat Steps 4 through 7 as necessary until the instrument is level and the optical plummet is centered directly over the established point.

Step 9   Turn on the instrument to activate the instrument's automatic leveling compensator. If the instrument is out of the automatic compensation range, it will indicate this condition via a message on the instrument display screen. If the instrument is out of its automatic compensation range, it is necessary to repeat the manual leveling process.

WARNING!

Do not sight the total station telescope on the sun or strong light sources. Doing so can cause irreparable damage to the eyes. It will also damage the instrument.

NOTE

Failure to properly focus the telescope crosshairs will cause parallax. Parallax occurs when there is an apparent movement of the crosshairs on the rod or object being viewed as the eye moves. If this occurs, layout errors can result. You can easily check for parallax by looking at the rod or object being viewed and moving your head slightly while looking at the crosshairs. If they stay on the same spot, no parallax exists.

Step 10   Sight the telescope exactly on a distant rod, prism, or other target, then turn the telescope eyepiece to focus the telescope so that the crosshairs and target are sharply defined. Check the telescope for parallax by moving your head slightly while looking through the eyepiece. There must be no relative movement between the crosshairs and the target. Repeat this step to refocus the telescope, if necessary.

10.3.0 Initializing the Total Station for Measurements

To get accurate measurements, the total station must be set up directly over the setup point as previously described. Next, the operator must initialize the instrument by entering certain reference data and parameter values needed to obtain accurate measurement results. This information typically includes the following:

- Units of measurement
- Constants
- Reference elevation
- Initial coordinates
- Reference direction

CAUTION

Before initializing and using a total station, allow sufficient time for it to adapt itself to the site ambient temperature. Sudden changes in temperature to the instrument or prism may result in a reduction in measuring distance range. This situation might occur when an instrument is removed from a heated truck in preparation for its use under cold weather conditions. Precautions should also be taken not to subject the instrument to extreme heat longer than necessary.

Initializing a total station must be done according to the manufacturer's instructions for the instrument being used. The following provides a general procedure for initializing a total station:

Step 1   Turn the instrument on. Check the battery power status. If the battery icon indicates a low battery condition, replace it with a fully charged battery. Ensure that the automatic leveling compensator is activated.

Step 2   Select the measurement units used for angle measurements (decimal degrees or decimal gons) and for distance measurements (feet, feet and inches, or meters). Select atmospheric correction units for the temperature (°C or °F) and pressure (in Hg or hPa).

Step 3   Input an atmospheric correction factor value for the prevailing ambient air temperature and barometric pressure. The correction factor value in ppm indicates how much a distance measurement made

Instructor's Notes:

_____

_____

_____

under ambient air conditions will differ from one made under the instrument calibration conditions. Typical calibration conditions used with instruments are 68°F (20°C) and 29.92 in Hg (760 mm Hg).

The atmospheric correction factor in ppm is determined using measured values for the ambient temperature and pressure of the air surrounding the instrument. Some instruments have built-in sensors that detect the atmospheric conditions and automatically correct readings for these natural errors. With most current total stations, the measured temperature and pressure values are entered via the keyboard, and the instrument automatically adjusts the ppm correction. With some instruments, an atmospheric correction chart (*Figure 37*) supplied with the instrument is used. Using the measured values for temperature and pressure, the correction factor is found using the chart, and then the factor is entered via the keyboard into the instrument.

Step 4   Input the correct prism constant value (0 mm, –18 mm, –30 mm, –40 mm) for the prism being used.

Step 5   Input the elevation of the setup point, commonly known as the occupied point.

Step 6   Input the height of the instrument and of the prism. These values are needed to accurately determine the elevation of the points where readings are taken.

Explain that the required correction factor must be entered into the total station to compensate for atmospheric changes.

Show Transparency 33 (Figure 37). Explain how the correction factor is found using an atmospheric correction chart.

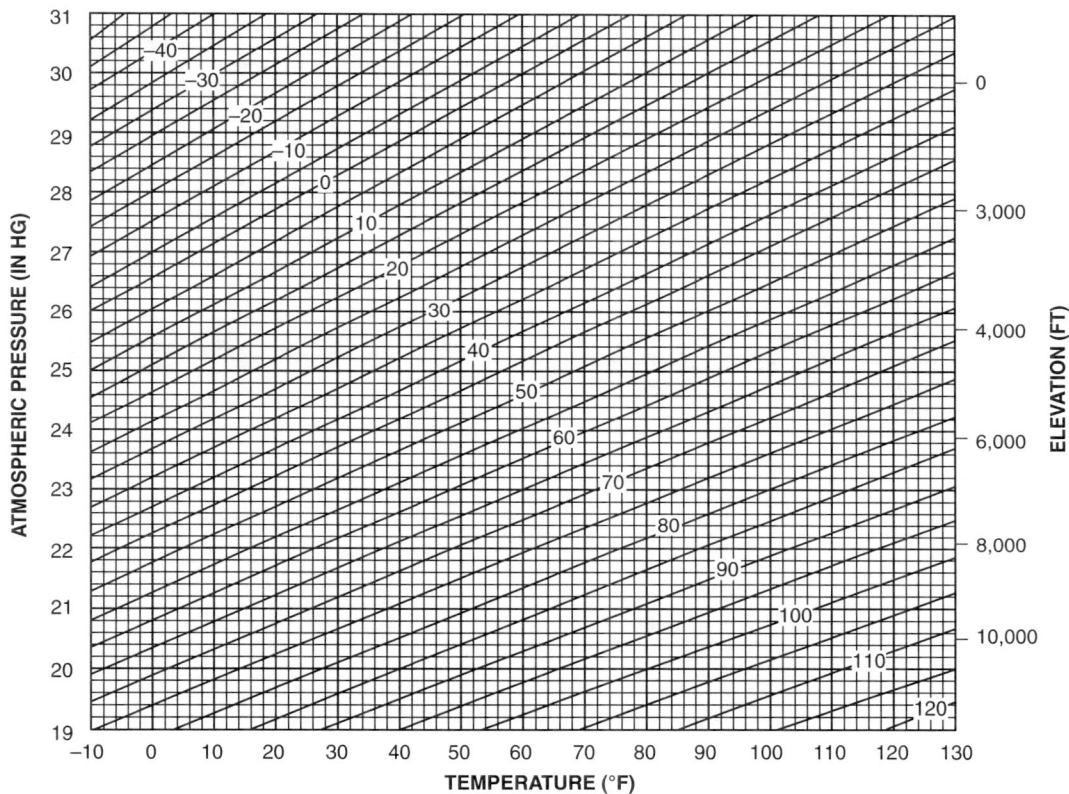

Figure 37 ◆ Atmospheric correction chart.

504F37.EPS

See the Teaching Tip for Sections 2.0.0, 5.0.0, 10.0.0, and 14.0.0 at the end of this module.

Show trainees how to check level using a total station, if available.

Have trainees practice checking level using a total station, if available. This laboratory corresponds to Performance Task 1.

Assign reading of Sections 11.0.0 and 12.0.0.

Ensure that you have everything required for teaching this session.

Discuss the purpose of field checks.

Show Transparency 34 (Figure 38). Discuss the geometric relationships for angular measurements.

Remind trainees that the manufacturer's instructions must be followed for each instrument.

Discuss how to perform a plate level check.

> **NOTE**
> Failure to enter a required correction factor to compensate for atmospheric changes will cause a total station to read too short or too long a distance, depending on the specific atmospheric conditions. This is important because the speed of light through air is not constant and depends on the atmospheric conditions. At lower temperatures, light slows down because the density of air increases. At higher temperatures, light speeds up because the density of air decreases. Similarly, light slows down as the barometric pressure increases and speeds up as the barometric pressure decreases.

Step 7 Select a point numbering scheme to identify occupied and sighted stations. For example, select 10, 100, 1,000, or some other number.

Step 8 Input the point number and the description of the occupied point. All occupied and backsighted points should be described thoroughly.

Step 9 Input the coordinates of the occupied point.

Step 10 Input the coordinates or directions to the backsight. Coordinates are usually used when using points from a previous alignment. A direction is typically used if it is a new alignment.

Step 11 At this point, the total station should be ready for making measurements. Make subsequent angle and/or distance measurements using the total station as directed in the manufacturer's instructions.

11.0.0 ◆ ALIGNMENT INSTRUMENT FIELD CHECKS

Total stations, as well as all other surveying and leveling instruments, should be field checked for correct calibration and adjustment if you are using the instrument for the first time, or if you suspect the instrument is out of adjustment.

11.1.0 Geometry of Angle Measuring Instruments

Transits, theodolites, and total stations used to make angular measurements consist essentially of an optical line of sight, which is perpendicular to and supported on a horizontal axis. As shown in *Figure 38*, the line of sight of the instrument is perpendicular to the horizontal axis, and the horizontal axis of the instrument is perpendicular to a vertical axis, about which it can rotate. Spirit levels mounted on the base of the instrument alidade are used to make the vertical axis coincide with the direction of gravity. These geometric relationships must be maintained in the instrument; otherwise, the instrument will be out of calibration and any angles measured or laid out with it will be incorrect.

11.2.0 Instrument Field Checks

This section describes field checks that apply to most all instruments. These include levels, transits, theodolites, and total stations. The checks given here are generic in nature. All checks for a specific make and model of instrument must be made as directed in the manufacturer's instructions for that specific instrument. If the manufacturer does not recommend that a field adjustment be performed for a particular adjustment situation, the instrument should be sent to a calibration/repair facility for adjustment.

11.2.1 Plate Level Check

The purpose of the plate level check is to make sure that the vertical axis of the instrument is actually vertical. Adjustment is required if the axis of the plate level is not perpendicular to the vertical

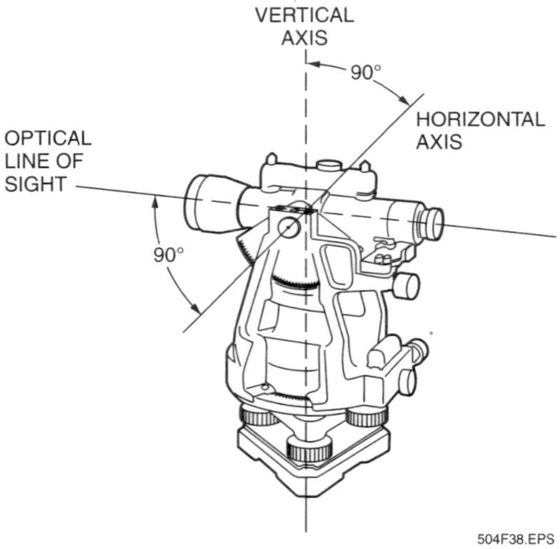

Figure 38 ◆ Geometry of an angle measurement instrument.

Instructor's Notes:

4.32 MILLWRIGHT ◆ LEVEL FIVE

axis. To determine whether the instrument requires adjustment, level the instrument carefully as follows:

Step 1   Rotate the instrument alidade so that the plate level vial is parallel with the centers of leveling screws A and B as shown in *Figure 39*. Adjust leveling screws A and B only to center the bubble in the plate level.

Step 2   Rotate the instrument alidade 180 degrees around its vertical axis and check the position of the plate level bubble. If the bubble remains centered, the plate level is adjusted correctly. If the bubble is displaced from the centered position, proceed with Steps 3 and 4.

Step 3   Adjust the plate level as directed in the manufacturer's instructions to move the position of the bubble back toward the center of the vial for a distance equal to one-half of the displacement (*Figure 40*). This adjustment is typically done using a capstan screw provided for this purpose. Next, center the bubble again using leveling screws A and B.

Step 4   Rotate the instrument alidade 180 degrees around its vertical axis again and check for bubble movement. If the bubble is displaced, repeat the adjustment again as necessary until the position of the plate level bubble does not change. This indicates the plate level is adjusted properly.

11.2.2 Circular Level Check

The only purpose served by the circular level is to aid in rough leveling of the instrument during its initial setup. After the instrument has been properly leveled, a perfectly centered bubble in the circular level indicates that it is in adjustment. If the bubble is displaced, this indicates that an adjustment is required. Adjustment of the circular level should be done only after the instrument has been carefully leveled using the plate level only. Next, the circular level adjustment screws on the bottom of the circular level are adjusted as needed to center the bubble in the circular level.

11.2.3 Optical Plummet Check

The alignment of the optical plummet's axis relative to the vertical axis of the instrument should be checked periodically for accuracy. An optical plummet can get out of adjustment, causing the instrument to be set over erroneous points. The alignment of an optical plummet can be checked by placing its crosshairs over a reference point on the ground with the instrument at 0 degrees (*Figure 41*). Next, the instrument is rotated 180 degrees, and the position of the crosshairs checked again. If the crosshairs are superimposed on the reference point at both positions, the plummet is aligned. If not, the plummet should be adjusted following the instrument manufacturer's instructions. Typically, this is done by turning the adjustment screws on the optical plummet

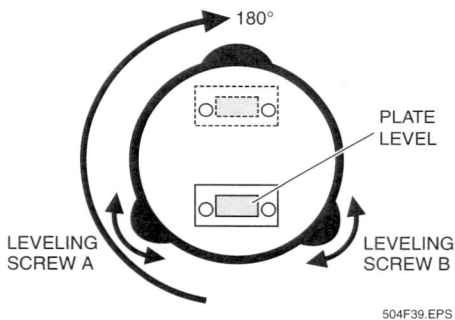

Figure 39 ◆ Plate level test positions.

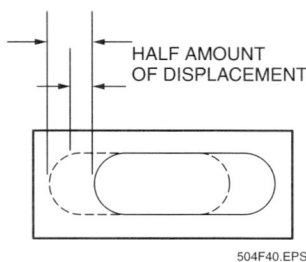

Figure 40 ◆ Plate level adjustment.

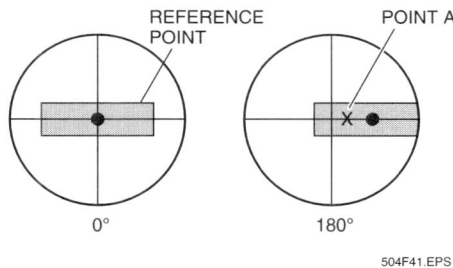

Figure 41 ◆ Optical plummet check.

so that its crosshairs are positioned over a point (Point A) midway between the original points sighted for the 0-degree and 180-degree positions. This procedure should be repeated as necessary until the crosshairs remain superimposed on the reference point when the instrument is at the 0-degree and 180-degree positions.

11.2.4 Crosshair Tests

Crosshair tests are performed to make sure that the vertical and horizontal crosshairs are plumb and level, respectively. Both tests are easy to perform. The object of the horizontal crosshair test is to make sure that the instrument's horizontal crosshair is in a plane that is perpendicular to the vertical axis of the instrument. With a properly adjusted instrument, you should be able to place any part of the horizontal crosshair on the object or point being viewed with the telescope and still get an accurate reading. First, carefully level the instrument, then sight the horizontal crosshair on a distant target or other well-defined point (*Figure 42*). Once the crosshair is placed on the point, turn the instrument's horizontal tangent screw so that the instrument slowly rotates about its vertical axis. The crosshair should stay fixed on the point as the instrument is rotated. If any part of the crosshair moves above or below the reference point, the instrument needs adjustment as directed in the manufacturer's instructions.

The object of the vertical crosshair test is to ensure that the instrument's vertical crosshair is perpendicular to the horizontal axis of the instrument. With a properly adjusted and leveled instrument, sight the vertical crosshair on a plumb bob string at rest. For a properly adjusted instrument, the vertical crosshair should coincide with the string. If any part of the crosshair does not coincide with the string (tilts to the left or right), the instrument needs adjustment as directed in the manufacturer's instructions.

NOTE

If either crosshair is out of adjustment, the instrument can still be used until the actual adjustment is made. This is done by using only the part of the crosshair that is nearest to the intersection point of the crosshairs.

11.2.5 Line-of-Sight Check (Levels and Transits)

The line-of-sight check, commonly called a **peg test**, determines if the instrument's telescope line of sight is horizontal. This means that the line of sight is parallel to the barrel of the telescope and the axis of the telescope bubble tube. The line of sight is checked as follows:

Step 1 In a fairly level and clear area, place two points at a distance of about 200 feet apart.

Step 2 Set up and level the instrument at a point exactly midway between the two points (*Figure 43A*). While working with a rod person, take several elevation rod readings at both locations (points A and B). Average the set of backsight readings and set of foresight readings, then subtract the averages to determine the actual difference in elevation between the two points. Record this difference.

NOTE

With the instrument placed midway between the two points, any error that may be caused by a line-of-sight problem will result in an identical amount of error in both the rod readings. Because the errors are identical, the calculated difference in elevation between the two points is the true difference in elevation.

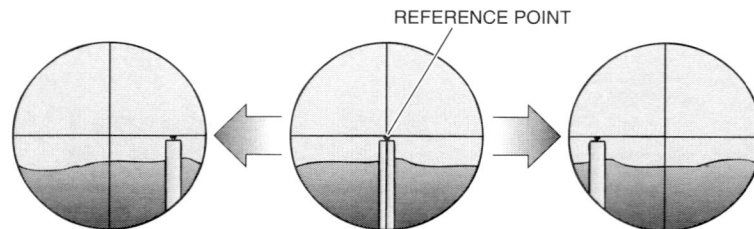

IF THE HORIZONTAL CROSSHAIR MOVES OFF THE REFERENCE POINT, THE RETICLE NEEDS ADJUSTMENT

504F42.EPS

Figure 42 ◆ Horizontal crosshair check.

Instructor's Notes:

_____

_____

_____

Step 3 Move and set up the instrument as close as possible (within a foot) of the stake at point A (*Figure 43B*).

Step 4 While the rod person holds a rod plumb on point A, sight backwards through the objective lens of the telescope at a pencil point that is being held and moved slowly up and down the rod by the rod person. When the pencil point is exactly centered in your view, read the rod and record the backsight elevation value.

Step 5 Rotate the telescope and take a rod reading on point B 200 feet away in a normal manner. Record the elevation as a foresight and subtract from the elevation obtained in Step 4. Do this several times and record the average as the difference in elevation readings.

Step 6 Compare the difference in the elevation readings obtained in Steps 1 and 5. Ideally, the difference in elevations should be the same. This indicates no adjustment is required. Any difference greater than two or three thousandths of a foot should be considered significant enough to require adjustment of the instrument. This is done by adjusting the instrument telescope level vial per the manufacturer's instructions.

11.2.6 Line-of-Sight Check (Theodolites and Total Stations)

Horizontal and vertical collimation checks make sure the line of sight of the theodolite or total station telescope, as represented by the vertical and horizontal crosshairs and the barrel of the telescope, are aligned. These checks can be done in several different ways. One simple way is described here:

Step 1 Set up and initialize the instrument about 200 feet away from a clearly defined target point.

Review the steps for performing a line-of-sight check on theodolites and total stations.

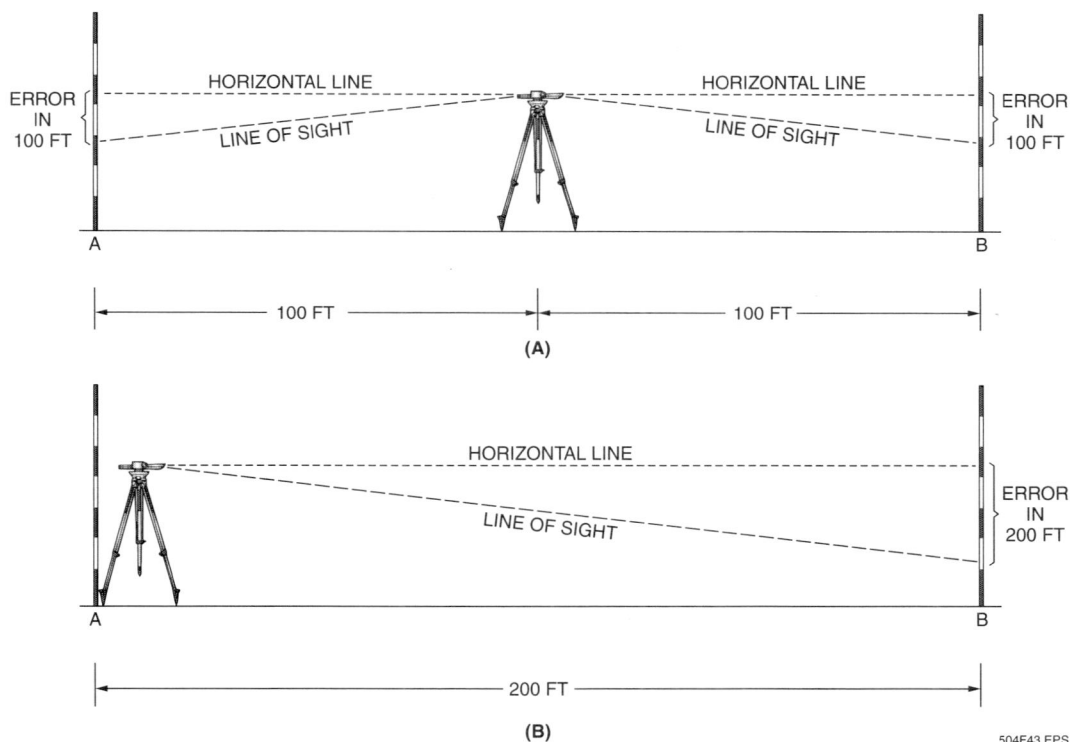

Figure 43 ◆ Peg test.

Step 2 With the telescope in the normal position, record the horizontal and vertical angles measured from the instrument to the target point.

Step 3 Invert the telescope, release the upper motion clamp if so equipped, and view the target point again. Measure and record the horizontal and vertical angles.

Step 4 Repeat Steps 2 and 3 several times. Separately average the values for the horizontal angles and the values for the vertical angles measured in Step 2. Do the same for the horizontal and vertical angle values measured in Step 3.

Step 5 Compare the average values for the horizontal and vertical angles measured in Step 2 with those measured in Step 3. With proper collimation, the difference between the horizontal angles should be exactly 180 degrees, with any difference being twice the horizontal collimation error. The sum of the vertical angle readings should be exactly 360 degrees, with any difference being twice the vertical collimation error. Small differences of one or two seconds are generally considered inconclusive as to whether the instrument is in or out of calibration. Larger differences indicate that the instrument should be calibrated before using it. The definition or magnitude of larger differences depends on the instrument and the type of measurements for which it will be used.

11.3.0 Laser Beam Level Check

A laser level's transmitter beam should be checked for proper calibration at regular intervals, particularly if using the instrument for the first time or if the unit has been handled roughly. Severe shock or vibration may have caused the instrument to be out of calibration. The laser transmitter needs calibration when the laser beam emitted from one side of the unit is above true level, and the beam emitted from the opposite side is below true level (*Figure 44*). When the instrument is correctly calibrated, it emits a 360-degree horizontal level plane beam. If the unit is turned 180 degrees or 90 degrees from its original position, the reading is within the manufacturer's specifications, typically within ±³⁄₃₂" per 100 feet of the original position. The laser transmitter calibration should be checked as described in the manufacturer's instructions. Manufacturers of different makes and models of instruments rec-

ommend checking their instruments in different ways. A general procedure for one method is given here:

> **WARNING!**
> Only trained and authorized personnel can operate a laser instrument. Inform all people within the operating range of the instrument that a laser instrument will be in use. Never look directly into a laser beam or point the beam into the eyes of others. Set up the laser transmitter so that the height of the emitted beam is above or below normal eye level.

Step 1 Set up and level the laser transmitter at a location that has a clear line of sight to an object at least 100 feet away. Attach a laser detector (receiver) unit on a level rod (*Figure 45*) or other calibrated rod and hold it plumb at the 100-foot location.

Step 2 Turn on the laser transmitter and rotate it so that the +X side is aimed at the laser detector.

Step 3 Move the receiver into the beam to get an on-grade reading. Mark and record the elevation, noting that it pertains to the +X axis.

Step 4 Rotate the laser transmitter 180 degrees so that the –X side is aimed at the laser detector. Mark and record the elevation, noting that it pertains to the –X axis.

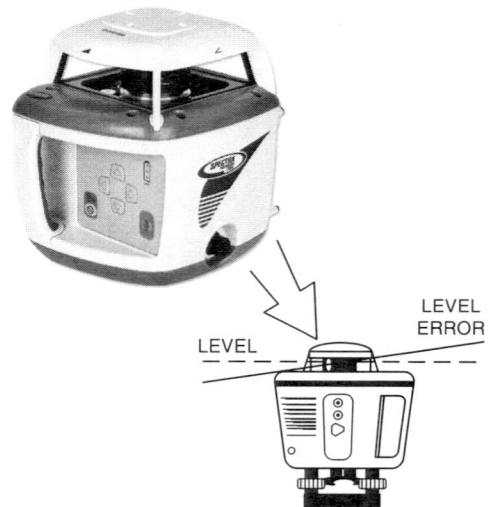

Figure 44 ◆ Laser transmitter emitted signals.

Instructor's Notes:
_____
_____
_____

Figure 45 ◆ Laser transmitter calibration setup.

Step 5 Compare the +X and –X elevation readings. If the difference between the readings is less than $\frac{3}{32}$", the X axis is within calibration. If within calibration, go to Step 9. If the difference in readings is more than $\frac{3}{32}$", X-axis calibration is required, as described in Steps 6 through 8.

Step 6 To correct for a calibration error, locate a new mark midway between the +X and –X marks. Move the receiver on the rod until its center-marking notch is aligned with the new midpoint mark.

Step 7 Adjust the X-axis calibration screw to obtain an on-grade laser beam reading at the midpoint line. Most detectors have move-up and move-down indicators that show the direction the beams need to be moved up or down relative to the on-grade point.

Step 8 Rotate the laser transmitter 180 degrees back to the original +X face. The on-grade reading should be on or within $\frac{3}{64}$" of the midpoint line. If not, repeat Steps 3 through 8.

Step 9 Rotate the laser 90 degrees. Repeat Steps 3 through 8 as required to check and adjust the beam emitted from the +Y and –Y sides of the unit.

12.0.0 ◆ TRIGONOMETRIC LEVELING

Trigonometric leveling is the process of measuring vertical angles and slope distances to determine the differences in elevation between points. Trigonometric leveling can be done using different methods, depending on the instruments available.

As shown in *Figure 46*, trigonometric leveling can be done using a transit or theodolite to measure the angle, then calculating the unknown height using right angle trigonometry. The procedure begins by first setting up and properly leveling the instrument over an established point at a known distance from the structure. The transit must also be set horizontal to use it as a level. Then, the height of the instrument (HI) is determined in the normal way by backsighting (BS) on a leveling rod or tape measure held on a benchmark (BM) of known elevation. The HI is calculated using the formula HI = BM + BS. For the example shown, the HI is 851.57 feet (845.23 feet + 6.34 feet). Following this, the vertical angle to the point on the structure is measured with the theodolite, and the angle is read on the vernier. For our example, the angle is 28°42'12". Using the values for the angle and the distance from the instrument to the building (125.00 feet), the height of the point on the building relative to the instrument (height X) can be calculated using the tangent function. For our example, height X is 68.44'. This height is then added to the HI to yield the actual elevation. For our example, the elevation is 920.01 feet (851.57 feet + 68.44 feet).

Today, a total station is commonly used for performing trigonometric leveling. In addition to measuring the slope distance and angles, these instruments also compute the horizontal distance and elevation. *Figure 47* shows an example of a total station being used to measure the elevation of a point on a second floor.

13.0.0 ◆ CHECKING HEIGHT

There is usually a rounded surface or a marked point on a horizontal plane surface on the work which is used as a reference height. Follow these steps to determine the difference in height between two objects:

Step 1 Mount the optical micrometer for vertical measurement.

Step 2 Set up the jig transit so that the horizontal distances are about the same from the transit to the reference mark and from the instrument to the part whose height is to be established. If the line of sight has a slight slope when the telescope is leveled, the error that this might cause is eliminated by making the horizontal distances equal (*Figure 48*).

Step 3 Place an optical tooling scale vertically on the reference mark, and direct the telescope toward the scale.

Describe how a calibration error is corrected.

Define trigonometric leveling. Point out that there are different methods, depending on the instruments available.

Show Transparency 41 (Figure 46). Explain how to perform trigonometric leveling using a theodolite.

Show Transparency 42 (Figure 47). Discuss the use of total stations to perform trigonometric leveling.

Assign reading of Sections 13.0.0 and 14.0.0.

Ensure that you have everything required for teaching this session.

Review the steps for finding the difference in height between two objects.

Show Transparency 43 (Figure 48). Explain how to make slope corrections.

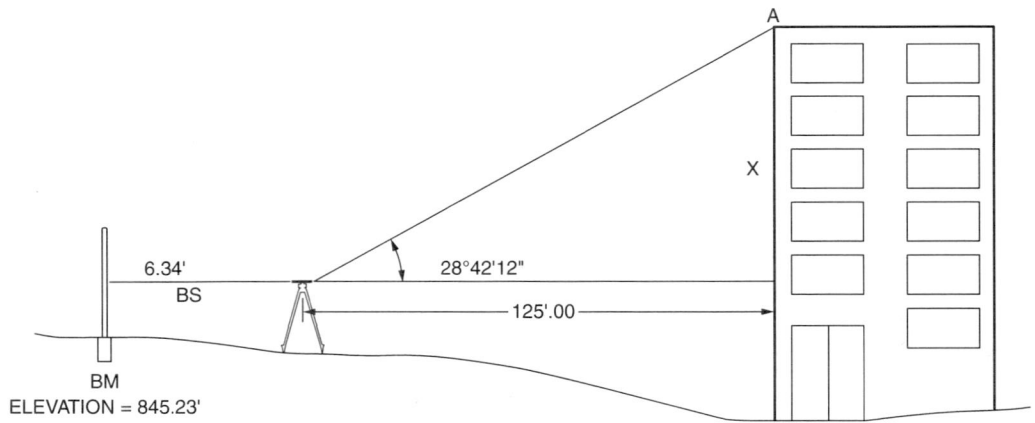

PROBLEM: Determine the elevation of A.

HEIGHT OF INSTRUMENT (HI) =
ELEVATION OF BENCH MARK (BM) + BACKSIGHT (BS)

HI = 845.23 + 6.34 = 851.57'

ANGLE = 28°42'12" = 28.70333333

$TAN\ 28.70333333° = \dfrac{X}{125}$

X = TAN 28.70333333° × 125'

X = 0.547559626 × 125'

X = 68.44495328 = 68.44'

ELEVATION OF A = 851.57' + 68.44'

$\boxed{= 920.01'}$

504F46.EPS

Figure 46 ◆ Trigonometric leveling using a theodolite.

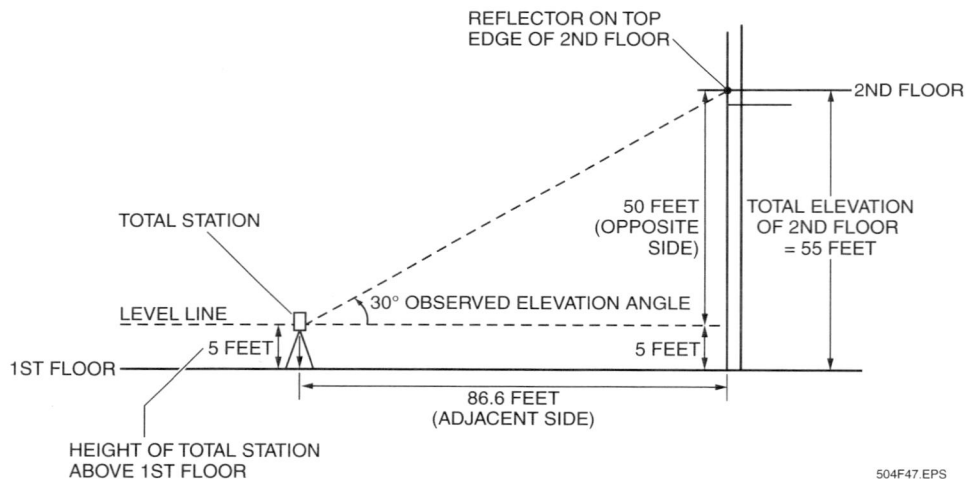

504F47.EPS

Figure 47 ◆ Total station being used to measure an elevation.

Instructor's Notes:

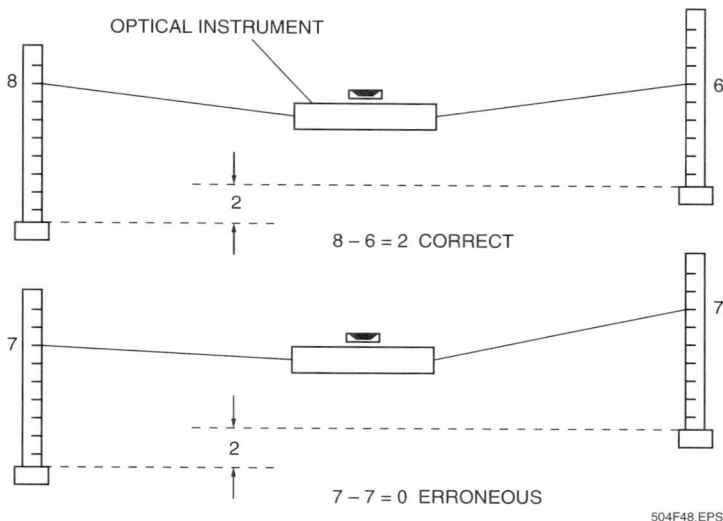

Figure 48 ◆ Slope correction.

504F48.EPS

Discuss the procedure for creating coincident lines of sight between two instruments.

Show Transparency 44 (Figure 49). Describe the different configurations that can occur relative to the lines of sight of two instruments.

Step 4 Set the micrometer at zero.

Step 5 Center the telescope level bubble using the elevation clamp and tangent screw.

Step 6 Turn the micrometer drum to move the line of sight toward the tenth division on the scale that shows the smaller reading.

Step 7 Read the scale using the micrometer.

Step 8 Check the bubble immediately to ensure that it is still centered.

Step 9 Compute the desired scale reading on the part.

Step 10 Set the micrometer for the desired reading, and direct the telescope toward the part.

Step 11 Center the bubble, and adjust the part so that when the optical tooling scale is held on it, the correct mark is on the line of sight.

13.1.0 Coincident Lines of Sight

When two theodolites are set at infinity focus and aimed at each other so that the reticles appear to coincide (*Figure 49*), the lines of sight are parallel, but they are not necessarily coincident.

When both instruments are focused at some point between them, the reticle of each instrument can be seen on the reticle of the other. When aimed so that the reticles coincide, the lines of sight meet at a point but are usually at an angle to each other. Therefore, the instruments must be aimed so that the reticles appear to coincide when focused at infinity and also when focused at a point between them. The reticle of each must be illuminated. Follow these steps to cause the lines of sight of two instruments to coincide:

Step 1 Aim each instrument at the center of the objective lens of the other.

Step 2 Mount a white card, with marks on each side for focusing, about halfway between the two instruments, and focus both instruments on the card without changing the aim of either.

Step 3 Remove the card, and look through the instruments. If the cross lines do not coincide, take up half the error with each instrument.

Step 4 Focus both instruments at infinity. If the reticle patterns do not coincide, take up half of the error with each instrument.

Step 5 Continue to focus alternately with the card and then with the infinity focus until the reticles coincide under both conditions.

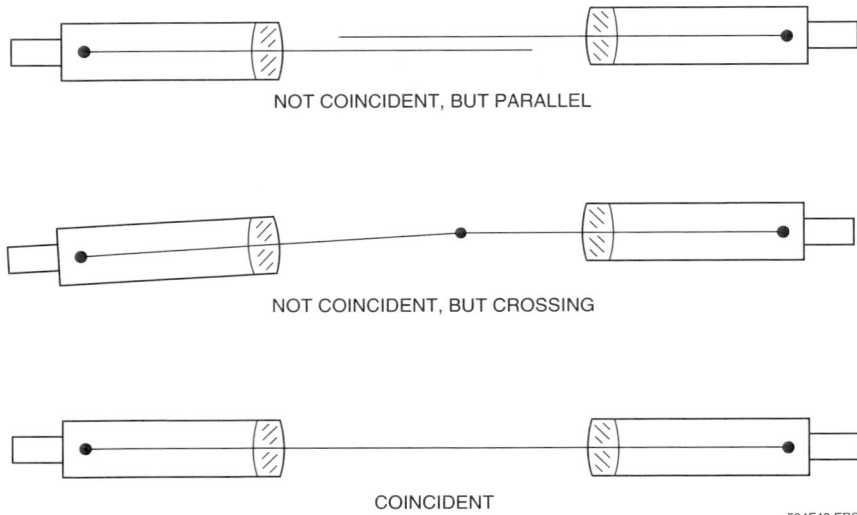

NOT COINCIDENT, BUT PARALLEL

NOT COINCIDENT, BUT CROSSING

COINCIDENT

504F49.EPS

Figure 49 ◆ Lines of sight.

Describe the bucking-in process.

Show Transparency 45 (Figure 50). Explain how to buck-in a line.

Explain how to properly care for, store, and transport optical instruments.

Describe the conditions to avoid with optical instruments.

13.2.0 Bucking-In

It is often necessary to place a theodolite in line with two targets or to satisfy two conditions simultaneously. This process is called **bucking-in** (*Figure 50*). The theodolite measures the exact error and reduces the time required to buck-in. Usually, a mechanical lateral adjuster is mounted on an instrument stand and the instrument is mounted on the lateral adjuster. Follow these steps to buck-in a line:

Step 1 Lock the adjuster so that its ways are approximately perpendicular to the final direction of the line of sight.

Step 2 Level the instrument.

Step 3 Aim at the far target.

Step 4 Take up any error on the near target, using the adjuster.

Step 5 Repeat the process until the instrument is nearly aligned.

Step 6 Relevel and perfect the alignment, using the optical micrometer and the azimuth tangent screw.

13.3.0 Care of Optical Instruments

Handle all instruments with great care. Never subject them to jolting or to vibration. Vibration loosens screws and often destroys all adjustments that depend on the use of the adjusting screws. Never store or transport an instrument within the shop or in shipment unless it is freely supported on resilient material.

The following are conditions to avoid with optical instruments:

- *Deflection* – Deflection is the bending or flexing of the instrument caused by external stress. The effect of deflection in instruments is sometimes as high as 50 to 100 times as great as the accuracy required. To avoid deflection, always support an instrument in its normal position during storage, use, or adjustment; allow an instrument to adjust to the surrounding air temperature before operating it; ensure that any attachments to be used with a telescope are in place when the instrument is adjusted; and always handle an instrument by its base to avoid a temporary set.

- *Temperature* – Some instruments, such as spirit levels, are extremely sensitive to ambient temperatures. For example, if the vial of a spirit level is unevenly heated so that the vapor pressure in the liquid is greater at one end than the other, the bubble does not come to rest at the highest point in the vial. A spirit level is useless when in the direct rays of the sun and unreliable when near an ordinary electric light or when touched with the finger.

Instructor's Notes:

_____

_____

_____

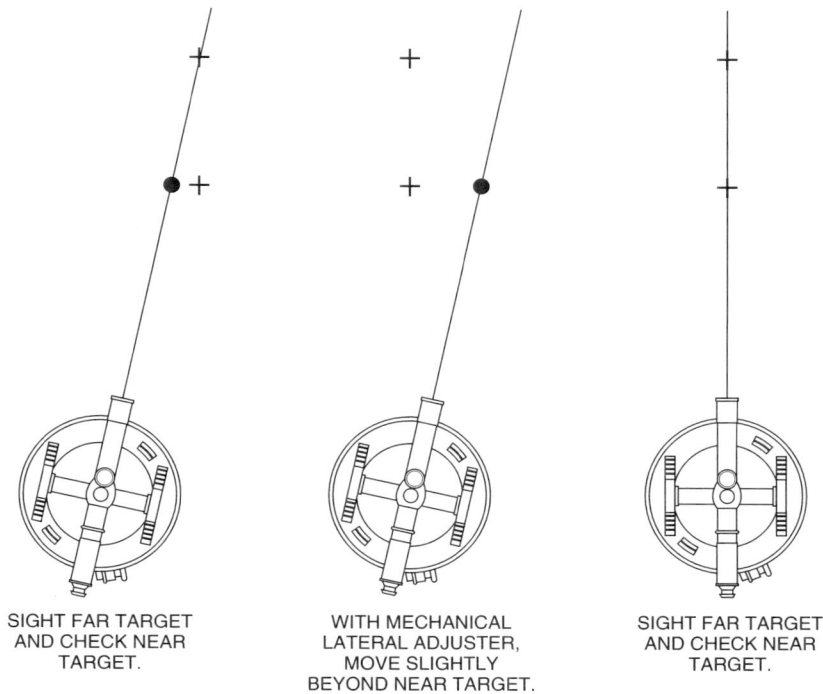

SIGHT FAR TARGET
AND CHECK NEAR
TARGET.

WITH MECHANICAL
LATERAL ADJUSTER,
MOVE SLIGHTLY
BEYOND NEAR TARGET.

SIGHT FAR TARGET
AND CHECK NEAR
TARGET.

504F50.EPS

Figure 50 ◆ Bucking-in.

- *Refraction* – A column of hot or cold air anywhere along a line of sight makes the image wobble and bend, thus destroying its accuracy.
- *Obstruction* – If any light rays from a target are prevented from reaching the entire area of an objective lens, the line of sight is unreliable.
- *Overlubrication* – The lubrication of an instrument is a shop operation. However, in use, the threads of the leveling screws can be lubricated with a very small quantity of fine instrument lubricant. This should be worked in by turning the screws up and down throughout their range and carefully wiping off the excess lubricant with a lint-free rag.
- *Dust* – Protect instruments from dust as much as possible. For example, the view through the telescope may be slightly dimmed if there is dust on the eyepiece lens nearest the eye. Clean this lens with a cotton swab dampened with lens cleaner.

14.0.0 ◆ USING OPTICAL LEVELS

The optical level is sometimes referred to as a spirit level or precision tilting level (*Figure 51*). These levels consist of a telescopic sight attached to a very sensitive spirit level. They are adjusted so that when the spirit level indicates level, the line of sight is perpendicular to gravity, or horizontal. This instrument is used to determine differences in height. It is necessary to establish one point from which the differences in height to all other points are measured. Measurements are made upward from top surfaces or downward from bottom surfaces.

When high accuracy is required, place the level so that the horizontal lengths of the sights to the two objects are approximately equal. This placement eliminates the effect of slight errors in instrument adjustment. To find the difference in height between two objects with great precision, make the horizontal distances from the instrument to the objects equal. *Figure 52* shows height difference.

Describe the proper way to clean dust from a lens.

Discuss the function of optical levels. Explain how adjustments are made.

Show Transparency 46 (Figure 51). Point out the components of this optical level.

Show Transparency 47 (Figure 52). Describe how to use an optical level to find the difference in height between two objects.

MODULE 15504-09 ◆ OPTICAL ALIGNMENT   4.41

Explain the procedure
for taking a sight using
an optical level.

Follow these steps to take a sight, using an optical level:

Step 1 Center the bubble of the circular level, using the four leveling screws. This aligns the **azimuth axis** very nearly in the direction of gravity.

Step 2 Free the azimuth clamp, and aim the telescope at a white or light-colored object.

Step 3 Look through the telescope, and turn the eyepiece focusing ring until the reticle pattern is sharp.

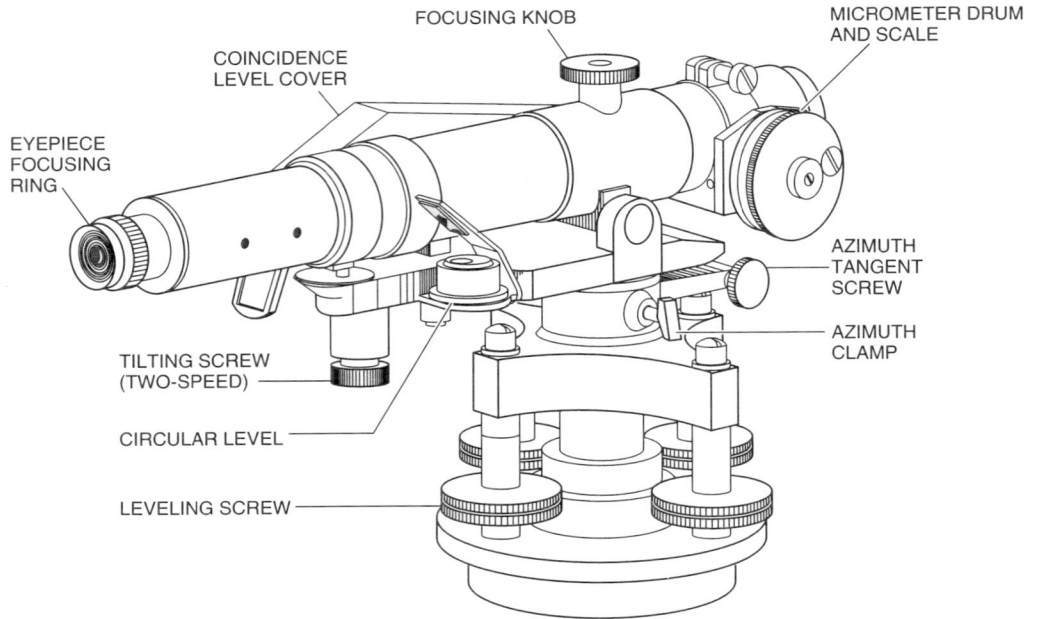

COINCIDENCE
LEVEL COVER

FOCUSING KNOB

MICROMETER DRUM
AND SCALE

EYEPIECE
FOCUSING
RING

AZIMUTH
TANGENT
SCREW

AZIMUTH
CLAMP

TILTING SCREW
(TWO-SPEED)

CIRCULAR LEVEL

LEVELING SCREW

504F51.EPS

Figure 51 ◆ Optical level.

Instructor's Notes:

_____

_____

_____

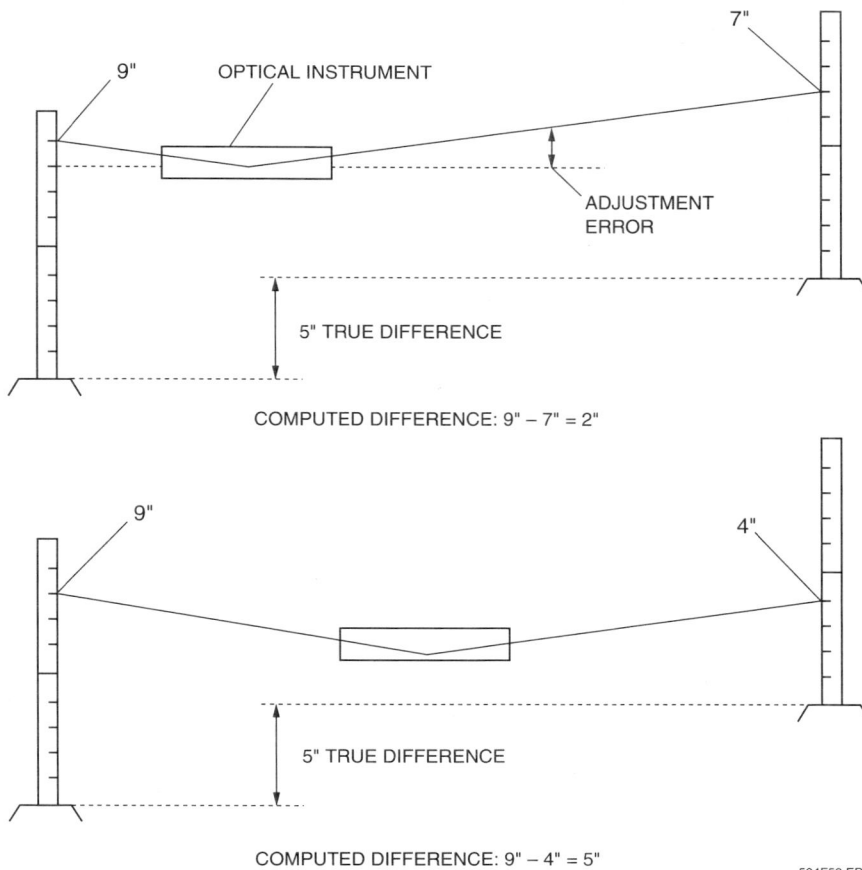

9" OPTICAL INSTRUMENT   7"

ADJUSTMENT
ERROR

5" TRUE DIFFERENCE

COMPUTED DIFFERENCE: 9" − 7" = 2"

9"   4"

5" TRUE DIFFERENCE

COMPUTED DIFFERENCE: 9" − 4" = 5"

504F52.EPS

Figure 52 ◆ Height difference.

Step 4 Write down the setting of the diopter scale for future use. The diopter scale is engraved on the eyepiece focusing ring.

Step 5 Position an optical tooling scale vertically with the zero end on a fixed object, and while looking over the telescope, aim the telescope toward the scale.

Step 6 Set the micrometer at zero.

Step 7 Sight through the telescope, and bring the scale into the field of view.

Step 8 Focus on the scale, using the focusing knob.

Step 9 Lock the azimuth clamp, and move the vertical line of the reticle so that it reads on the scale, using the azimuth tangent screw.

Step 10 Adjust the two ends of the coincidence level bubble, using the tilting screw, so that the ends coincide.

Step 11 Observe where the horizontal cross line falls on the tooling scale, and turn the optical micrometer so that the cross line moves to the nearest lower tenth of an inch.

Step 12 Check the coincidence level bubble to ensure that the lines are still coincident.

Discuss the features of the micrometer on an optical level.

Provide an optical level for trainees to examine, if available.

Show Transparency 48 (Figure 53). Explain how to read the erect and reversed scales.

See the Teaching Tip for Sections 2.0.0, 5.0.0, 10.0.0, and 14.0.0 at the end of this module.

Show trainees how to check level using a precision tilting level, if available.

Have trainees practice checking level using a precision tilting level, if available. This laboratory corresponds to Performance Task 1.

NOTE

The tilting screw on the level is a two-speed screw. Within the knurled head of the screw are two stops located one-third of a turn apart. Between stops, the screw operates in low gear at a ratio of 8:1. Beyond stops, it operates in high gear at a ratio of 1:1. To use the screw, turn it in either direction until it is just past the desired point; then reverse its direction. Low gear starts to function and continues to function in either direction until the final setting is made.

The micrometer of the optical level is graduated from zero at the middle to 0.100 inch at each end. The scale and the numbers are black in one direction and red in the other. When reading an optical tooling scale that is erect or above the object to be located, the line of sight must be lowered, and the red numbers are used. When the optical tooling scale is below the object, or reversed, the black numbers are used. A double vernier makes it possible to estimate to 0.0001 inch increments in whichever direction the drum has moved. Use that part of the vernier where the numbers increase in the same direction as those of the micrometer drum. *Figure 53* shows reading the scale.

Follow these steps to read the scale erect:

Step 1   Set the micrometer scale at zero, and read the optical tooling scale at the cross line. The reading shown is between 1.3 and 1.4 on *Figure 53*.

Step 2   Move the cross line to the least tenth, and add the micrometer reading in the red numbers. The final reading shown is between 1.3 and 1.4 inches.

Follow these steps to read the scale reversed:

Step 1   Set the micrometer scale at zero, and read the optical tooling scale at the cross line. This reading is shown as between 2.7 and 2.8 on *Figure 53*.

Step 2   Move the cross line to the least tenth, and add the micrometer reading in the black numbers. The final reading is between 2.7 and 2.8 inches.

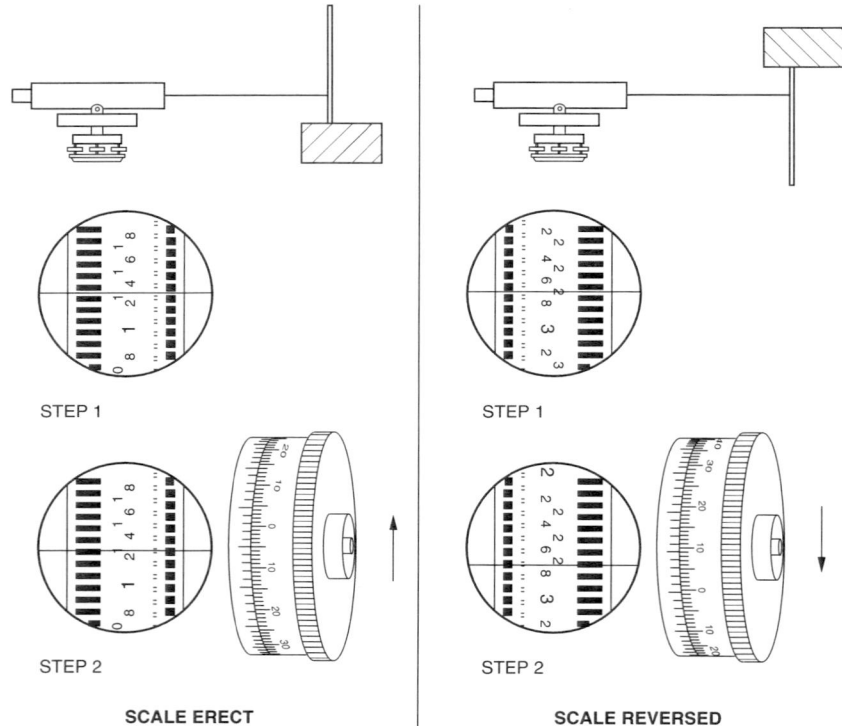

STEP 1

STEP 1

STEP 2

STEP 2

SCALE ERECT

SCALE REVERSED

504F53.EPS

Figure 53 ◆ Reading the scale.

4.44 MILLWRIGHT ◆ LEVEL FIVE

Instructor's Notes:

Review Questions

1. Optical alignment procedures are based on the use of one or more _____.
   a. sonic echoes
   b. lines of sight
   c. right angles
   d. circles

2. In a telescopic sight, there is a(n) _____ between the reticle and the objective lens.
   a. eyepiece
   b. mirror
   c. movable focusing lens
   d. reflective sheet

3. The change in apparent relative position when viewed from different angles with respect to the target is called _____.
   a. parallelism
   b. astigmatism
   c. focus shift
   d. parallax

4. Spirit levels provide reference to a _____ line, which is the direction of gravity.
   a. parallel
   b. diagonal
   c. horizontal
   d. vertical

5. Circular levels are used for _____ leveling.
   a. approximate
   b. vertical
   c. diagonal
   d. precision

6. The part of the instrument that displaces a line parallel to the instrument's line of sight is called an _____.
   a. optical filter
   b. objective lens
   c. optical micrometer
   d. auto-collimator

7. The _____ is an instrument used to check elevations.
   a. optical filter
   b. auto collimator
   c. builder's level
   d. tribrach

8. Tripod legs should have about a _____ foot spread.
   a. two
   b. three
   c. four
   d. five

9. One degree equals _____ of a circle.
   a. $\frac{1}{4}$
   b. $\frac{1}{90}$
   c. $\frac{1}{180}$
   d. $\frac{1}{360}$

10. Theodolites employ an optical line of sight that is _____ to a horizontal axis.
    a. adjacent
    b. vertical
    c. perpendicular
    d. opposite

11. To test the vertical crosshair, sight it on a _____.
    a. plumb bob string at rest
    b. building wall
    c. roofline
    d. person holding a level

12. The difference between the two types of electro-optical instruments is in the _____ of the signal transmitted.
    a. amplitude
    b. radar type
    c. wavelength
    d. power

13. Correction factors for atmospheric pressure and temperature are expressed as _____.
    a. inches
    b. millimeters
    c. microns
    d. ppm

14. The display of a total station typically is capable of displaying _____.
    a. a diagram of base alignment
    b. four lines of text
    c. two lines of text
    d. a picture of the slope triangle

Have trainees complete the Trade Terms Quick Quiz for Sections 1.0.0–14.0.0, and go over the answers prior to administering the Review Questions.

Have trainees complete the Review Questions, and go over the answers prior to administering the Module Examination.

MODULE 15504-09 ◆ OPTICAL ALIGNMENT   4.45

15. When rotating the two leveling screws on a total station, the plate level bubble will always follow the direction of the _____.
    a. right index finger
    b. right thumb
    c. left hand
    d. left thumb

16. The axis of the plate level must be _____.
    a. parallel with the vertical axis
    b. parallel with the nearest leg of the tripod
    c. perpendicular to the vertical axis
    d. perpendicular to the plate

17. The peg test determines if the instrument's line of sight is _____.
    a. parallel to the vertical axis
    b. perpendicular to the horizontal axis
    c. diagonal
    d. horizontal

18. A column of hot or cold air along the line of sight of an optical instrument, which destroys the accuracy, is called _____.
    a. reflection
    b. parallax
    c. refraction
    d. direction

19. When high accuracy with a level is required, place the level so that the _____ are approximately equal.
    a. heights of the objects
    b. horizontal distances to the objects
    c. distances to the ground and to the objects
    d. parallel sight lines

20. The tilting screw on a precision tilting level is a _____ screw.
    a. reticle
    b. left-handed
    c. vernier
    d. two-speed

Instructor's Notes:
_____
_____
_____

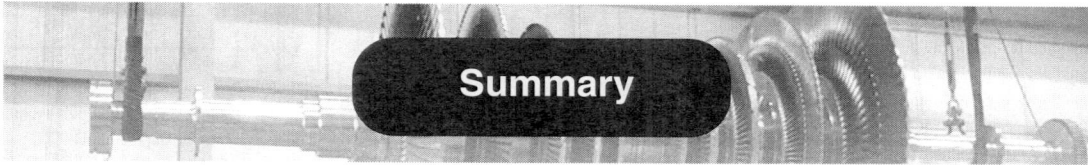

Optical alignment and leveling instruments provide a consistent method of accurately installing equipment. Each type of instrument has a specific use and requires that the millwright practice in order to become proficient.

Shortcuts produce inconsistent results and should be avoided. Because the instruments are designed for precision and have fine adjustments, they should be cared for, cleaned, and stored properly at all times.

Notes

Summarize the major concepts presented in the module.

Administer the Module Examination. Be sure to record the results of the Examination on Craft Training Report Form 200, and submit the results to the Training Program Sponsor.

Administer the Performance Tests, and fill out Performance Profile Sheets for each trainee. If desired, trainee proficiency noted during laboratory sessions may be used to complete the Performance Test. Record the results on Craft Training Report Form 200, and submit the results to the Training Program Sponsor.

MODULE 15504-09 ◆ OPTICAL ALIGNMENT   4.47

Aim: To regulate the direction of a sighting device.

Alidade: The upper part of an instrument that turns in azimuth with the sighting device.

Azimuth: The direction in a horizontal plane.

Azimuth axis: The vertical axis, or the axis of the bearing and spindle that confines rotation to a horizontal plane.

Buck-in: To place an instrument so that the line of sight satisfies two requirements simultaneously, such as aiming at two targets. It is usually accomplished by trial and error.

Circular level: The round level attached to the alidade.

Clamp: The device used to connect the part to be aimed with the stationary part of an instrument so that the tangent screw can operate.

Degree: An angle equal to $\frac{1}{360}$ of a full circle.

Diopter: The measurement of the power of a lens, equal to the reciprocal of its focal length in meters.

Electro-optical instruments: EDMIs that transmit modulated infrared or visible light signals in short wavelengths of about 0.4 micrometers or microns (μm) to 1.2 μm.

Elevation axis: The horizontal axis, or the axis of the bearing and the journal of the telescope axle that confines rotation to a vertical plane.

Elevation: The direction of a line of sight in a vertical plane.

Focus: To move the optical parts so that a sharp image is seen, or the point at which the image is perfectly clear.

Horizontal: Perpendicular to the direction of gravity.

Infrared light: An electromagnetic signal having frequencies below the visible portion of the frequency spectrum. The infrared light frequencies lie between those of light and radio waves with wavelengths ranging from 0.7 μm to 1.2 μm.

Minute: An angle equal to $\frac{1}{60}$ of a degree.

Objective lens: The lens at the front end of a telescope and therefore nearest the object sighted.

Parallax: The change in the apparent relative orientations of objects when viewed from different positions.

Peg test: A procedure used to check for an out-of-adjustment bubble vial on levels and other instruments.

Plate level: A fairly sensitive tubular level mounted on the plate of a jig transit, used to place the azimuth axis in the direction of gravity.

Position: The location of an instrument in a horizontal plane about which the instrument turns when it is being leveled. Once a jig transit is leveled, it is the point at which the vertical and horizontal axes intersect.

Reference line: A line of sight from which measurements are made.

Reticle: A series of fine lines that are placed in the focus of the objective lens of an optical instrument to quantify the measurements of angles or distances.

Second: An angle equal to $\frac{1}{60}$ of a minute, or $\frac{1}{3600}$ of a degree.

Standards: Uprights that support the telescope axle bearings of a jig transit.

Station: The distance given in inches and decimal parts of an inch measured parallel to a chosen center line from a single, chosen point.

Tangent screw: A hand-operated screw that is used to change the direction of the line of sight either in azimuth or in elevation.

Telescope axle: The horizontal axle that supports the telescope of a jig transit.

Telescope direct: The normal position of a jig transit telescope, as opposed to telescope reversed.

Telescope reversed: The position of a jig transit telescope when it is turned over, or transited, so that it is upside down to its normal position.

Telescopic sight: An optical system that consists of an objective lens and a focusing device and that forms an image on a cross-line reticle which is viewed through an eyepiece that magnifies the image and the cross lines together.

Vertical: In the direction of gravity.

Visible light: An electromagnetic signal having frequencies located in the part of the electromagnetic spectrum to which the eye is sensitive. The visible light frequencies have wavelengths ranging from 0.4 μm to 0.7 μm.

Instructor's Notes:

_____

_____

_____

Resources & Acknowledgments

Additional Resources

This module is intended to present thorough resources for task training. The following reference works are suggested for further study. These are optional materials for continued education rather than for task training.

Brunson Instrument Company
 http://www.brunson.us
Topcon Corporation
 http://www.topcon.com
Trimble Navigation Limited
 http://www.trimble.com

Figure Credits

Brunson Instrument Co., 504F08, 504F51 (photo)

Toolz, Inc., 504F10

Leica Geosystems, Inc., 504F11

Topcon Positioning Systems, Inc., 504F17 (photo), 504F30, 504F33–504F37, 504F39

Ed LePage, 504F19

Trimble Navigation Limited, 504F28, 504F44, 504F45

CST/berger, 504F31

Topaz Publications, Inc., 504F32

MODULE 15504-09 ◆ OPTICAL ALIGNMENT 4.49

MODULE 15504-09 — TEACHING TIPS

The following are suggested activities or instructional methods to help you teach the material in this module.

General

When you call on someone to answer a question, the rest of the class relaxes or even tunes out because they expect that the question and answer will take place only between you and the trainee you called on. Instead, use this technique to involve more trainees in answering questions and to keep them on their toes.

1. Ask trainees to define a term or explain a concept.
2. After one trainee has answered, ask a trainee seated nearby if the answer is right. Then ask whether a trainee in the back of the room agrees.
3. Ask trainees to explain why they think an answer is right or wrong.
4. Use the session to clear up incorrect ideas and encourage trainees to learn from their mistakes.

Sections 1.0.0 – 14.0.0 *Trade Terms*

This Quick Quiz will familiarize trainees with trade terms commonly used in relation to optical alignment. You will need photocopies of the quiz provided on the following page. Trainees will need pencils. If you allow trainees to use the Trainee Guide, decrease the amount of time you give them to complete the quiz, and remind them to bring their books to class.

1. Make a photocopy of the quiz for each trainee.
2. Give trainees between 5 and 10 minutes to complete the quiz.
3. Go over the answers to the quiz.
4. Ask trainees if they have questions.

Answers to Quick Quiz

1. e
2. g
3. f
4. c
5. h
6. b
7. a
8. d
9. i
10. j

Quick Quiz   *Trade Terms*

For each description listed, identify the term that the text best describes. Write the corresponding letter in the blank provided.

_____ 1. An electromagnetic signal that has frequencies below the visible portion of the frequency spectrum and light frequencies lying between those of light and radio waves is called _____.

_____ 2. The change in the apparent relative orientations of objects when viewed from different positions is referred to as _____.

_____ 3. An angle equal to $\frac{1}{60}$ of a degree is a(n) _____.

_____ 4. The direction in a horizontal plane is called the _____.

_____ 5. The series of fine lines placed in the focus of the objective lens of an optical instrument to quantify the measurement of angles or distances is the _____.

_____ 6. To place an instrument so that the line of sight satisfies two requirements simultaneously, such as aiming at two targets, is called _____.

_____ 7. The upper part of an instrument that turns in azimuth with the sighting device is termed the _____.

_____ 8. The distance, given in inches and decimal parts of an inch, measured parallel to a chosen center line from a single selected point is called the _____.

_____ 9. A procedure used to check for an out-of adjustment bubble vial on levels and other instruments is the _____.

_____ 10. Instruments that transmit modulated infrared or visible light signals in short wavelengths of about 0.4 micrometers or microns (μm) to 1.2 μm are called _____.

a. alidade
b. bucking-in
c. azimuth
d. station
e. infrared
f. minute
g. parallax
h. reticle
i. peg test
j. electro-optical instruments (EDMIs)

Section 2.0.0   *Establishing Line of Sight*

This Quick Quiz will familiarize trainees with terms related to establishing line of sight. You will need photocopies of the quiz provided on the following page. Trainees will need pencils. If you allow trainees to use the Trainee Guide, decrease the amount of time you give them to complete the quiz, and remind them to bring their books to class.

1. Make a photocopy of the quiz for each trainee.
2. Give trainees between 5 and 10 minutes to complete the quiz.
3. Go over the answers to the quiz.
4. Ask trainees if they have questions.

Answers to Quick Quiz

1. j
2. h
3. k
4. b
5. a
6. d
7. c
8. f
9. e
10. l
11. g
12. i

Quick Quiz   *Establishing Line of Sight*

For each description listed, identify the term that the text best describes. Write the corresponding letter in the blank provided.

_____ 1. Nearly every optical alignment procedure is based on the use of one or more _____.

_____ 2. A tube consisting of an objective lens near the front end, a reticle with cross lines or a similar pattern near the back end, and an eyepiece mounted behind the reticle is referred to as a(n) _____.

_____ 3. When the optical parts are moved so that a sharp image is seen, the image is said to be in _____.

_____ 4. Objects in front of the telescope are focused on the reticle by moving the _____.

_____ 5. The measurement of the power of a lens, which is equal to the reciprocal of its focal length in meters, is the _____.

_____ 6. The horizontal axle that supports the telescope of a jig transit is called the _____ axle.

_____ 7. Accurate right angles are formed by _____.

_____ 8. In optical alignment, _____ means in the direction of gravity.

_____ 9. A type of level that can be centered more accurately than the spirit level is the _____ level.

_____ 10. Instruments that are graduated in units of 0.100 inch and used to measure short distances from lines of sight are called _____.

_____ 11. An instrument used to establish elevations, but which cannot be used to measure vertical angles, is the _____.

_____ 12. When setting up on dirt, make sure that the points of the _____ are well into the ground.

    a.   diopter
    b.   focusing lens
    c.   auto-collimation
    d.   telescope
    e.   coincidence
    f.   vertical
    g.   builder's level
    h.   telescopic sight
    i.   tripod
    j.   lines of sight
    k.   focus
    l.   optical tooling scales

Checking Level

For successful completion of Performance Task 1, trainees will check level using a theodolite, a precision tilting level, a total station, or an auto level. If you only have one of these instruments available, use that instrument to check the level on some part of a building, such as a window sill or a machine base. If you have more than one instrument, check level on several targets at the same time with different groups of trainees. Make sure that complete setups are performed each time a trainee checks level. A juncture of any kind, or even a nail in the floor, can be used as a location point for an optical or string plummet.

1. Show trainees how to check level using the appropriate instrument(s).
2. Set up the instrument for each trainee, and have each trainee check level.
3. Observe trainees, and evaluate their efforts at checking level.
4. Answer any questions trainees might have.

MODULE 15504-09 — ANSWERS TO REVIEW QUESTIONS

Answer		Section
1.	b	2.0.0
2.	c	2.0.0
3.	d	2.0.0
4.	d	2.2.0
5.	a	2.2.0
6.	c	2.2.0
7.	c	2.3.0
8.	b	2.4.3
9.	d	3.1.0
10.	c	4.2.1
11.	a	4.2.3
12.	c	6.2.0
13.	d	7.0.0
14.	b	10.1.0
15.	d	10.2.0
16.	c	11.2.1
17.	d	11.2.5
18.	c	13.3.0
19.	b	14.0.0
20.	d	14.0.0

The NCCER makes every effort to keep these textbooks up-to-date and free of technical errors. We appreciate your help in this process. If you have an idea for improving this textbook, or if you find an error, a typographical mistake, or an inaccuracy in NCCER's Contren® textbooks, please write us, using this form or a photocopy. Be sure to include the exact module number, page number, a detailed description, and the correction, if applicable. Your input will be brought to the attention of the Technical Review Committee. Thank you for your assistance.

Instructors – If you found that additional materials were necessary in order to teach this module effectively, please let us know so that we may include them in the Equipment/Materials list in the Annotated Instructor's Guide.

Write: Product Development and Revision
National Center for Construction Education and Research
3600 NW 43rd St., Bldg. G, Gainesville, FL 32606

Fax: 352-334-0932

E-mail: curriculum@nccer.org

Craft _____ Module Name _____

Copyright Date _____ Module Number _____ Page Number(s) _____

Description _____

_____

_____

_____

(Optional) Correction _____

_____

_____

(Optional) Your Name and Address _____

_____

_____

Turbines

NCCER STANDARDIZED CRAFT TRAINING PROGRAM

The National Center for Construction Education and Research (NCCER) provides a standardized national program of accredited craft training. Key features of the program include instructor certification, competency-based training, and performance testing. The program provides trainees, instructors, and companies with a standard form of recognition through a National Craft Training Registry. The program is described in full in the *Guidelines for Accreditation*, published by NCCER. For more information on standardized craft training, contact the NCCER by writing us at 3600 NW 43rd St., Bldg. G, Gainesville, FL 32606; calling 352-334-0911; or emailing info@nccer.org. More information may be found at our website, www.nccer.org.

HOW TO USE THIS ANNOTATED INSTRUCTOR'S GUIDE

Each page presents two sections of information. The larger section displays each page exactly as it appears in the Trainee Module. The narrow column ties suggested trainee and instructor actions to each page and provides icons (detailed below) to call your attention to material, safety, audiovisual, or testing requirements. The bottom of each page includes space for your notes.

The **Audiovisual** icon indicates an appropriate time to show a transparency or other audiovisual aid.

The **Classroom** icon prompts you to define a term, stress a point, ask trainees to explain a concept, or give examples.

The **Demonstration** icon directs you to show trainees how to perform tasks.

The **Examination** icon tells you to administer the written module examination.

The **Homework** icon is placed where you may wish to assign reading for the next class, assign a project, or advise trainees to prepare for an examination.

The **Laboratory** icon is used when trainees are to practice performing tasks.

The **Materials** icon is a reminder for you to gather materials needed for classes, labs, and testing.

The **Performance Testing** icon tells you to administer a performance test or a portion thereof.

The **Safety** icon is used to emphasize safety issues. It is often keyed to *Caution* and *Warning!* statements in the Trainee Module.

The **Teaching Tip** icon indicates additional guidance is available, such as how to conduct an exercise, get the most educational value from a field trip, or encourage class participation. Teaching Tips may expand on a feature (*Think About It*, *Did You Know?*) or provide *Quick Quizzes* or similar exercises. You will be referred to the Teaching Tips section at the back of the module if there is additional material.

The **Combination** icon indicates that the laboratory listed corresponds with a performance task. If desired, you can note the proficiency of the trainees during the laboratory, and use it to satisfy performance testing requirements.

PREPARATION

Before teaching this module, you should review the Objectives, Performance Tasks, Materials and Equipment List, and Module Outline. Be sure to allow ample time to prepare your own training or lesson plan and gather all required materials and equipment.

MODULE OVERVIEW

This module describes steam, gas, and hydraulic turbines. Trainees will become familiar with turbine components as they learn the principles by which turbines operate.

PREREQUISITES

Prior to training with this module, it is recommended that the trainee shall have successfully completed *Core Curriculum; Millwright Level One; Millwright Level Two; Millwright Level Three; Millwright Level Four;* and *Millwright Level Five*, Modules 15501-09 through 15504-09.

OBJECTIVES

Upon completion of this module, the trainee will be able to do the following:

1. Identify and explain impulse and reaction blades.
2. Identify and explain types of turbines.
3. Identify and explain steam turbine components.
4. Identify and explain gas turbine components.
5. Explain types of water turbines.

PERFORMANCE TASKS

This is a knowledge-based module. There are no Performance Tasks.

MATERIALS AND EQUIPMENT LIST

Overhead projector and screen

Transparencies

Blank acetate sheets

Transparency pens

Whiteboard/chalkboard

Markers/chalk

Pencils and scratch paper

Appropriate personal protective equipment

Photos or videos/DVDs showing the components and/or operation of steam, gas, and water turbines

TV/VCR/DVD player

Examples or pictures of impulse and reaction blades

Examples, photos, videos, or DVDs of Pelton wheels, Francis Wheels, and Kaplan wheels

Examples or pictures of turbine components:
  Compressor parts
  Combustor parts
  Turbine section parts
  Auxiliary support systems
  Bearings

Copies of the Quick Quiz*

Module Examinations**

*Located at the back of this module.

**Located in the Test Booklet.

SAFETY CONSIDERATIONS

Ensure that the trainees are equipped with appropriate personal protective equipment and know how to use it properly.

ADDITIONAL RESOURCES

This module is intended to present thorough resources for task training. The following reference works are suggested for both instructors and motivated trainees interested in further study. These are optional materials for continued education rather than for task training.

Environmental Protection Agency
www.epa.gov/CHP/documents/tech_turbines.pdf

General Electric Company
www.gepower.com/home/index.htm

Siemens Corporation
www.powergeneration.siemens.com/home

http://mysite.du.edu/~jcalvert/tech/fluids/turbine.htm

TEACHING TIME FOR THIS MODULE

An outline for use in developing your lesson plan is presented below. Note that each Roman numeral in the outline equates to one session of instruction. Each session has a suggested time period of 2½ hours. This includes 10 minutes at the beginning of each session for administrative tasks and one 10-minute break during the session. Approximately 20 hours are suggested to cover *Turbines*. You will need to adjust the time required for hands-on activity and testing based on your class size and resources.

Topic	Planned Time
Session I. Introduction; Turbine Operating Principles; Types of Turbines	
A. Introduction	_____
B. Turbine Operating Principles	_____
1. Impulse Turbines	_____
2. Reaction Turbines	_____
C. Types of Turbines	_____
1. Steam Turbines	_____
2. Gas Turbines	_____
3. Hydroelectric Turbines	_____
Session II. Steam Turbine Casing Parts	
A. Inlet Parts	_____
B. Blade Supports	_____
C. Extraction and Exhaust	_____
D. Safety	_____
E. Bearings	_____
F. Seals	_____
Session III. Steam Turbine Rotor Parts	
A. Turbine Rotor	_____
B. Blades	_____
C. Turning Gear and Coupling	_____
D. Lube Oil Gear Pump	_____
Session IV. Steam Turbine Auxiliary Systems	
A. Lubrication and Jacking Oil Equipment	_____
B. Turbine Control and Protection System	_____
C. Electrohydraulic Governor Equipment	_____
D. Seal Steam System	_____
E. Steam Distribution System	_____
F. Condensate System	_____

Session V. Gas Turbine Compressor Parts
   A. Inlet Guide Vanes
   B. Rotor and Stator Blades
   C. Discharge Section
   D. Bleed-Off Lines

Session VI. Gas Turbine Combustor and Turbine Section Components
   A. Combustor Components
      1. Combustion Chambers
      2. Fuel Nozzles
      3. Igniters and Crossfire Tubes
      4. Flame Detectors
      5. Transition Pieces
   B. Turbine Section Components
      1. Nozzles
      2. Rotor Blades
      3. Exhaust Silencer

Session VII. Gas Turbine Auxiliary Support Systems
   A. Starting Systems
   B. Lube Oil System
   C. Fuel System
   D. Water Cooling System

Session VIII. Review and Testing
   A. Module Review
   B. Module Examination
      1. Trainees must score 70% or higher to receive recognition from NCCER.
      2. Record the testing results on Craft Training Report Form 200, and submit the results to the Training Program Sponsor.

Millwright Level Five

15505-09

Turbines

15505-09
Turbines

Topics to be presented in this module include:

Overview

Turbines are machines that are used for power generation. This module provides an overview of steam, gas, and hydraulic turbines, and their components. Basic operating principles and common applications for impulse turbines and reaction turbines are presented.

Instructor's Notes:

_____

_____

_____

Objectives

When you have completed this module, you will be able to do the following:

1. Identify and explain impulse and reaction blades.
2. Identify and explain types of turbines.
3. Identify and explain steam turbine components.
4. Identify and explain gas turbine components.
5. Explain types of water turbines.

Trade Terms

Combustion
Combustion chamber
Compressor
Condensate

Governor
Kinetic energy
Mechanical energy
Turbine

Required Trainee Materials

1. Pencil and paper
2. Appropriate personal protective equipment

Prerequisites

Before you begin this module, it is recommended that you successfully complete *Core Curriculum*; *Millwright Level One; Millwright Level Two; Millwright Level Three; Millwright Level Four;* and *Millwright Level Five*, Modules 15501-09 through 15504-09.

This course map shows all of the modules in the fifth level of the Millwright curriculum. The suggested training order begins at the bottom and proceeds up. Skill levels increase as you advance on the course map. The local Training Program Sponsor may adjust the training order.

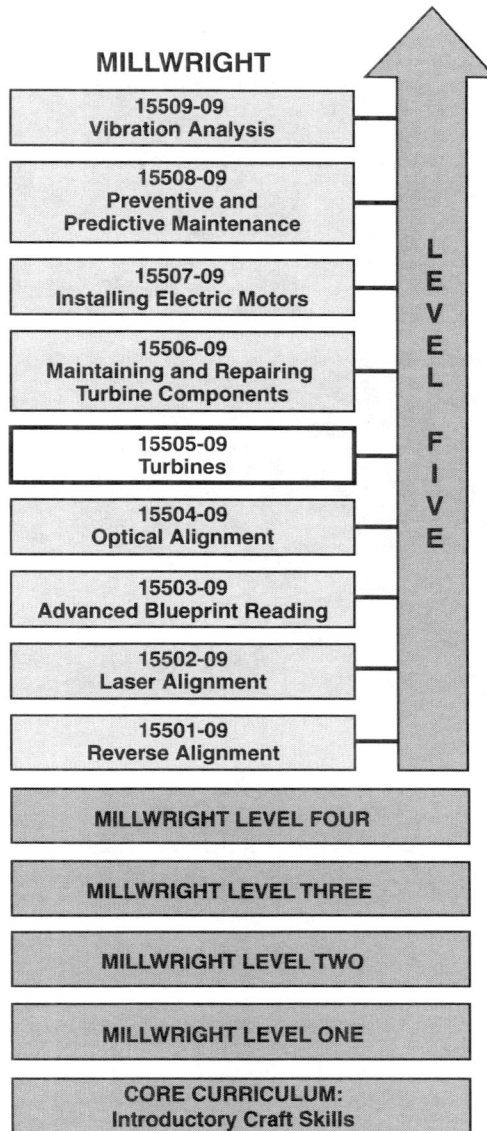

MILLWRIGHT

| 15509-09 |
| Vibration Analysis |
| 15508-09 |
| Preventive and Predictive Maintenance |
| 15507-09 |
| Installing Electric Motors |
| 15506-09 |
| Maintaining and Repairing Turbine Components |
| 15505-09 |
| Turbines |
| 15504-09 |
| Optical Alignment |
| 15503-09 |
| Advanced Blueprint Reading |
| 15502-09 |
| Laser Alignment |
| 15501-09 |
| Reverse Alignment |

LEVEL FIVE

MILLWRIGHT LEVEL FOUR

MILLWRIGHT LEVEL THREE

MILLWRIGHT LEVEL TWO

MILLWRIGHT LEVEL ONE

CORE CURRICULUM:
Introductory Craft Skills

505CMAP.EPS

Ensure that you have everything required to teach the course. Check the Materials and Equipment List at the front of this module.

See the general Teaching Tip at the end of this module.

Explain that terms shown in bold are defined in the Glossary at the back of this module.

Show Transparency 1, Objectives. Review the goals of the module, and explain what will be expected of the trainees.

Review the modules covered in Level Five, and explain how this module fits in.

1.0.0 ◆ INTRODUCTION

A **turbine** (*Figure 1*) is a machine that converts **kinetic energy** into **mechanical energy**. A turbine receives energy from a moving fluid or gas and converts this energy into the rotary movement of a shaft. When the turbine shaft is directly connected or geared to another machine, that machine is driven by the turbine to produce useful work.

The primary mechanical feature of a turbine is the rotating element known as the rotor. The rotor is a shaft and has a series of vanes, blades, or buckets supported around it. This arrangement of

Figure 1 ◆ Single-flow steam turbine.

505F01.EPS

vanes, blades, or buckets is called a wheel. The blades of each wheel are angled around the rotor so that when the moving fluid or gas moves through the blades, the energy of the fluid is transferred to the rotor, causing the rotor to rotate. The casing surrounding the rotor has stationary vanes, or stators, that direct the fluid through the blades at the optimum angle.

The amount of steam, fluid, or gas entering the turbine can be controlled so that the horsepower of the turbine shaft can be varied. This is useful when the turbine shaft is connected to an electric generator because the turbine can provide a constant supply of electricity even under varying load conditions.

2.0.0 ◆ TURBINE OPERATING PRINCIPLES

A turbine can be powered by water, steam, gas, or air. There are two basic classifications of turbines, depending on its operating principles: impulse turbines or reaction turbines. These two basic designs are often combined.

2.1.0 Impulse Turbines

Impulse turbines (*Figure 2*) use the impulse of a moving fluid or gas on the rotor blades to turn the rotor. One or more high-velocity jets force water,

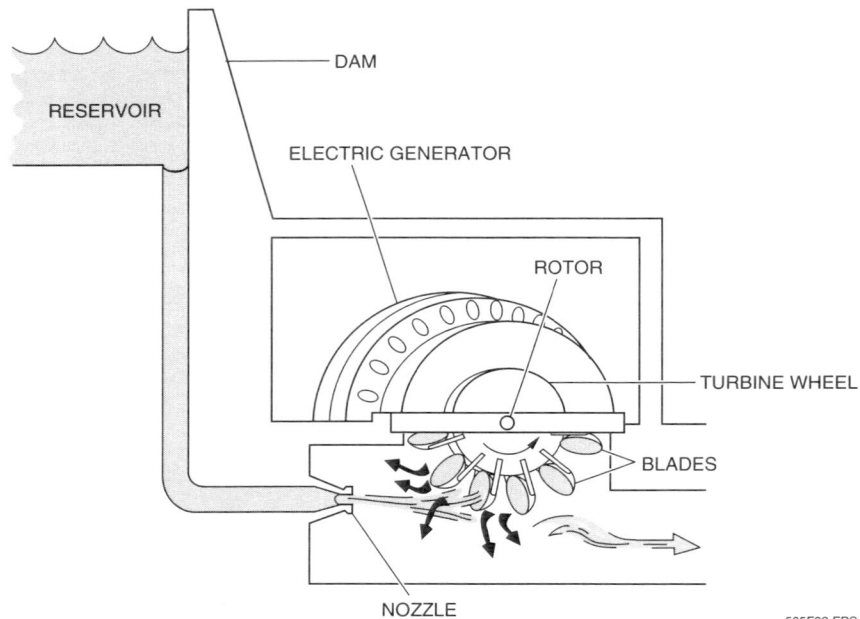

Figure 2 ◆ Hydroelectric impulse turbine.

505F02.EPS

Instructor's Notes:

_____

_____

_____

gas, or steam onto the rotor blades. In the impulse turbine, the flow of fluid or gas contacts a small portion of the blades on each wheel at any given instant. Since the fluid does not fill all passages of the wheel, impulse turbines are known as constant-pressure turbines because the pressure of the fluid changes minimally when the fluid passes from the front of the turbine wheel to the back of the turbine wheel.

Kinetic energy from a jet of water that sprays on the blades of the turbine wheel turns the hydro-electric impulse turbine. Many turbines in operation today are multistage impulse turbines (*Figure 3*). Multistaging does not change the principle of impulse operation, but it does greatly increase the efficiency of operation.

2.2.0 Reaction Turbines

In reaction turbines (*Figure 4*), the fluid or gas pressure is distributed across all the blades in the turbine wheel at once. Since the fluid or gas is admitted to the turbine wheels around their entire circumferences, the wheels may have a smaller diameter than the wheels in an impulse turbine

having the same power depending on the head. A portion of the power of a reaction turbine comes from the actual velocity of the fluid or gas, while the rest of the power comes from the pressure drop acting between the front and the back of the turbine blades. A windmill is an example of a basic reaction turbine.

3.0.0 ◆ TYPES OF TURBINES

Industrial turbines are custom-designed to meet specific operating needs. They are simple and powerful machines that are classified according to the fluid or gas that they use. The three most common types are the steam turbine, the gas turbine, and the hydraulic (water) turbine.

3.1.0 Steam Turbines

Steam turbines (*Figure 5*) are used primarily in fossil-fuel and nuclear power plants to drive electric generators for producing electric power to consumers. They are also used to drive electric generators in large industrial plants to provide the

Explain why multi-staging increases efficiency.

Compare reaction turbines to impulse turbines.

Show Transparency 3 (Figure 4). Describe the operation of a reaction turbine.

Introduce steam, gas, and water turbines.

Briefly describe common applications for steam turbines.

Show Transparency 4 (Figure 5).

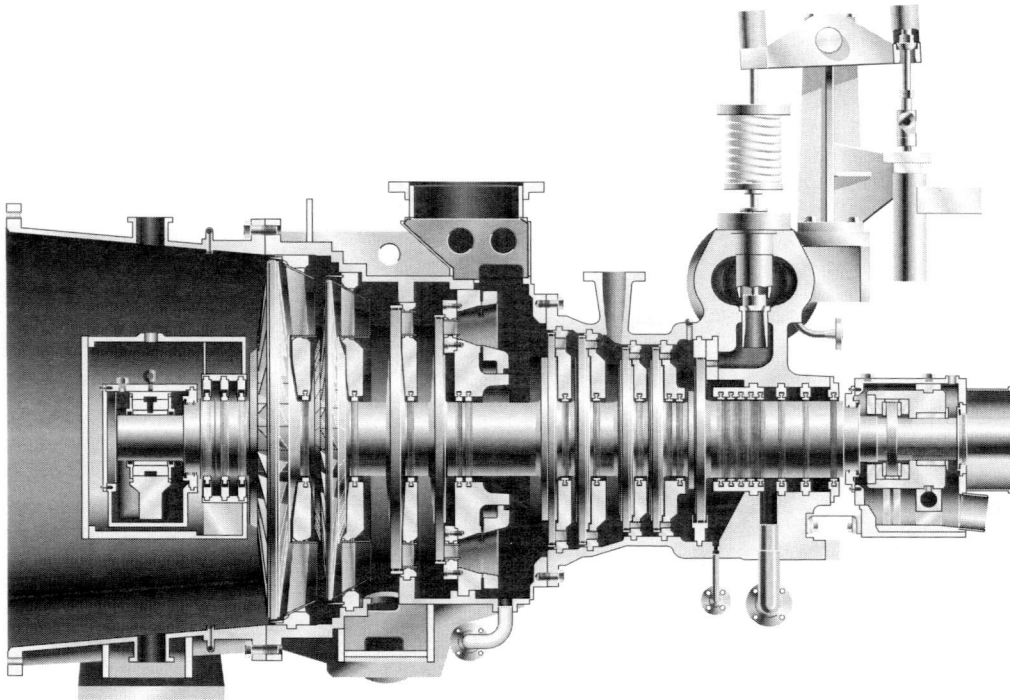

505F03.EPS

Figure 3 ◆ Basic multistage impulse turbine.

MODULE 15505-09 ◆ TURBINES 5.3

Compare condensing and noncondensing turbines.

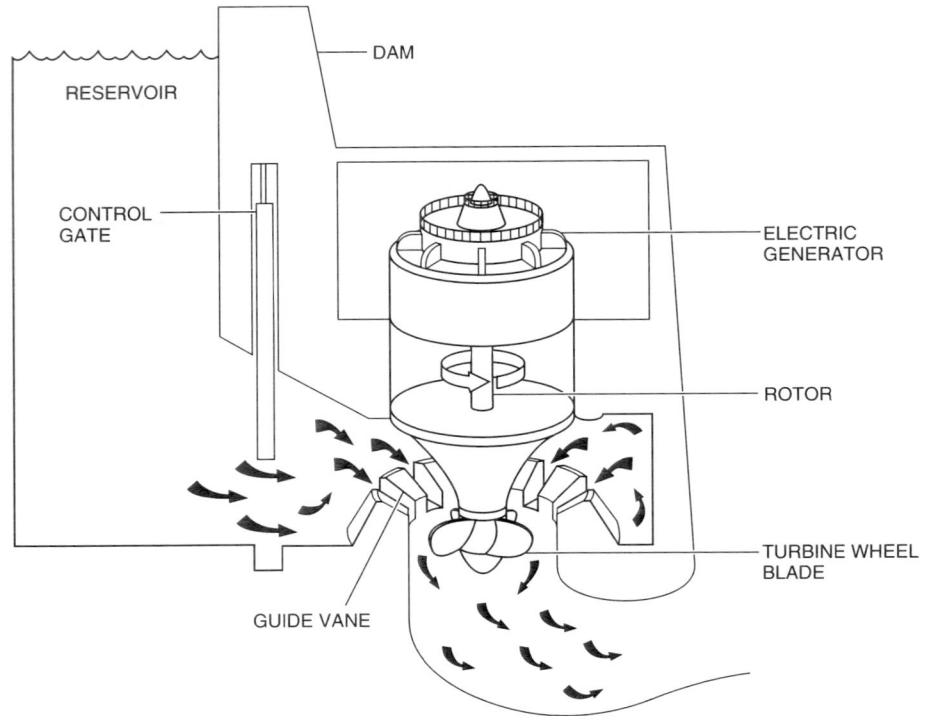

Figure 4 ◆ Hydroelectric reaction turbine.

plant's electricity, and they are used to drive the propellers of large ships and submarines. The steam needed to operate the turbine is produced in a boiler and piped under pressure to the steam turbine. Because of the velocity of the steam, the rotational speed of the turbine can exceed 10,000 rpm.

> **NOTE**
>
> Steam turbines are usually operated at pressures from 600 psi to 3,600 psi. Most run from 600°F to 1,200°F. At these levels, even a very small leak can cause serious injury or death. Follow all safety procedures at all times.

Modern industrial turbines can have as many as 50 or more stages set on a horizontal axle. Each stage consists of a turbine wheel of rotating blades and a set of stationary blades. The curved wheel blades and the stationary blades are shaped so that the spaces between them act as nozzles that aim the steam and increase the velocity of the steam before it enters the next stage. As the steam travels through the turbine, it expands because of pressure and temperature drops. Therefore, to get the maximum horsepower in the turbine, each stage is larger than the one that precedes it.

Steam turbines may be either condensing or noncondensing, depending on how the steam exiting the turbine is used. In the condensing turbine, the steam exiting the turbine enters a condenser to convert the steam back to water. This creates a vacuum which helps draw the steam through the turbine. The water is then pumped back to the boiler to be converted into steam again.

The steam coming from a noncondensing turbine is not converted into water but is sent to another unit or system in the plant operating at a lower pressure. In a noncondensing turbine, steam can be extracted from the turbine at extraction points to obtain desired pressure flow. Controlling valves at each extraction point regulate the amount of steam extracted from the turbine.

Instructor's Notes:

_____

_____

_____

_____

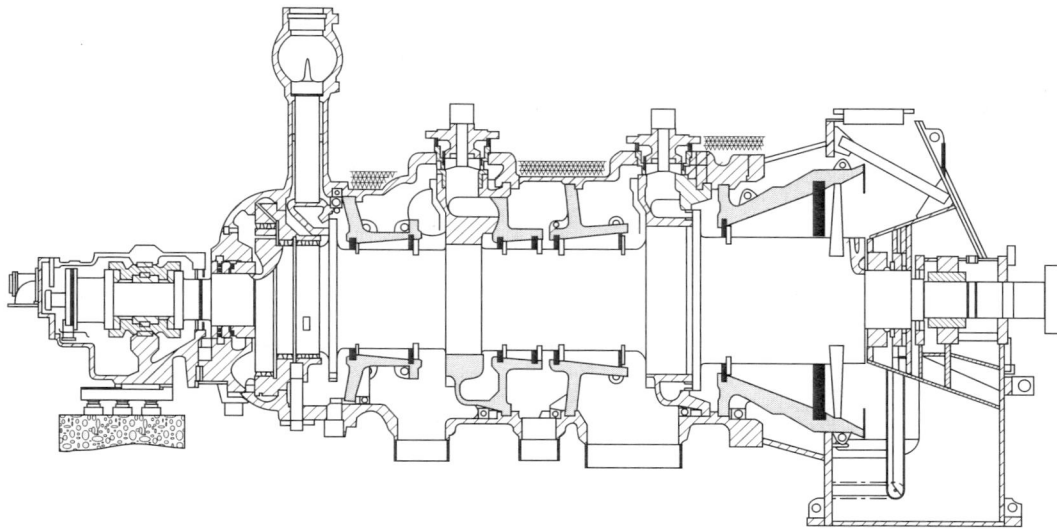

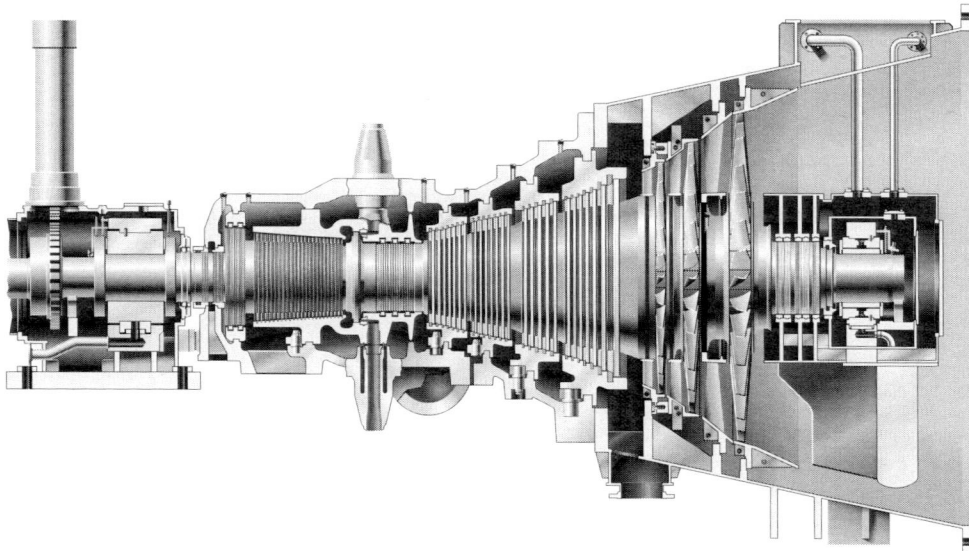

Figure 5 ◆ Industrial steam turbine.

505F05.EPS

3.2.0 Gas Turbines

Gas turbines (*Figure 6*) are the most common types of turbines. They are internal **combustion** turbines used to power electric generators and ship propellers. They are also an important part of jet aircraft engines. Some industrial gas turbines are used to provide additional power during peak periods when plant steam turbines are not able to meet the demand, or to provide emergency power during a power outage when no outside source of electricity is available. The basic gas turbine has three main components: a **compressor**, a group of combustors, and a turbine.

The compressor consists of hundreds of blades that turn like a fan to pull air into the gas turbine and then draw the air into the combustors. In the combustors, fuel is mixed with the incoming air and is then ignited. The burning gases move out of the combustors and enter the turbine. Once in the turbine, the burning gases move through the turbine blades the same way that steam moves through a steam turbine, causing the turbine to rotate at high speeds.

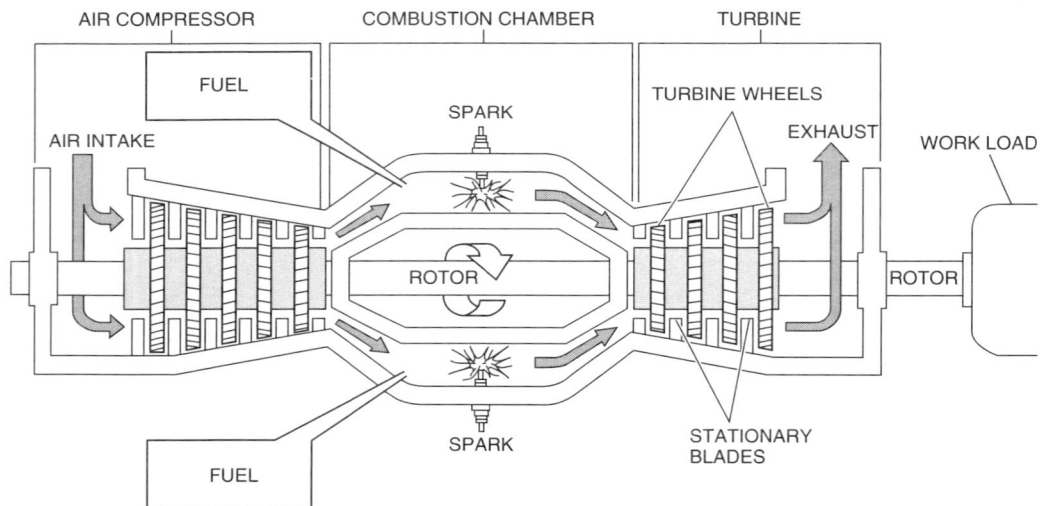

Figure 6 ◆ Basic components of a gas turbine.

505F06.EPS

Instructor's Notes:

_____

_____

_____

Gas turbine exhaust gases are approximately 800°F to 1,000°F when they leave the turbine. There are ways to capture some of the heat lost in a gas turbine. A typical compressor consumes about one half of the energy output of the gas turbine. Combined circle systems use the excess heat in many ways.

3.3.0 Hydroelectric Turbines

Hydroelectric, or water, turbines are used primarily to drive electric generators at power plants. Hydroelectric turbines are driven by water falling from waterfalls or by water stored behind dams. The type of hydroelectric turbine used at a plant depends on the head available. The head is the distance that the water falls before striking the turbine and can range from 8 feet to more than 1,000 feet. The water is conveyed to the turbine from a river flow or through a pressure conduit known as a penstock. Water turbines currently being built are one of the following types:

- Pelton wheel
- Francis wheel
- Kaplan wheel

Pelton wheels (*Figure 7*) consist of a wheel with multiple arms ending in double bowl shapes joined in the middle. The water jets strike the bowls where they are joined. The Pelton wheel is the only impulse water turbine presently being manufactured. The Pelton wheel must have a head of at least 800 feet to operate efficiently. Impulse turbines can be mounted on a horizontal or vertical shaft, and there may be several jet nozzles acting on a single turbine wheel.

Most hydroelectric turbines are reaction-type turbines. Reaction turbines include the Francis wheel (*Figure 8*), which is much like a squirrel cage blower—that is, a cylinder with angled slits. Kaplan turbines (*Figure 9*) are a more recent development, and consist of a propeller shape at the end of a tube. The reaction-turbine wheel operates under water. The turbine wheel is surrounded by guide vanes, and water coming from the reservoir is directed around the entire periphery of the turbine. The guide vanes can be used to regulate the flow of water through the turbine. The rotor is turned by the weight or pressure of the water flowing through the turbine wheels.

Many hydroelectric power plants use reversible pump-turbine units. The runners look like a Francis runner, but are much smaller. These units can act as a turbine to drive an electric generator or as a pump to force water back up to the upper water reservoir to be used again in the turbine. During peak periods of demand for electricity, water flows

505F07.EPS

Figure 7 ◆ Pelton wheels.

from an upper reservoir, through the turbine, and into a lower reservoir. At other times, the generator serves as an electric motor to turn the turbine in reverse and pump the water from the lower reservoir back up into the upper reservoir to be held until it is needed again.

505F08.EPS

Figure 8 ◆ Francis wheels.

Introduce different types of hydroelectric turbines.

Explain the configuration of a Pelton wheel turbine. Refer to Figure 7.

Explain that a Francis wheel is like a squirrel cage blower. Refer to Figure 8.

Describe how a Kaplan wheel operates. Refer to Figure 9.

If available, show examples, photos, videos, or DVDs of Pelton wheels, Francis Wheels, and Kaplan wheels in operation.

Discuss pump-turbine units and how their function changes according to demand for electricity.

MODULE 15505-09 ◆ TURBINES 5.7

Assign reading of
Sections 4.0.0–4.1.13.

Ensure that you have
everything required for
teaching this session.

Name the three
categories of steam
turbine components.

List the parts of a
turbine casing.

Describe steam
chests, supply lines,
and inlet control
valves.

Show Transparencies 6
and 7 (Figures 10 and
11). Point out the parts
of the turbine casing.

Figure 9 ◆ Kaplan turbines.

505F09.EPS

4.0.0 ◆ STEAM TURBINE PARTS

The major working parts and components of a steam turbine can be divided into the following categories:

- Casing parts
- Rotor parts
- Auxiliary system components

4.1.0 Turbine Casing Parts

The turbine casing is the outer shell of the steam turbine. In most cases, there are multiple casings in a turbine. The casing is a thick, metal casting with insulation covering the outer surface. The casing is made in upper and lower halves, divided horizontally so that the machine can be opened for maintenance and repair work. Turbine casing parts are parts that are often attached directly to the casing. They include the following:

- Steam chest
- Supply lines
- Inlet control valves and pipes
- Nozzle blocks
- Blade carriers
- Stationary blades
- Extraction and exhaust ports
- Grounding brushes
- Rupture disc
- Insulation and lagging
- Bearings
- Equalizing seals
- Gland seals

Figures 10 and *11* show turbine casing parts and components.

4.1.1 Steam Chest

The steam chest is the short header pipe or part of the casing across the smaller end of the turbine to which the steam inlet control valves are attached. Steam is piped from the boiler to the steam chest and from the steam chest to the inlet pipes of the turbine. There are control valves mounted in the steam chest to control the amount of steam entering the turbine.

4.1.2 Supply Lines

The supply lines are steam pipes that carry the steam to the turbine. A high-pressure supply line is connected to the steam chest to allow high-pressure steam to enter the turbine. Older models may have a lower-pressure line that can connect with the turbine at a point near an intermediate-pressure extraction point. Such a line can admit intermediate-pressure steam during boiler outages, so that the turbine can be run at reduced capacity.

4.1.3 Inlet Control Valves and Pipes

The inlet control valves may be on or off the base, and are variable-position valves that regulate the amount of steam admitted into each section of the turbine or that stop the flow of steam entirely. The opening of the control valves determines the horsepower of the turbine and the back-pressure level. The control valves also control the extraction steam-flow rates. The inlet pipes are smaller pipes that connect the steam chest and valves to the turbine.

Instructor's Notes:

_____

_____

_____

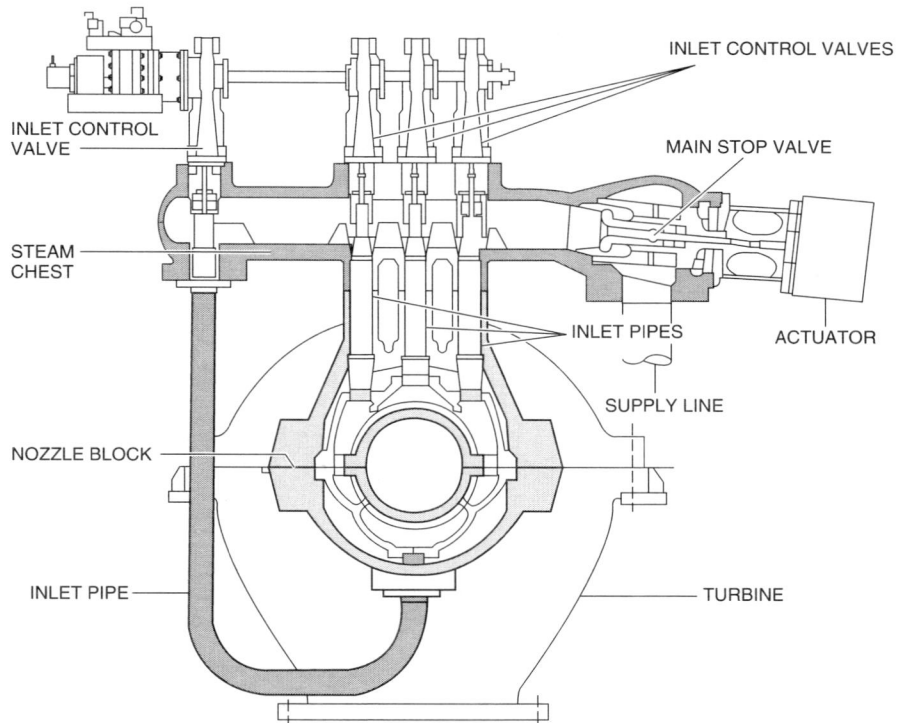

Figure 10 ◆ Turbine casing parts.

Labels in figure:
INLET CONTROL VALVES
INLET CONTROL VALVE
MAIN STOP VALVE
STEAM CHEST
INLET PIPES
ACTUATOR
SUPPLY LINE
NOZZLE BLOCK
INLET PIPE
TURBINE
505F10.EPS

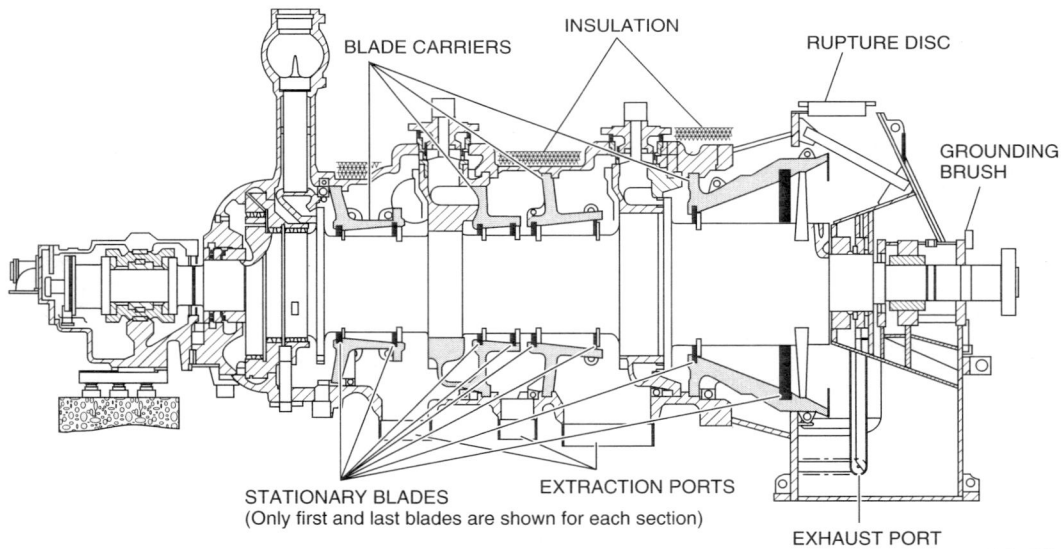

Figure 11 ◆ Turbine casing components.

Labels in figure:
BLADE CARRIERS
INSULATION
RUPTURE DISC
GROUNDING BRUSH
STATIONARY BLADES
(Only first and last blades are shown for each section)
EXTRACTION PORTS
EXHAUST PORT
505F11.EPS

Describe the function of nozzle blocks, blade carriers, and stationary blades.

Explain the purpose of extraction valves.

Provide pictures or examples of turbine components for trainees to examine.

Describe grounding brushes and explain why it is important to conduct shaft current away from the rotor.

Explain that the rupture disc provides emergency release of excess pressure.

Caution trainees to stay away from the area in front of the rupture disc.

Explain why the steam turbine is covered with insulation and note how it is attached.

Explain how different types of bearings are used to reduce friction.

Show Transparency 8 (Figure 12). Note the position of the equalizing seal.

4.1.4 Nozzle Blocks

The nozzle blocks, also called nozzle plates, or rings, or box, are chambered bulkheads in the turbine that may separate the high-pressure and low-pressure sections from the intermediate-pressure section. Nozzle ports in each chamber direct steam at a high velocity and at the correct angle into the turbine wheels in the next section of the turbine. The nozzle block has a set of chambers and nozzles for each inlet valve. They may also be used to redirect steam.

4.1.5 Blade Carriers

The blade carriers, also called blade rings or diaphragms, are special alloy cylinders mounted inside the turbine casing. Fixed vanes are attached to the inner surface of the carriers to direct the steam into the wheels at the proper direction for optimum energy transfer. The blade carriers are attached to the casing in a manner that allows them to expand both radially and axially independently as the turbine is warmed up.

4.1.6 Stationary Blades

The stationary blades are special alloy vanes that are attached to the blade carriers and direct the steam flow at the required angel to the wheel blades.

4.1.7 Extraction and Exhaust Ports

The extraction and exhaust ports are openings in the turbine casing that allow steam to be removed from the turbine. Extraction ports located at different sections in the turbine allow steam at different pressures to be removed for use elsewhere in the plant. Extraction valves associated with the extraction ports regulate the amount of steam removed. The exhaust trunk feeds all remaining steam directly into the condenser to be sent back to the boiler.

4.1.8 Grounding Brushes

The grounding brushes are carbon elements or graded copper straps mounted in an insulated casing, with springs that hold the brushes against the rotor. A grounding wire runs from the brushes to conduct shaft current away from the rotor because static electric discharges can cause pitting and corrosion to the turning parts and bearings.

4.1.9 Rupture Disc

The rupture disc is a thin, soft metal diaphragm across a flanged opening at the top of the low-pressure section of the turbine. When the pressure in the turbine exceeds the maximum high limit, the rupture disc bursts, allowing the steam to escape.

4.1.10 Insulation and Lagging

The entire steam turbine is covered with insulating material to prevent heat loss and dangerous burns and to maintain even expansion and contraction. The insulation may be attached with lag screws or it may be in the form of removable blankets.

4.1.11 Bearings

The bearings are used to reduce friction between the shaft housing and the rotating shaft. The bearing surfaces are babbited, and the bearings are hydrodynamic in design. A groove between the babbited surfaces retains oil that forms a film to separate the metallic running surfaces. This lubrication reduces the coefficient of friction and minimizes wear. Two types of bearings are used: a thrust end bearing and a journal bearing.

The thrust end bearing holds the rotor in place and must be able to withstand the axial force, along the rotor, of the steam passing through the turbine. The steam exerts pressure on the rotor, pushing the shaft in the direction of the steam flow. The thrust bearing is sometimes installed on the high-pressure end of the shaft and helps absorb the force to hold the shaft rigidly in place while allowing free rotational movement.

A journal bearing is a babbited liner that fits into a machined fit in the support casing to hold the shaft in place. Journal bearings are made in two halves, which can be replaced without removing the turbine rotor. The journal bearings serve as radial support for the turbine shaft.

Both types of bearings use an air or oil seal system to prevent lube oil from escaping into the turbine.

4.1.12 Equalizing Seals

The equalizing seals are cylinders that encircle the shaft at the front of each section. The equalizing seals divide the steam path and provide back surfaces for the steam pressure to act against. Labyrinth seals between the equalizing seals and the rotor and turbine housing prevent steam leakage between the sections. *Figure 12* shows the steam path.

Instructor's Notes:

_____

_____

_____

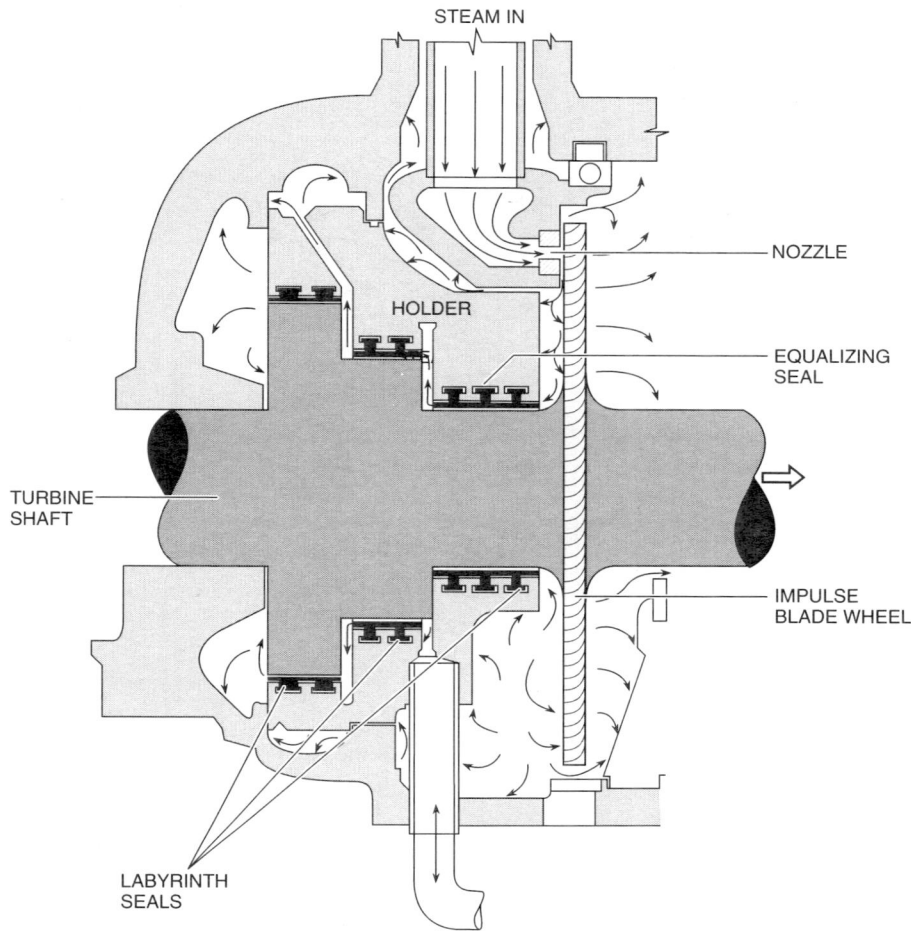

STEAM IN

NOZZLE

HOLDER

EQUALIZING
SEAL

TURBINE
SHAFT

IMPULSE
BLADE WHEEL

LABYRINTH
SEALS

505F12.EPS

Figure 12 ◆ Steam path at high-pressure equalizing seal.

Explain the function of gland seals.

Assign reading of Sections 4.2.0–4.2.6.

Ensure that you have everything required for teaching this session.

Explain the function of the turbine rotor, and introduce its basic parts.

4.1.13 Gland Seals

Gland seals are rings in a split housing that use a closed steam supply and spring tension to hold them close to the shaft at a prescribed distance to prevent leakage of steam.

4.2.0 Turbine Rotor Parts

The turbine rotor parts (*Figure 13*) are all parts in the turbine that are in contact at some time with a rotor or central shaft. These parts include the following:

- Turbine rotor
- Impulse wheels
- Reaction blades
- Turning gear
- Shaft coupling
- Lube oil gear pump

4.2.1 Turbine Rotor

The turbine rotor is the primary turning component of the turbine. The rotor shaft is connected directly to the generator rotor shaft and performs the work required to produce electricity.

MODULE 15505-09 ◆ TURBINES 5.11

Compare the function-
ality of impulse and
reaction blades.

Show Transparency 10
and 11 (Figures 14 and
15).

Discuss the effects of
the pressure drop
across reaction blades.

Show Transparency 12
(Figure 16). Note the
position of the blade
shrouds.

Explain the functions
of the turning gear and
shaft coupling.

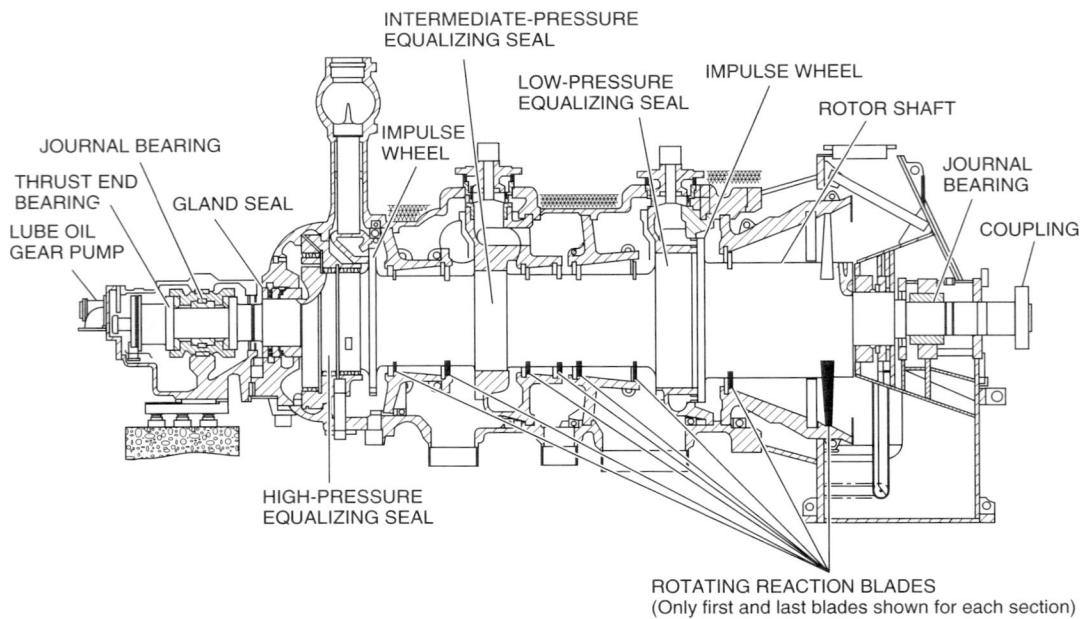

Figure 13 ◆ Turbine rotor parts.

4.2.2 Impulse Wheels

In compound steam turbines containing impulse and reaction wheels, the impulse wheels are located at the beginning of each section of the turbine. The impulse wheels are bladed discs mounted on the shaft past the nozzle block and the crossover block. In the nozzles, the steam pressure drops and the steam velocity is increased. The impulse wheel blades are positioned to the nozzles so that the steam changes directions as it passes across the impulse wheel. The steam does not actually drive directly against the blades. The change of direction causes the wheel to turn with minimal pressure drop in the steam passing through, but there is a reduction in steam velocity. *Figure 14* shows the impulse blade effect.

4.2.3 Reaction Blades

The rotating reaction blades are special alloy blades that are positioned in relation to stationary vanes so that the steam force works directly against them. The force of the steam pressure pushes into the blades, causing the steam to divert in the opposite direction. This causes a more turbulent steam flow and results in a pressure drop as the component rotates. *Figure 15* shows this effect.

All the remaining blades in each section of the turbine are reaction blades. The remaining sets of rotating and stationary blades increase in size throughout the rest of the turbine due to the expansion in the steam volume as it travels through the turbine. Since a pressure drop occurs across the moving reaction blades, there is a tendency for steam to leak around the ends of the blades (*Figure 16*). This is prevented by integral bucket or blade shrouds.

4.2.4 Turning Gear

The turning gear is a motor-driven reduction gear. Most are located on the low-pressure end of the turbine shaft. The turning gear is used to periodically turn the shaft of the turbine during start-up and shutdown of the turbine. This prevents thermal distortion of the shaft, which is common at startup and shutdown.

4.2.5 Shaft Coupling

The shaft coupling is a flange that is normally located on the end of the shaft at the low-pressure end of the turbine. In some applications, it is located on the high-pressure end. The coupling

Instructor's Notes:

_____

_____

_____

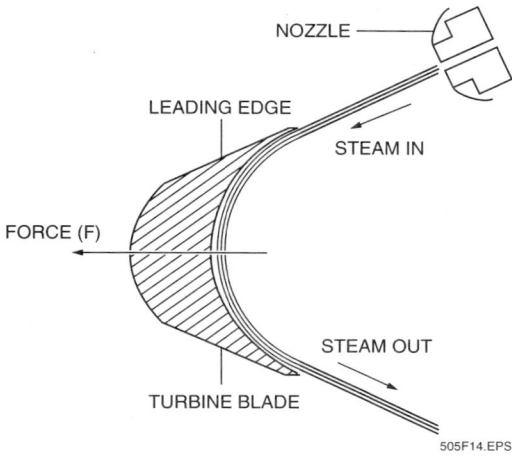

Figure 14 ◆ Principle of impulse blade effect.

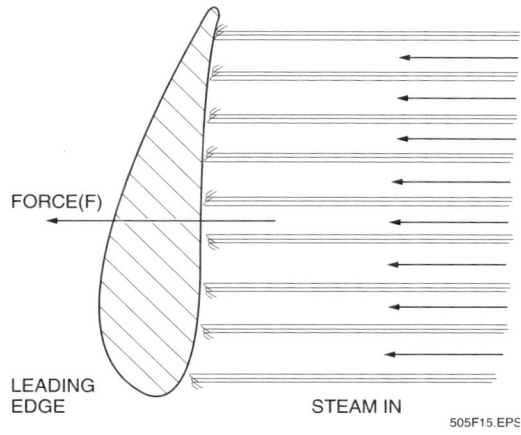

Figure 15 ◆ Reaction blade effect.

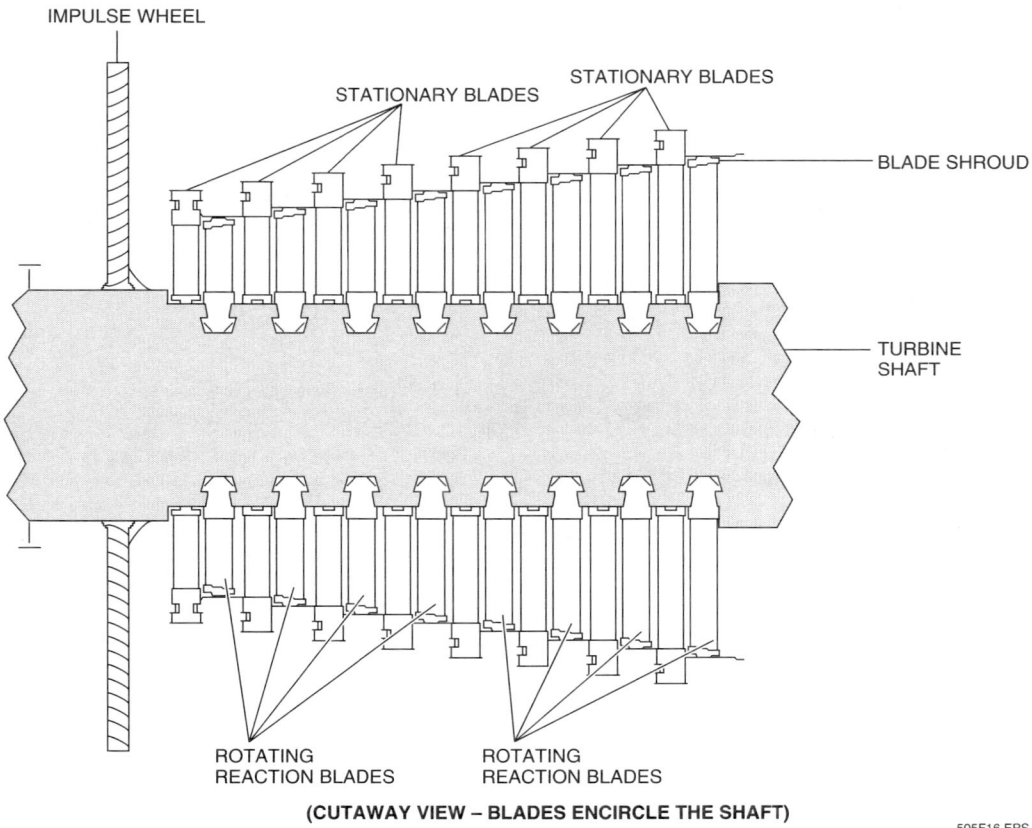

(CUTAWAY VIEW – BLADES ENCIRCLE THE SHAFT)

505F16.EPS

Figure 16 ◆ Low-pressure blading.

provides a positive lock with the driven shaft so that the rotation of the turbine shaft can be transmitted to the driven shaft.

4.2.6 Lube Oil Gear Pump

The lube oil gear pump is a shaft-driven pump usually on the high-pressure end. This pump provides lubricating oil to the bearings and is part of the lube oil system.

4.3.0 Turbine Auxiliary System Components

Several auxiliary systems are required to operate the turbine safely and efficiently for an extended service life. These systems include the following equipment:

- Electrohydraulic **governor** equipment
- Lubrication and jacking oil system equipment
- Turbine control and protection system equipment
- Seal steam system
- Steam distribution system
- **Condensate** drain and injection system

4.3.1 Lubrication and Jacking Oil System Equipment

The lubrication and jacking oil (lift oil) system (*Figure 17*) is a closed piping system that consists of main and backup pumps, a reservoir, oil coolers and oil heaters, oil filters, a centrifugal purifier, a vapor extractor, valves, and pipes. This system provides oil to the bearings in the turbine and generator. The lubrication oil system components provide oil to form a film between the metallic surfaces. This reduces friction and limits wear and heat on the surfaces.

The jacking and/or lifting oil components provide oil at a much higher pressure during startup and shutdown. The high-pressure oil actually lifts the rotor shaft from the bearing surfaces.

4.3.2 Turbine Control and Protection System Equipment

The turbine control and protection system is a closed piping system that may have a common or separate reservoir. It consists of pumps, one or more reservoirs, valves, and actuators. The turbine control and protection system provides hydraulic oil, under operating pressure, to the control valves. On a steam system, there is a separate system.

4.3.3 Electrohydraulic Governor Equipment

The electrohydraulic governor is an electronic controller apparatus that regulates the hydraulic servos of the control valves. This piece of equipment governs horsepower, regulates extraction and back pressure, controls the startup load rate, and synchronizes the shaft speed and frequency output. It is never used, however, as a shutdown device.

The electronic components of most governor systems consist of separate speed and extraction valve controllers, valve positioners, speed and pressure sensors, angle transmitters, valve coordinators to stabilize the responses of the separate controllers, limiters with fixed-value threshold elements, a starting and loading device, and the frequency controller.

The hydraulic components consist of the valve actuators, including the hydraulic rotary drives and the valve drive camshafts.

4.3.4 Seal Steam System

In the case of steam turbines, the seal steam system is a system of pipes, valves, a desuperheater, a condenser, an exhaust fan, and regulators that supply steam to the gland seals on the turbine shaft. The system then collects the steam and condenses it. The condensate is sent to the surface condenser.

4.3.5 Steam Distribution System

For steam systems, the steam distribution system is a system of pipes, valves, and headers that supply steam from the heat recovery steam generator to the turbine and other user points. This sends steam to user points in the rest of the plant.

4.3.6 Condensate Drain and Injection System

For steam systems, the condensate drain and injection system is a series of pipes, drains, valves, and nozzles that supplies condensate water to the turbine for cooling purposes and provides a means of draining condensate from the turbine. The condensate is injected into the back of the low-pressure section of the turbine. While the steam passing through the area helps cool the blades, additional cooling water is sprayed into the section to provide conductive cooling of the area. Under normal operation, the condensate flashes to steam upon contact and lowers the steam temperature in a manner similar to a desuperheater.

Instructor's Notes:

_____

_____

_____

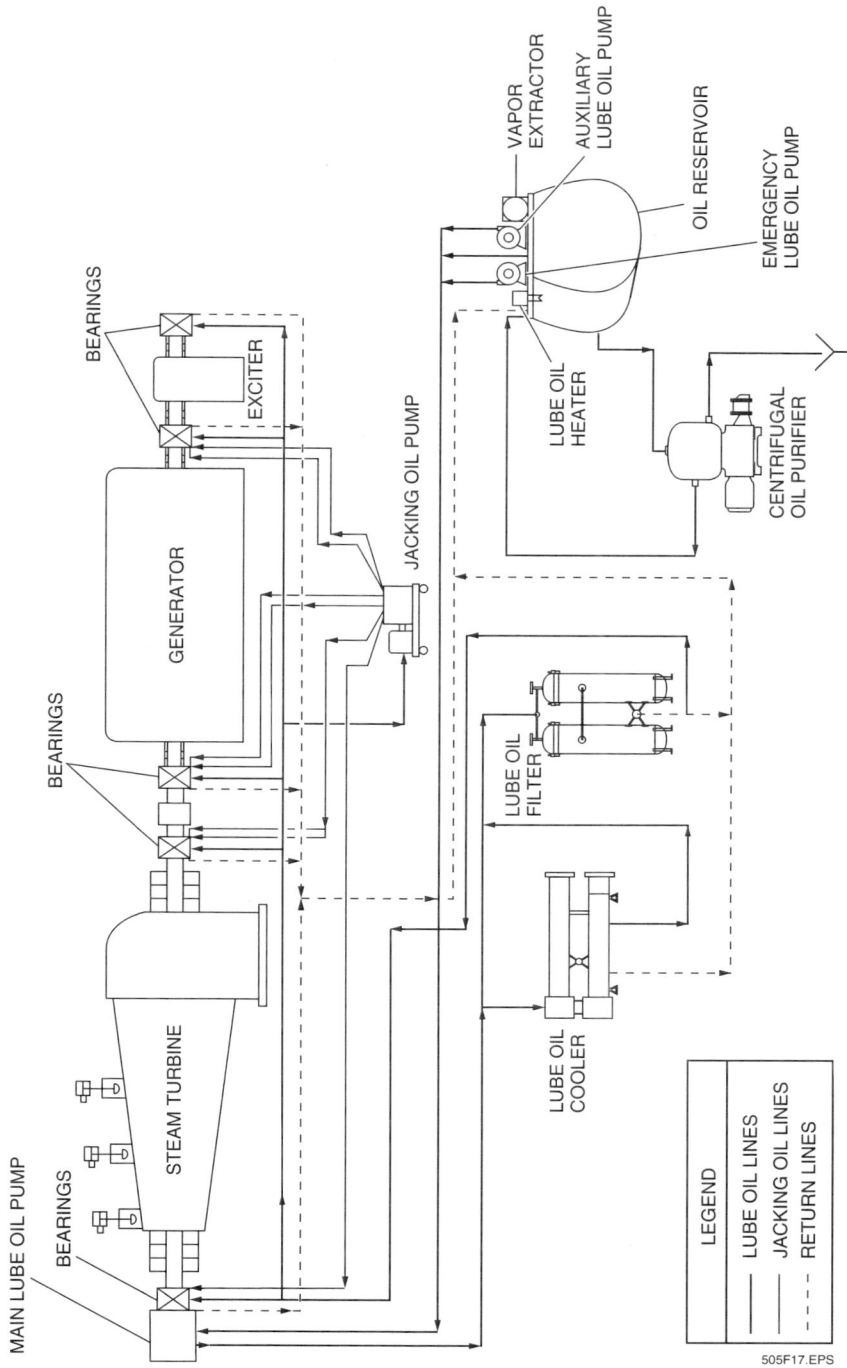

Figure 17 ◆ Lubrication and jacking oil system.

LEGEND

LUBE OIL LINES
JACKING OIL LINES
RETURN LINES

505F17.EPS

Assign reading of
Sections 5.0.0–5.1.5.

Ensure that you have
everything required for
teaching this session.

Compare gas and
steam turbines.

Show Transparency 14
(Figure 18). Point out
the major sections of
the gas turbine.

Describe the operation
of gas turbine com-
pressors.

Show Transparency 15
(Figure 19). Point out
the parts of the com-
pressor section.

Describe how the inlet
guide vanes adjust for
different phases of
turbine operation.

Explain the relationship
between the rotor
blades and the stator
blades.

Point out the succes-
sive stages of rotor and
stator blades.

Describe the function
of the diffuser section.

Discuss the use of
bleed-off lines to
prevent compressor
surge.

Residual steam in the turbine during shutdown can condense as the turbine cools. Some of the injected condensate may also collect in the turbine near shutdown. Turbine damage and corrosion is often caused by moisture in the turbine. To prevent condensate from remaining in the turbine, drains are provided to facilitate the removal of liquid before startup.

5.0.0 ◆ GAS TURBINE PARTS

Gas turbines burn their own fuel and expose the turbine parts to temperatures above 2,400°F. For this reason, the working parts of a gas turbine require different designs than those of steam turbines. *Figure 18* shows a typical gas turbine layout. The four major sections of a gas turbine are as follows:

- Compressor
- Combustion systems
- Turbine section
- Auxiliary support systems

5.1.0 Compressor Parts

Gas turbine compressors draw in air from the atmosphere and reduce the volume of the air, thus increasing its pressure. The compressors then force the air into the combustors. The compressed air in the combustors causes more fuel to be burned, which produces more power in the turbine. Most power plant turbines use axial-flow compressors, meaning that the air flows in the direction of the axis or rotor. The following are the major parts of a gas turbine compressor (see *Figure 19*):

- Inlet guide vanes
- Rotor blades
- Stator blades
- Discharge section
- Bleed-off lines

5.1.1 Inlet Guide Vanes

The inlet guide vanes are attached to the inlet compressor casing before the compressor rotor. Their purpose is to feed the air at a specific angle that is efficient into the first row of rotor blades. These guide vanes can be turned at different angles to control the amount of air entering the compressor. At low speeds during startup, the vanes are kept partially closed to prevent too much air from entering the compressor. As the turbine reaches operating speed, the vanes slowly open wider to allow more air into the compressor. This also provides surge protection during a shutdown or trip.

5.1.2 Rotor Blades or Vanes

Rotor blades are mounted on discs around the rotor throughout the compressor. The moving rotor blades accept the air from the inlet guide vanes and force the air to slide along the faces of the rotor blades. The rotor blades also push the air sideways in the direction in which the blades are rotating. These two forces cause the air to leave the rotor blades at the proper angle to enter the stator blades.

5.1.3 Stator Blades or Vanes

The stator blades are stationary blades mounted to the inside of the compressor casing between each row of rotor blades. The stator blades direct the airflow to the next set of rotor blades. A row of stator blades and a row of rotor blades make up a stage in the compressor. A typical compressor is made up of ten or more stages. The compressor decreases the size of the blades and air space in each stage, while decreasing the casing size and increasing the blade angle, to compress the air and increase the air pressure.

5.1.4 Discharge Section

Each stage of rotor and stator blades increases the air pressure by forcing the air into smaller and smaller spaces. The final section of the compressor is known as the diffuser section. The diffuser section extends beyond the final stage of blades. In the diffuser section, the air slows down, causing the air pressure to increase to its final pressure before entering the combustors. Most typical gas turbine compressors increase the air pressure to 10 or 14 times the normal atmospheric pressure.

5.1.5 Bleed-Off Lines

Air pressure sometimes increases too rapidly in a compressor and causes a problem known as compressor surge, which can result in serious compressor problems. This happens most frequently during startup. Compressor surge occurs when the air pressure forces the air to reverse direction and rush back through the compressor and out the inlet.

To prevent the problem of compressor surge, bleed-off lines remove excess air from the middle and rear sections of the turbine. Pressure-relief valves located in the bleed-off lines open when the air pressure exceeds a set point level and allow some of the air to escape through the lines that carry the air to the turbine exhaust. When the air pressure decreases to a safe operating pressure, the valves close to return the compressor to normal operation.

5.16 MILLWRIGHT ◆ LEVEL FIVE

Instructor's Notes:
_____
_____
_____
_____

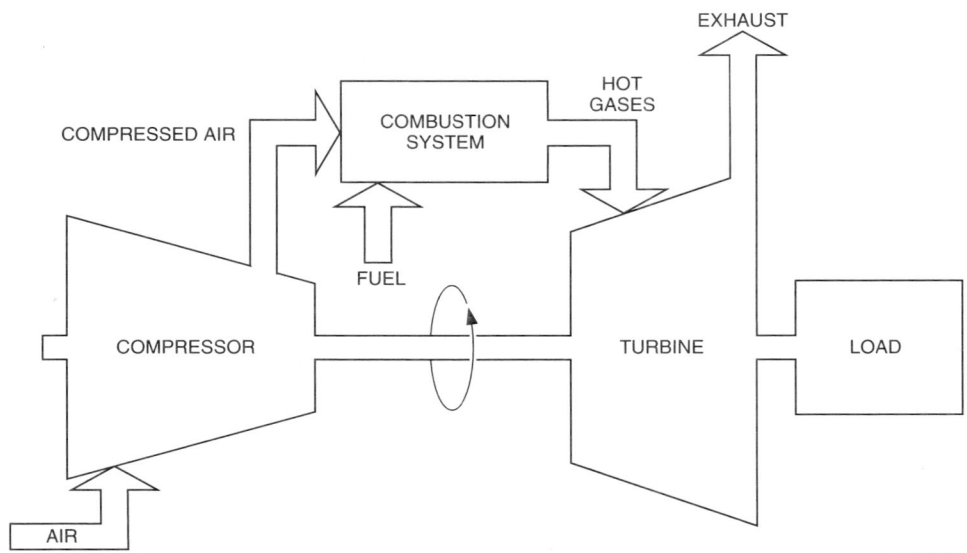

Figure 18 ◆ Gas turbine block layout.

505F18.EPS

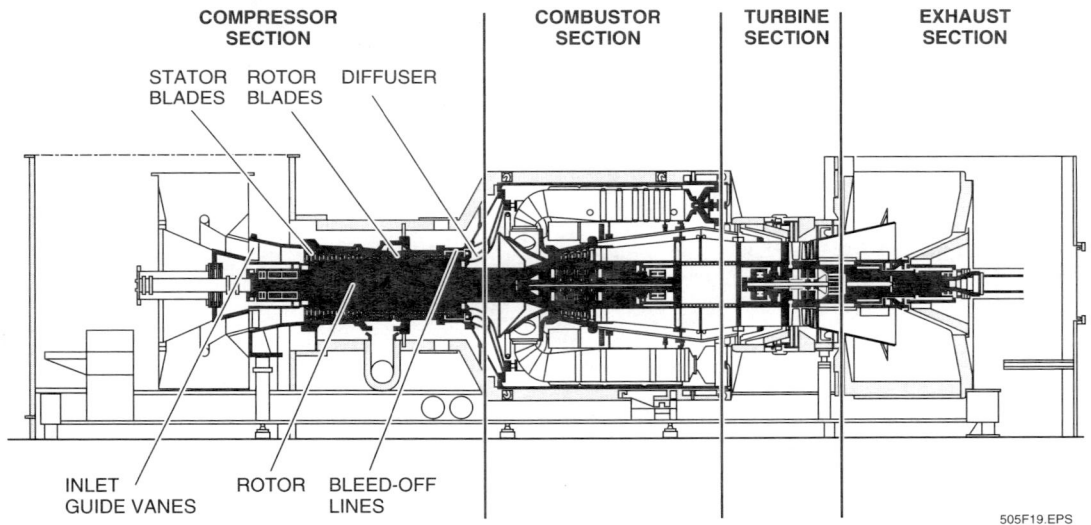

Figure 19 ◆ Compressor parts.

505F19.EPS

Assign reading of
Sections 5.2.0–5.3.3.

Ensure that you have
everything required for
teaching this session.

Discuss the types of
combustors used in
gas turbines.

Show Transparencies
16–18 (Figures 20–22).
Describe each type of
combustor.

Introduce the parts of
the combustor section.

Show Transparency 19
(Figure 23).

Describe reverse air-
flow.

5.2.0 Combustors

Gas turbines can use many different kinds of fuels in the same basic combustion system. The combustion section mixes compressed air from the compressor with the selected fuel and ignites the fuel with a spark ignition in the combustion can. The gases from the burning fuel and air drive the turbine. The following are the three basic types of combustors that are used in gas turbines:

- **Combustion chambers**
- Can-annular combustors
- Annular combustors

Combustion chambers (*Figure 20*) are found mainly on gas turbines designed specifically for use in power plants. They are the easiest to maintain because the individual combustors can be removed for inspection and repair.

Can-annular combustors (*Figure 21*) were originally designed as jet engines. The individual combustors in this arrangement are surrounded by a ring-shaped manifold, known as a common annular, that directs the compressor air into the combustors. Can-annular combustors join together at their ends into a common chamber that leads to the turbine.

Annular combustors (*Figure 22*) do not have individual cans. An annular combustor consists of a single, circular combustion chamber. It is essentially two cylinders, one inside the other. These are primarily aero derivatives.

5.2.1 Combustor Parts

The combustor section produces a great deal of energy for its size. The amount of heat released inside the combustor can raise the combustion gas temperature to about 2,600°F. Since the burning gas at this temperature could seriously damage the walls of the combustor, the metal must be cooled. The following are the parts of the combustor section:

- Combustion chambers
- Fuel nozzles
- Igniters and crossfire tubes
- Flame detectors
- Transition pieces

Combustion chambers – A can combustor uses a reverse airflow design (*Figure 23*). This type of combustion chamber is mounted around the discharge section of the compressor and bolted to the compressor discharge section bulkhead. Reverse airflow means that the air coming from the compressor enters the combustor at the same end at

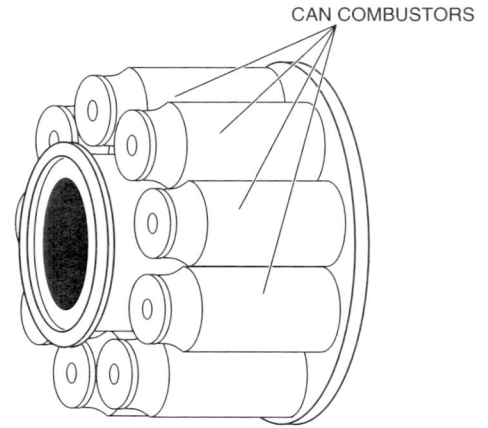

Figure 20 ◆ Basic combustion chamber.

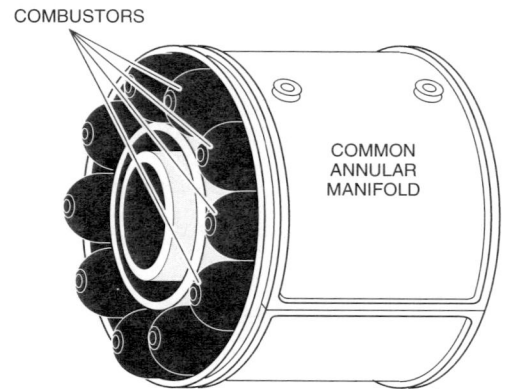

Figure 21 ◆ Basic can-annular combustor.

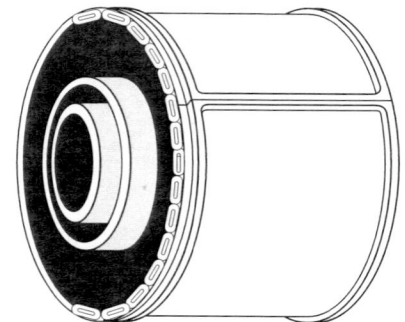

Figure 22 ◆ Basic annular combustor.

5.18   MILLWRIGHT ◆ LEVEL FIVE

Instructor's Notes:

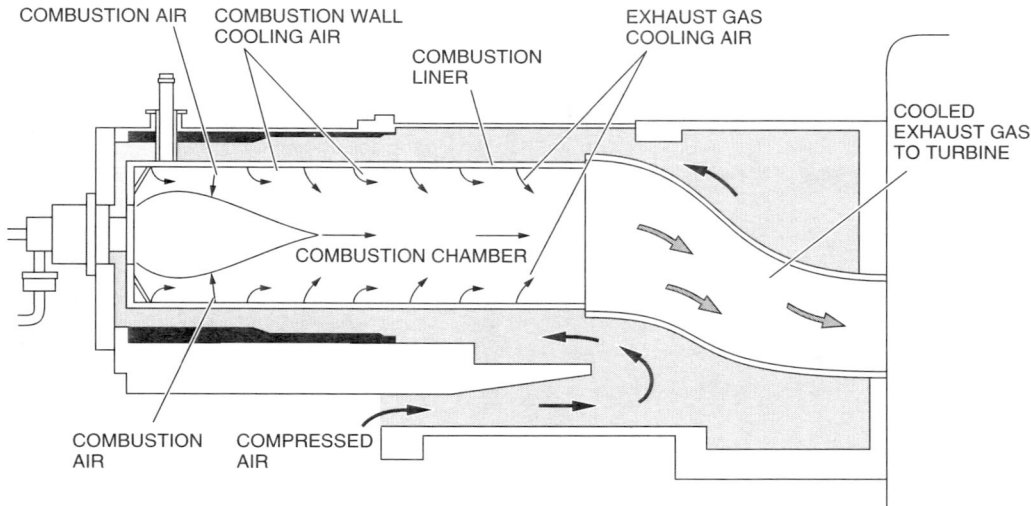

Figure 23 ♦ Reverse airflow combustor.

which the combustion gases flow out. The compressed air leaves the discharge end of the compressor and enters the forward section of the turbine shell. At this point, the air reverses direction and enters the combustors from the turbine end. The air flows through the combustors and enters the combustor liner through holes and slits. A portion of the air reaches the head of the combustor chamber and enters the liner cap and nozzle. This air mixes with the incoming fuel to provide oxygen for combustion.

Most can-annular combustion chambers are positioned so that the air flows in a straight path through the chambers (*Figure 24*). Air flows through slits in the chamber walls to cool the burning gases and through the larger holes in the chamber walls to provide oxygen for combustion. The air entering the larger holes reverses direction and circulates back into the fuel spray. This creates the turbulence needed for more efficient mixing of fuel and air.

Fuel nozzles – Fuel sprays into the combustor through a fuel nozzle that mixes the fuel with compressed air. There are many types of fuel nozzle designs used in gas turbines today.

Once operating speed is reached, the fuel pressure increases. The higher fuel pressure opens the check valve, allowing fuel into the outer fuel channels and the inner fuel channels at the same time. This produces a narrow, high-speed spray to help keep the flame away from the combustor walls.

Igniters and crossfire tubes – Combustion of the fuel and air mixture in the combustors is initiated by igniters that operate in the same manner as spark plugs. The igniters, usually installed in only two combustion chambers located next to each other, receive power from the ignition transformers. The combustion chambers are interconnected by cross-fire tubes. The chambers without igniters are fired by the flame from the fired chambers through the interconnecting cross-fire tubes.

After the gas turbine combustion begins, it is continuous. The igniters are fired only long enough to ensure that a stable flame has been established. Once the flame is established, it continues to burn without the need for additional sparks from the igniters.

Flame detectors – Two flame detectors or scanners are usually located on each side of two adjacent combustors on the opposite side of the turbine from the igniters. The flame detectors verify the presence of flame in the combustors. If combustion does not take place, or if the flame goes out, the flame detector sends a signal to the fuel system governor to stop the flow of fuel and shut down the turbine. *Figure 25* shows the ignition system components.

Transition pieces – The transition pieces (*Figure 26*) carry the hot combustion gases from the combustion chambers to the first-stage nozzle of the turbine. Before the compressor discharge air flows into the combustion chamber, it must first pass

Discuss airflow through a can-annular combustor.

Show Transparency 20 (Figure 24).

Explain fuel nozzle operation.

Discuss igniters, crossfire tubes, and flame detectors.

Show Transparency 21 (Figure 25). Point out the ignition system components.

Explain the function of transition pieces in cooling the exhaust gases.

Show Transparency 22 (Figure 26).

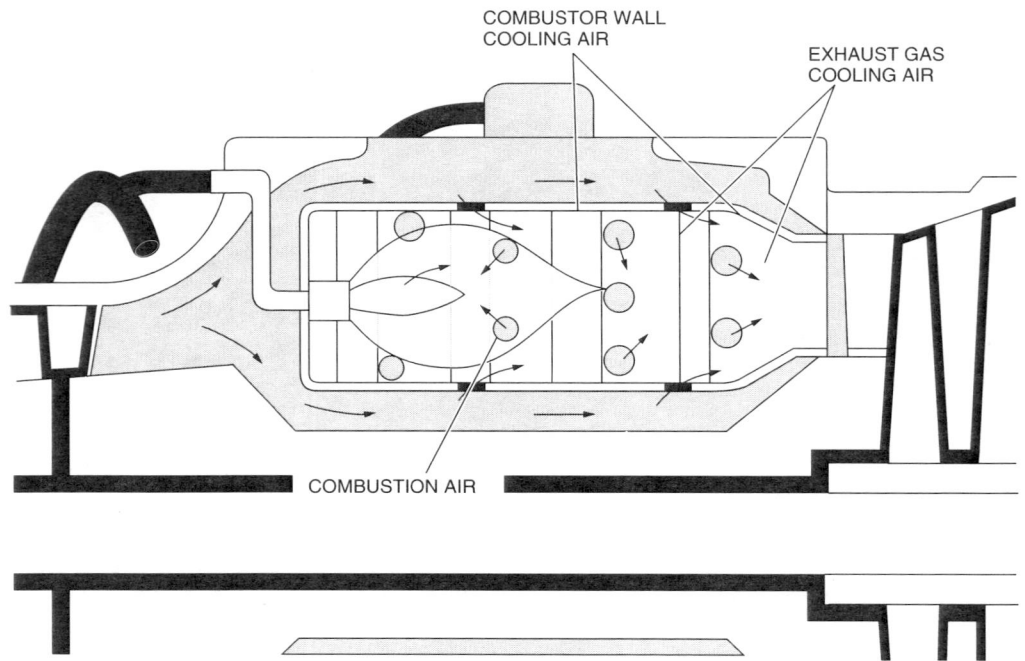

COMBUSTOR WALL COOLING AIR

EXHAUST GAS COOLING AIR

COMBUSTION AIR

505F24.EPS

Figure 24 ◆ Airflow through a can-annular combustor.

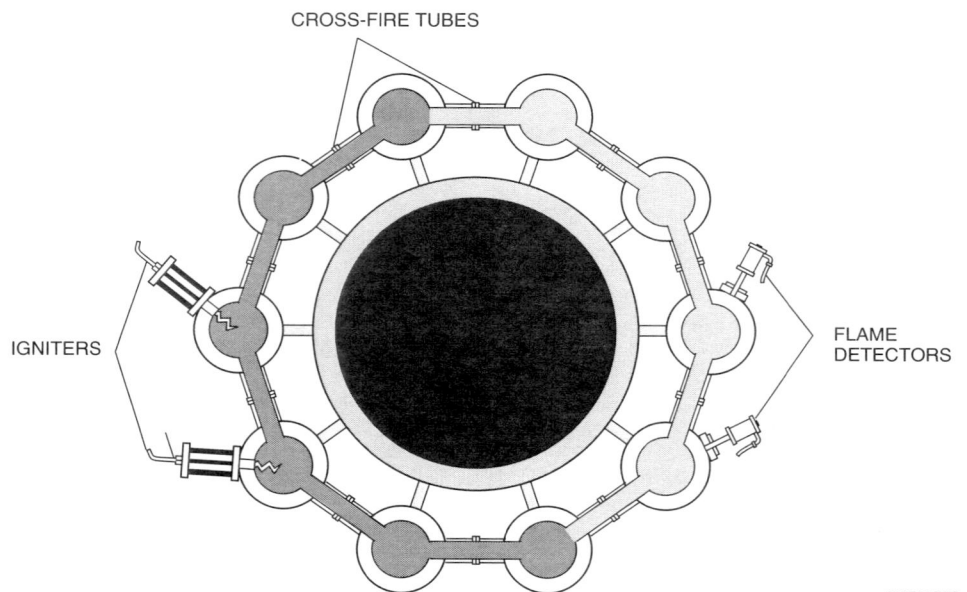

CROSS-FIRE TUBES

IGNITERS

FLAME DETECTORS

505F25.EPS

Figure 25 ◆ Ignition system components.

Instructor's Notes:

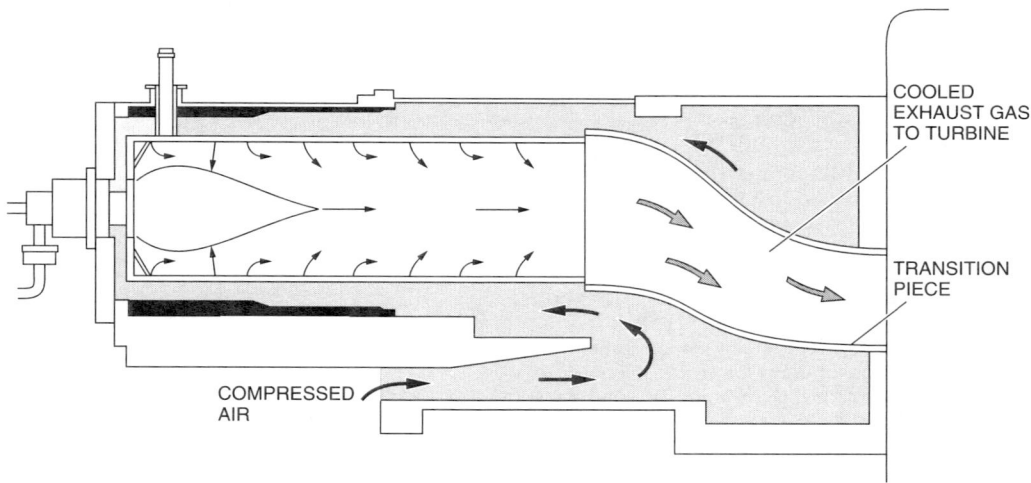

Figure 26 ◆ Transition piece.

505F26.EPS

Describe the major components of the turbine section.

Describe the flow of combustion gases through the nozzles and into the rotor blades.

Show Transparency 23 (Figure 27).

around the transition pieces. This causes a heat exchange between the extremely hot combustion gases leaving the combustion chamber and the cool compressor discharge air entering the combustion chamber. The air entering the combustion chamber is preheated. This produces a balance between the hot and cold air to protect the inner walls of the combustion chambers.

5.3.0 Turbine Section Parts

The turbine section of a gas turbine is similar to that of the steam turbine. The pressurized, hot gases from the combustors are directed at the turbine blades and produce the rotating force in the turbine shaft. Both gas and steam turbines extract energy from a moving fluid as the fluid goes from an area of higher pressure to an area of lower pressure. The following are the major parts of the turbine section:

• Nozzles
• Rotor blades
• Exhaust silencer

5.3.1 Nozzles

Combustion gases enter the turbine through a row of stationary blades known as nozzles. The nozzles direct the hot gases into the moving rotor blades. As in the compressor, a row of nozzles and rotor blades is known as a stage (*Figure 27*). A row of nozzles at the beginning of each stage redirects the

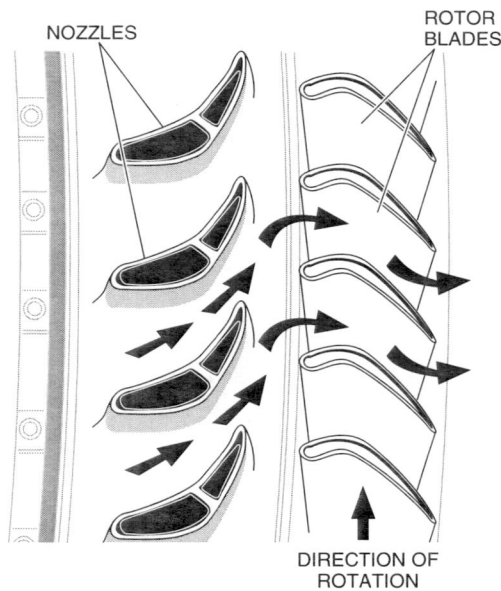

Figure 27 ◆ Nozzles and rotor blades.

505F27.EPS

gases to the next row of rotor blades. Gas turbines generally have multiple stages, and each stage is larger than the preceding stage to extract the maximum power from the gases.

MODULE 15505-09 ◆ TURBINES 5.21

5.3.2 Rotor Blades

The moving blades in a turbine are known as rotor blades. The rotor blades are attached to a disc made of special alloy steel. These discs are bolted to the turbine shaft. The combustion gases moving through the turbine cause the rotor blades to move and therefore rotate the turbine shaft, which is connected to a load such as a generator. The turbine shaft may be either straight or cone-shaped (*Figure 28*). Many turbine shafts are hollow to allow cool compressor air to flow through the middle and cool the shaft discs and blades.

The two basic rotor blade types are impulse blades and reaction blades. The impulse blades are cup-shaped, and the reaction blade is similar to the shape of an aircraft wing (*Figure 29*).

Most gas turbine rotor blades are a combination of impulse blades and reaction blades. The base of the rotor blade is shaped like an impulse blade, and the tip of the rotor blade is shaped like a reaction blade (*Figure 30*). The combination form of the blade is used because the tip of the blade has to cover a greater distance on each rotation, and therefore it moves faster than the base of the blade. The streamlined, reaction-shaped tip moves more smoothly through the fast-moving combustion gas than does the cup-like, impulse-shaped base.

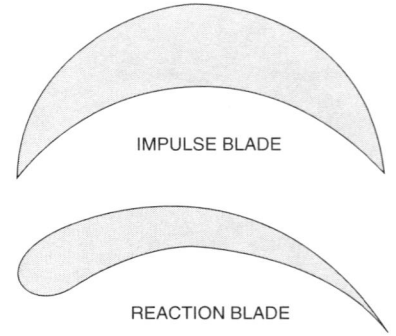

Figure 29 ◆ Cross section of turbine blades.

5.3.3 Exhaust Silencer

The hot combustion gases exit the turbine through the exhaust stack silencer (*Figure 31*). The silencer consists of a series of baffles to quiet the sound of the turbine. The silencer is actually a large muffler, somewhat like a car muffler. As the exhaust gases travel through the silencer, the offset baffles force the gases to change direction, reducing the sound with each change of direction. In addition to the silencer, the exhaust diffuser distributes (diffuses) heat to the atmosphere through its skin. The

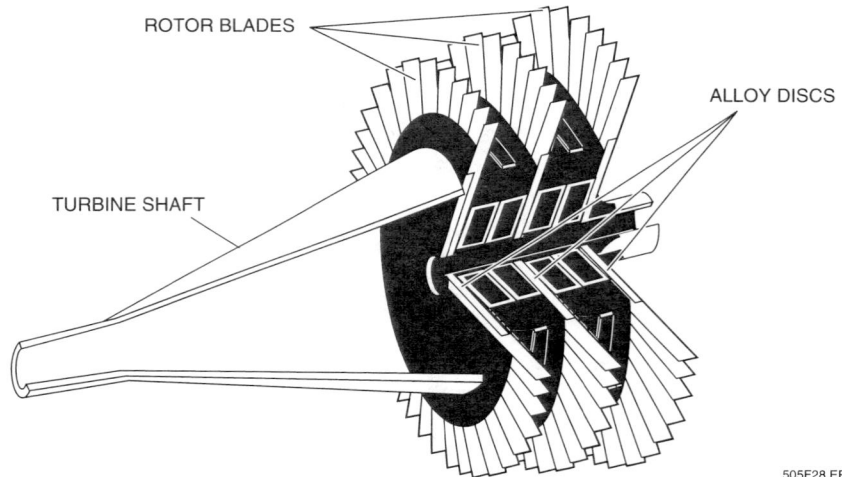

Figure 28 ◆ Rotor blades on a cone-shaped turbine shaft.

5.22   MILLWRIGHT ◆ LEVEL FIVE

Instructor's Notes:

_____

_____

_____

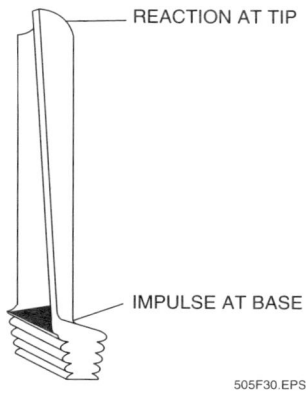

Figure 30 ◆ Combination rotor blade.

exhaust plenum is a surround or duct to contain and transfer the exhaust to the silencer or to a later HRSG stage.

5.4.0 Auxiliary Support Systems

There are many auxiliary support systems involved in the operation of a gas turbine. The major auxiliary support systems include the following:

- Starting systems
- Lube oil system
- Fuel system
- Water cooling system

5.4.1 Starting Systems

Large industrial gas turbines need powerful, heavy-duty starters to be able to turn the heavy compressor and turbine rotor for several minutes until self-sustaining speed is obtained. All starters use a clutch, or trip coil, to disengage the starter motor when the turbine reaches a self-sustaining speed. This allows the turbine to accelerate to operating speed without damaging the starter motor. Most industrial gas turbines use air motors, diesel engines, or electric motors for starters. However, the type of starter varies from one customer to the next.

Air motor starters are generally used for smaller gas turbines. They are small and rugged and can attain very high speeds. The air to drive these motors can be stored in tanks at a pressure of about 400 psi or in bottles at a pressure of about 3,000 psi. In both cases, a regulator reduces the air pressure to between 50 and 100 psi before the air enters the motor.

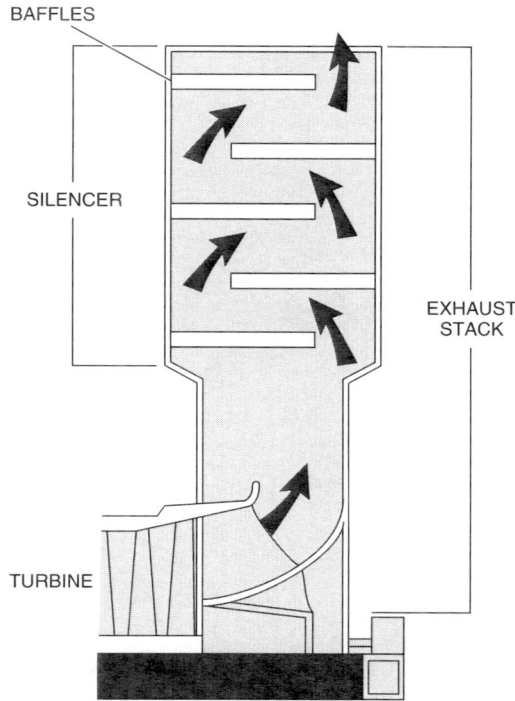

Figure 31 ◆ An exhaust silencer.

Diesel engines are used most often to start large gas turbines during black starts. These engines require no outside source of electrical power and can provide the power needed to bring large industrial turbines to self-sustaining speeds.

Electric motors are used to start the largest gas turbines because they can deliver the most power at the fastest rate of all the other starters, and they require the least maintenance. Electric motors usually start gas turbines that are used for peak power loads. They are normally useless during black starts because there is no outside electrical power available unless the plant has emergency generators driven by diesel engines or smaller gas turbines to provide power to the electric starters.

All starters require a torque converter and a gearbox.

5.4.2 Lube Oil System

The lube oil system (*Figure 32*) is one of the most important auxiliary systems on the gas turbine. The lube oil system is designed to provide filtered lubricant at the proper temperature and pressure

Assign reading of Sections 5.4.0–5.4.4.

Ensure that you have everything required for teaching this session.

Describe the major auxiliary support systems for gas turbines.

Discuss the advantages and disadvantages of different types of starters.

Explain the importance of the lube oil system.

Explain that a break-
down in the lubrication
system could cause
serious damage to the
turbine.

Show Transparency 28
(Figure 32). Point out
how the oil circulates
through the system.

Discuss the need for
an oil cooler.

Compare light and
heavy fuel oil systems.

Show Transparencies
29 and 30 (Figures 33
and 34).

Describe natural gas
turbine fuel systems.

Show Transparency 31
(Figure 35).

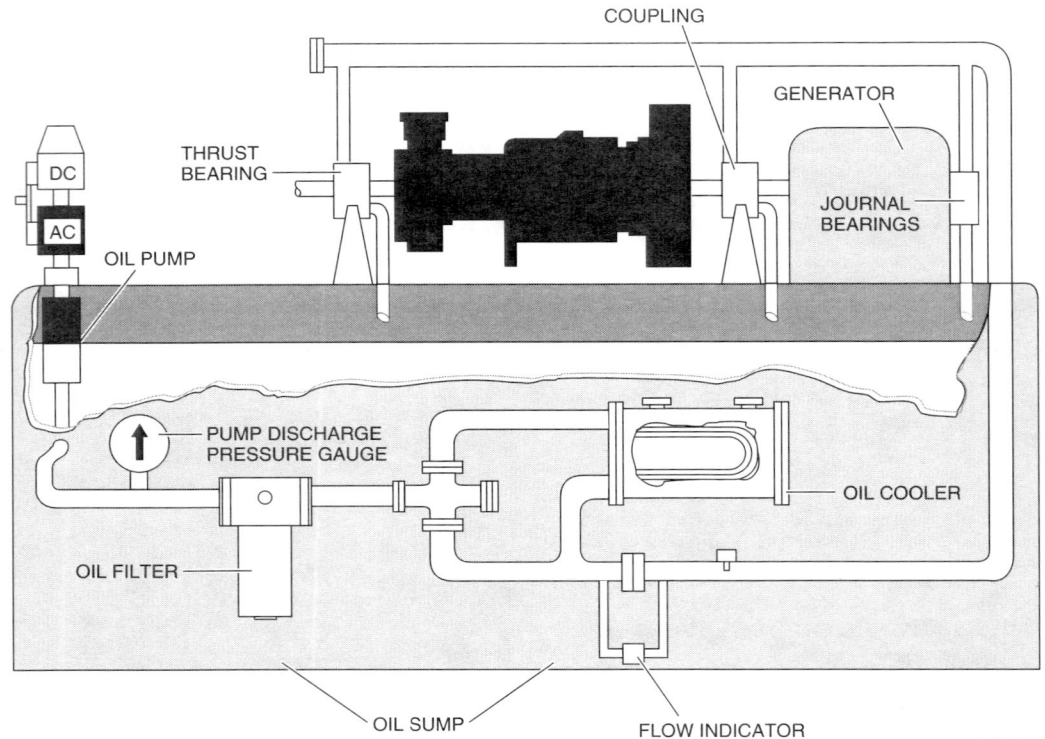

Figure 32 ◆ Lube oil system.

for operation of the turbine and its associated equipment. The lube oil system lubricates, cools, and protects the vital components, including the bearings, couplings, gears, generator bearings, and hydrogen seals. If the lube oil system fails, serious damage occurs very quickly. The lube oil system typically consists of an oil sump, oil pump, oil filter, and an oil cooler.

The oil sump is a large tank used to store the oil. It is usually built into an accessory base from the one on which the turbine is installed. The pump, which can be driven by an AC motor, circulates the oil through the system. The pump sends the oil through an oil filter to remove contaminants, and then through an oil cooler that keeps the oil between 80°F and 120°F. The oil cooler is a heat exchanger that uses water or air to remove heat from the oil. The cooled oil flows on the bearings, gears, and other lubricating points and is then drained back into the sump. Many flow gauges and temperature gauges are installed in the system to monitor the operation of the system at all times.

5.4.3 Fuel System

The fuel system must feed fuel to all the fuel nozzles on the turbine at the same rate to ensure that each combustor produces the same amount of hot gas. The type of fuel system varies depending on the type of fuel being used.

A typical light fuel oil system (*Figure 33*) consists of individual free-wheeling flow dividers for each combustor used. The flow dividers are constructed like gear pumps, but the fuel flowing through them makes them turn. These are all connected to the same shaft, so they all turn at the same rate, called their frequency match.

When using heavy fuel oils, the same system is used except that the flow dividers are powered by a hydraulic motor (*Figure 34*). The hydraulic motor turns all the flow dividers at the same rate to keep the flow of oil to each combustor constant.

When using natural gas as the turbine fuel, the pump in the fuel line is eliminated because the gas flows under pressure into the plant (*Figure 35*). A control valve regulates the flow of gas into the combustors.

Instructor's Notes:

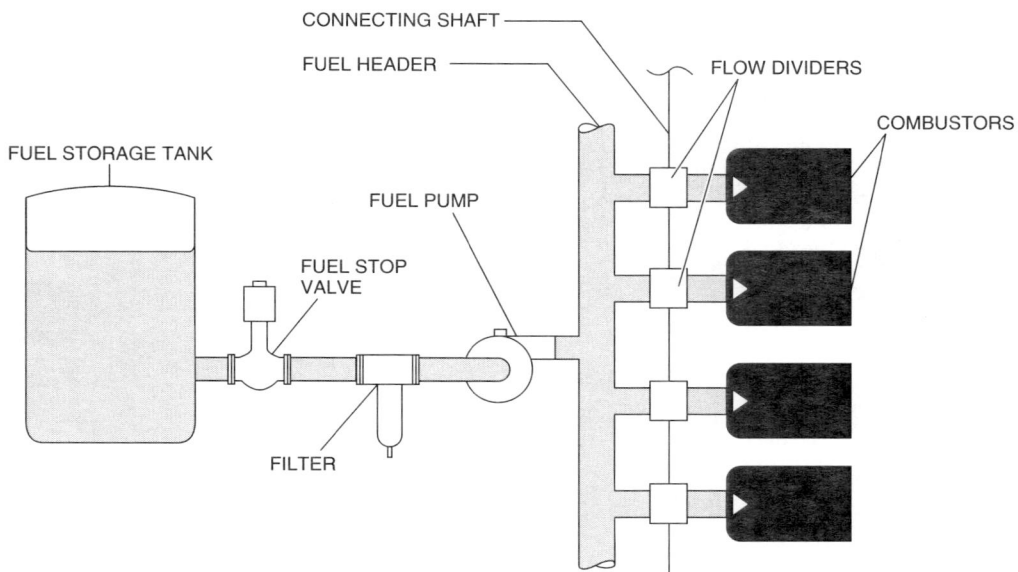

Figure 33 ◆ Light fuel oil system.

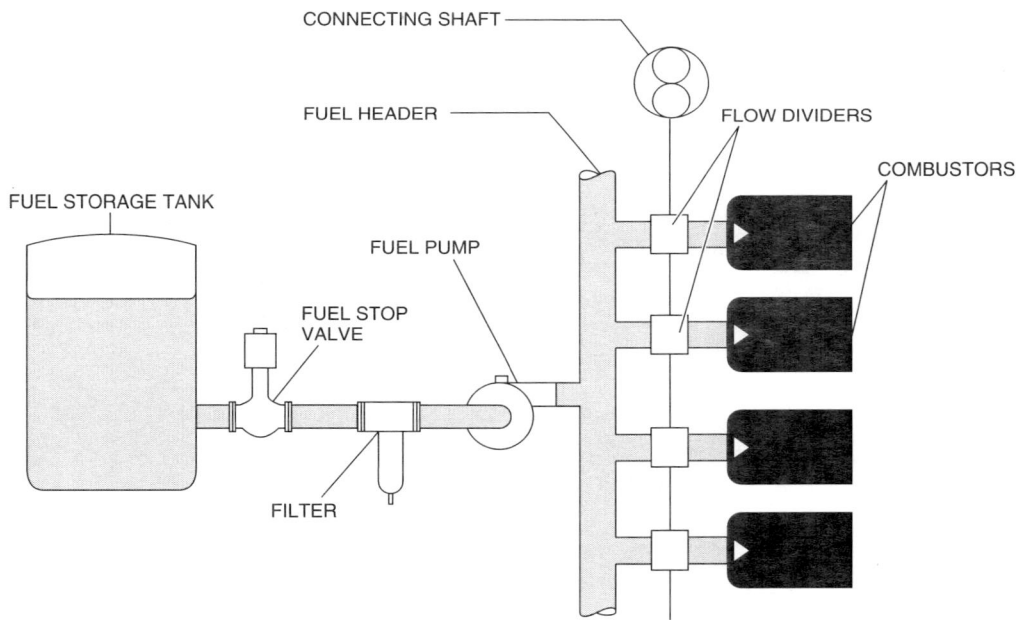

Figure 34 ◆ Heavy fuel oil system.

Describe the parts of a
water cooling system.

Show Transparency 32
(Figure 36).

See the Teaching Tips
for Sections 1.0.0–5.0.0
at the end of this
module.

5.4.4 Water Cooling System

Three main areas are cooled by the water cooling system (*Figure 36*): the lube oil cooler, the turbine support leg, and the diesel starter engine. The cooling water is stored in a large storage tank. A cooling water circulating pump, which can be driven by an AC motor, a DC motor, or by gears from the turbine, circulates the cooling water through the system. After the water travels through the components to be cooled, it goes to a cooling tower or other system before going back to the storage tank.

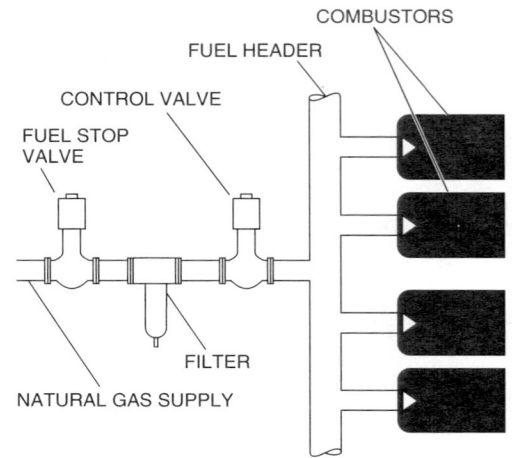

505F35.EPS

Figure 35 ◆ Natural gas fuel system.

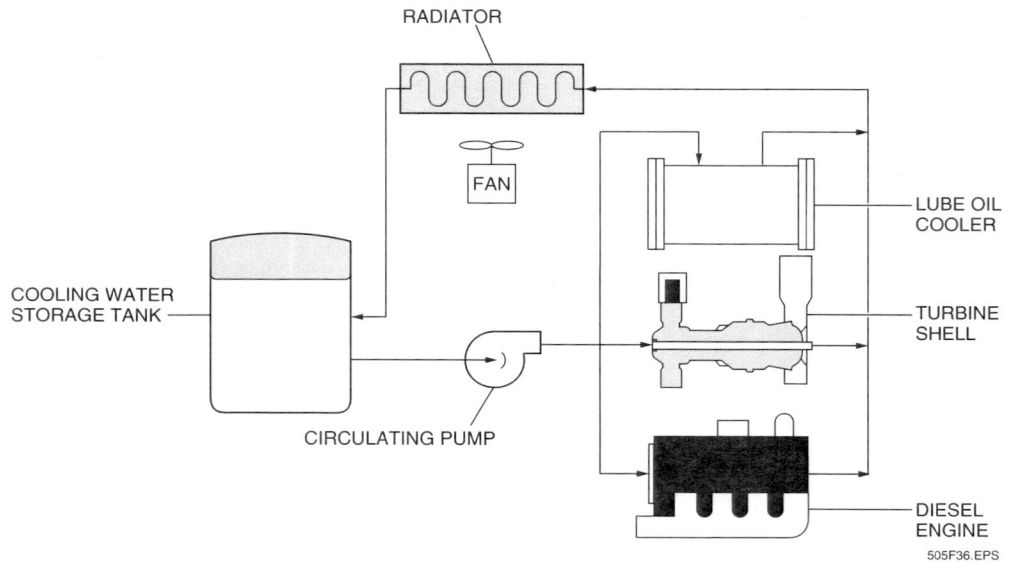

505F36.EPS

Figure 36 ◆ Water cooling system.

Instructor's Notes:

1. In the impulse turbine, the flow of fluid or gas contacts _____ at any given instant.
   a. the entire wheel
   b. all the blades
   c. alternate blades
   d. a small portion of the blades

2. A windmill is an example of a(n) _____ turbine.
   a. impulse
   b. reaction
   c. multistage
   d. condensing

3. The most common turbines in use today are _____ turbines.
   a. gas
   b. hydraulic
   c. steam
   d. water

4. In the condensing turbine, the steam exiting the turbine enters a condenser to convert the steam _____.
   a. to low-pressure steam
   b. back to water
   c. to high-pressure steam
   d. to superheated steam

5. When they leave the turbine, gas turbine exhaust gases are approximately _____.
   a. 250°F
   b. 425°F
   c. 560°F
   d. 1,000°F

6. To operate efficiently, the Pelton wheel must have a head of at least _____ feet.
   a. 300
   b. 500
   c. 800
   d. 1,000

7. Most hydraulic turbines are _____-type turbines.
   a. impulse
   b. reaction
   c. condensing
   d. noncondensing

8. A journal bearing in the steam turbine is a babbited liner that fits into a machine fit in the support casing to _____.
   a. absorb the force on the shaft
   b. hold the shaft in place radially
   c. balance the axial thrust
   d. prevent steam leakage

9. The electrohydraulic governor is an electronic controller apparatus that regulates the _____.
   a. steam pressure
   b. lube oil system
   c. vacuum distribution
   d. hydraulic positioners

10. Gas turbines burn their own fuel and expose the turbine parts to temperatures above _____.
    a. 500°F
    b. 1,000°F
    c. 1,500°F
    d. 2,400°F

11. The final section of the gas turbine compressor is known as the _____.
    a. discharge
    b. transition piece
    c. combustion chamber
    d. outlet section

12. A can combustor uses a _____ airflow design.
    a. straight-path
    b. circulating
    c. reverse
    d. center-channel nozzle

13. Combustion chambers are interconnected by _____.
    a. igniters
    b. spark plugs
    c. flame detectors
    d. crossfire tubes

MODULE 15505-09 ◆ TURBINES 5.27

14. During black starts, large gas turbines most often use _____ starters.
    a. air motor
    b. diesel engine
    c. electric motor
    d. battery-driven

15. Which of the following fuels does *not* require a pump to regulate the flow?
    a. Natural gas
    b. Light fuel
    c. Heavy fuel
    d. Diesel

Instructor's Notes:

_____
_____
_____
_____

Summary

The principal function of a turbine is to rotate a shaft by moving steam, water, or gas through a series of blades attached to the shaft to drive a generator or other piece of equipment. As a millwright, you will be expected to identify the different types of turbines, know their basic components, and understand the operating principles of different types of turbines. Always follow all safety guidelines that pertain to rotating equipment when working in the vicinity of operating turbines.

Summarize the major concepts presented in the module.

Administer the Module Examination. Be sure to record the results of the Examination on Craft Training Report Form 200, and submit the results to the Training Program Sponsor.

Notes

Trade Terms Introduced in This Module

Combustion: The burning of a gas, liquid, or solid in which the fuel is oxidized and emits a great amount of heat.

Combustion chamber: Any enclosed space in which a fuel, such as oil, coal, kerosene, or natural gas, is burned to provide heat.

Compressor: A machine used for increasing the pressure of a gas or vapor.

Condensate: The liquid by-product of cooling steam.

Governor: A device used to provide automatic control of speed or power of a prime mover but never as a method for shutting down a turbine.

Kinetic energy: The energy of movement.

Mechanical energy: Energy in the form of mechanical power.

Turbine: A machine that generates rotary, mechanical power from the energy in a stream of fluid or gas.

Turbine generator: An electric generator driven by a steam, hydraulic, or gas turbine.

Instructor's Notes:

_____

_____

_____

Resources & Acknowledgments

Additional Resources

This module is intended to present thorough resources for task training. The following reference works are suggested for further study. These are optional materials for continued education rather than for task training.

Environmental Protection Agency
  www.epa.gov/CHP/documents/
  tech_turbines.pdf

General Electric Company
  www.gepower.com/home/index.htm

Siemens Corporation
  www.powergeneration.siemens.com/home

http://mysite.du.edu/~jcalvert/tech/fluids/
  turbine.htm

Figure Credits

Courtesy of ThermoDudes, LLC, 505F01

GE Oil & Gas, 505F03, 505F05 (photo)

Ed LePage, 505F06 (photo)

By courtesy of Voith Siemens Hydro, 505F07

CANADIAN HYDRO COMPONENTS LTD.,
  505F08, 505F09

MODULE 15505-09 — TEACHING TIPS

The following are suggested activities or instructional methods to help you teach the material in this module.

General

When you call on someone to answer a question, the rest of the class relaxes or even tunes out because they expect that the question and answer will take place only between you and the trainee you called on. Instead, use this technique to involve more trainees in answering questions and to keep them on their toes.

1. Ask trainees to define a term or explain a concept.
2. After one trainee has answered, ask a trainee seated nearby if the answer is right. Then ask whether a trainee in the back of the room agrees.
3. Ask trainees to explain why they think an answer is right or wrong.
4. Use the session to clear up incorrect ideas and encourage trainees to learn from their mistakes.

**Sections
1.0.0–5.0.0**

Quick Quiz

This Quick Quiz will familiarize trainees with terms commonly used in relation to turbines and their components. You will need photocopies of the quiz provided on the following page. Trainees will need pencils. If you allow trainees to use the Trainee Guide, decrease the amount of time you give them to complete the quiz, and remind them to bring their books to class.

1. Make a photocopy of the quiz for each trainee.
2. Give trainees between 5 and 10 minutes to complete the quiz.
3. Go over the answers to the quiz.
4. Ask trainees if they have questions.

Answers to Quick Quiz

1. d
2. f
3. h
4. c
5. k
6. g
7. a
8. e
9. m
10. i
11. l
12. j
13. n
14. o
15. b

Quick Quiz   *Turbine Components*

For each description listed, identify the term that the text best describes. Write the corresponding letter in the blank provided.

_____ 1. Chambered bulkheads in the turbine that may separate the high-pressure and low-pressure sections from the intermediate-pressure section are called the _____.

_____ 2. _____ is used to periodically turn the shaft of the turbine during startup and shutdown.

_____ 3. Stationary blades that direct airflow in the compressor are called _____.

_____ 4. The _____ allows emergency release of excess pressure.

_____ 5. The combustion gases from the combustor are directed to the turbine by the _____.

_____ 6. The pressure drop helps to drive the _____.

_____ 7. The _____ are split rings that use a closed steam supply and spring tension to hold them close to the shaft to prevent leakage of steam.

_____ 8. The _____ prevent static discharges that can cause pitting and corrosion in the moving parts.

_____ 9. The most common starter for black starts is the _____.

_____ 10. A cluster of combustors inside a common ring is a(n) _____ combustor.

_____ 11. The device that determines whether a combustor is firing is the _____.

_____ 12. The part of the ignition system that ignites most of the chambers is the _____ tube.

_____ 13. The _____ synchronize the supply of fuel to the combustors.

_____ 14. Lube oil is stored in the _____.

_____ 15. Fixed vanes are attached to the inner surface of the _____ to direct the steam into the wheels at the proper direction for optimum energy transfer.

   a. gland seals
   b. blade carrier
   c. rupture disc
   d. nozzle blocks
   e. grounding brushes
   f. turning gear
   g. reaction blades
   h. stator blades
   i. can-annular
   j. crossfire
   k. transition piece
   l. flame detector
   m. diesel engine
   n. flow dividers
   o. sump

Identifying Pieces of Turbine Equipment

This exercise will familiarize trainees with various types of turbine components. Trainees will need appropriate personal protective equipment, pencils, and paper. You will need to arrange to obtain different types of turbine components and have them set up at several workstations. Allow 30–45 minutes for this exercise.

Alternatively, obtain manufacturers' literature, photographs, or illustrations of various types of turbine components. Discuss each type of component with trainees. Ask trainees to identify each type. If desired, have trainees identify the functions of each component.

1. Show trainees various types of turbine components, and discuss their functions.
2. Set up one or two different types of turbine components at each workstation, depending on the space available.
3. Have trainees circulate to all of the stations and identify each type of turbine component. If desired, have trainees identify the function of each component.
4. Answer any questions trainees might have.

MODULE 15505-09 — ANSWERS TO REVIEW QUESTIONS

	Answer	Section
1.	d	2.1.0
2.	b	2.2.0
3.	a	3.2.0
4.	b	3.1.0
5.	d	3.2.0
6.	c	3.3.0
7.	b	3.3.0
8.	b	4.1.11
9.	d	4.3.3
10.	d	5.0.0
11.	a	5.1.4
12.	c	5.2.1
13.	d	5.2.1
14.	b	5.4.1
15.	a	5.4.3

CONTREN® LEARNING SERIES — USER UPDATE

The NCCER makes every effort to keep these textbooks up-to-date and free of technical errors. We appreciate your help in this process. If you have an idea for improving this textbook, or if you find an error, a typographical mistake, or an inaccuracy in NCCER's Contren® textbooks, please write us, using this form or a photocopy. Be sure to include the exact module number, page number, a detailed description, and the correction, if applicable. Your input will be brought to the attention of the Technical Review Committee. Thank you for your assistance.

Instructors – If you found that additional materials were necessary in order to teach this module effectively, please let us know so that we may include them in the Equipment/Materials list in the Annotated Instructor's Guide.

Write: Product Development and Revision
National Center for Construction Education and Research
3600 NW 43rd St., Bldg. G, Gainesville, FL 32606

Fax: 352-334-0932

E-mail: curriculum@nccer.org

Craft _____ Module Name _____

Copyright Date _____ Module Number _____ Page Number(s) _____

Description
_____
_____
_____
_____

(Optional) Correction
_____
_____
_____

(Optional) Your Name and Address
_____
_____
_____

Maintaining and Repairing Turbine Components

NCCER STANDARDIZED CRAFT TRAINING PROGRAM

The National Center for Construction Education and Research (NCCER) provides a standardized national program of accredited craft training. Key features of the program include instructor certification, competency-based training, and performance testing. The program provides trainees, instructors, and companies with a standard form of recognition through a National Craft Training Registry. The program is described in full in the *Guidelines for Accreditation*, published by NCCER. For more information on standardized craft training, contact the NCCER by writing us at 3600 NW 43rd St., Bldg. G, Gainesville, FL 32606; calling 352-334-0911; or emailing info@nccer.org. More information may be found at our website, www.nccer.org.

HOW TO USE THIS ANNOTATED INSTRUCTOR'S GUIDE

Each page presents two sections of information. The larger section displays each page exactly as it appears in the Trainee Module. The narrow column ties suggested trainee and instructor actions to each page and provides icons (detailed below) to call your attention to material, safety, audiovisual, or testing requirements. The bottom of each page includes space for your notes.

The **Audiovisual** icon indicates an appropriate time to show a transparency or other audiovisual aid.

The **Classroom** icon prompts you to define a term, stress a point, ask trainees to explain a concept, or give examples.

The **Demonstration** icon directs you to show trainees how to perform tasks.

The **Examination** icon tells you to administer the written module examination.

The **Homework** icon is placed where you may wish to assign reading for the next class, assign a project, or advise trainees to prepare for an examination.

The **Laboratory** icon is used when trainees are to practice performing tasks.

The **Materials** icon is a reminder for you to gather materials needed for classes, labs, and testing.

The **Performance Testing** icon tells you to administer a performance test or a portion thereof.

The **Safety** icon is used to emphasize safety issues. It is often keyed to *Caution* and *Warning!* statements in the Trainee Module.

The **Teaching Tip** icon indicates additional guidance is available, such as how to conduct an exercise, get the most educational value from a field trip, or encourage class participation. Teaching Tips may expand on a feature (*Think About It*, *Did You Know?*) or provide *Quick Quizzes* or similar exercises. You will be referred to the Teaching Tips section at the back of the module if there is additional material.

The **Combination** icon indicates that the laboratory listed corresponds with a performance task. If desired, you can note the proficiency of the trainees during the laboratory, and use it to satisfy performance testing requirements.

PREPARATION

Before teaching this module, you should review the Objectives, Performance Tasks, Materials and Equipment List, and Module Outline. Be sure to allow ample time to prepare your own training or lesson plan and gather all required materials and equipment.

MODULE OVERVIEW

This module covers basic turbine components, typical problems encountered when working with turbines, and guidelines for maintaining and repairing various types of turbines. Techniques for gaining access to components and replacing them are also covered.

PREREQUISITES

Please refer to the Course Map in the Trainee Module. Prior to training with this module, it is recommended that the trainee shall have successfully completed *Core Curriculum; Millwright Level One; Millwright Level Two; Millwright Level Three; Millwright Level Four;* and *Millwright Level Five*, Modules 15501-09 through 15505-09.

OBJECTIVES

Upon completion of this module, the trainee will be able to do the following:

1. Inspect sealing glands and carbon rings.
2. Replace nozzle rings and reversing blade assemblies.
3. Inspect governor systems.
4. Replace rotor bearings.
5. Adjust overspeed trip mechanisms.
6. Inspect rotor assemblies.

PERFORMANCE TASKS

Under the supervision of the instructor, the trainee should be able to do the following:

1. Identify six of the following pieces of turbine equipment:
   - Sealing glands
   - Carbon rings
   - Rotor bearings
   - Nozzle rings
   - Governor
   - Trip linkage
   - Rotor
   - Oil pump

MATERIALS AND EQUIPMENT LIST

Overhead projector and screen

Transparencies

Blank acetate sheets

Transparency pens

Whiteboard/chalkboard

Markers/chalk

Pencils and scratch paper

Appropriate personal protective equipment

Toolbox with common mechanic tools

Plastic sealing compound

Paste sealing compound

Antigalling compound

Silcone grease

Bearing puller

Dial indicator

Sleeve-type bearing driver

Torch

Hot oil or bearing heater

Shims

Compressed air

continued

Examples of the following turbine components:

Turbine casing

Sealing glands

Carbon rings

Rotor bearing

Nozzle rings

Governor

Trip linkage

Bearing pedestals and housings

Rotor assembly

Hydraulic jack or wooden blocks

Applicable rigging equipment

Hand-held grinder

Dry ice

Prussian blue

Photos or videos/DVDs showing the components and/or operation of large turbines

Copies of the Quick Quizzes*

Module Examinations**

Performance Profile Sheets**

* Located at the back of this module.

**Located in the Test Booklet.

SAFETY CONSIDERATIONS

Ensure that the trainees are equipped with appropriate personal protective equipment and know how to use it properly.

ADDITIONAL RESOURCES

This module is intended to present thorough resources for task training. The following reference works are suggested for both instructors and motivated trainees interested in further study. These are optional materials for continued education rather than for task training.

Environmental Protection Agency
www.epa.gov/CHP/documents/tech_turbines.pdf

General Electric Company
www.gepower.com/home/index.htm

Siemens Corporation
www.powergeneration.siemens.com

http://mysite.du.edu/~jcalvert/tech/fluids/turbine.htm

TEACHING TIME FOR THIS MODULE

An outline for use in developing your lesson plan is presented below. Note that each Roman numeral in the outline equates to one session of instruction. Each session has a suggested time period of 2½ hours. This includes 10 minutes at the beginning of each session for administrative tasks and one 10-minute break during the session. Approximately 15 hours are suggested to cover *Maintaining and Repairing Steam Turbine Components*. You will need to adjust the time required for hands-on activity and testing based on your class size and resources. Because laboratories often correspond to Performance Tasks, the proficiency of the trainees may be noted during these exercises for Performance Testing purposes.

Topic	Planned Time
Session I. Introduction; Maintaining and Repairing Turbine Casings, Sealing Glands, and Carbon Rings	
A. Introduction	_____
B. Maintaining and Repairing Turbine Casings	_____
C. Maintaining and Repairing Sealing Glands and Carbon Rings	_____
1. Disassembling Sealing Glands	_____
2. Replacing Carbon Rings	_____
3. Assembling Sealing Glands	_____

Session II. Maintaining Governor Systems; Replacing Nozzle Rings, Reversing Blade Assemblies, and Rotor Locating Bearings

A. Maintaining Governor Systems _____

    1. Removing and Replacing Governor Components _____

B. Replacing Nozzle Rings and Reversing Blade Assemblies _____

C. Replacing Rotor Locating Bearings _____

Session III. Replacing Bearing Pedestals and Housings; Maintaining Overspeed Trip Mechanisms, Part One

A. Replacing Bearing Pedestals and Housings _____

    1. Replacing Exhaust-End Bearing Pedestals _____

    2. Replacing Steam-End Bearing Housings _____

    3. Aligning Exhaust-End Bearing Pedestals and Steam-End Housings _____

B. Maintaining Overspeed Trip Mechanisms _____

    1. Disassembling/Assembling Overspeed Trip Mechanisms _____

    2. Replacing Plunger Assemblies and Trip Bodies _____

Session IV. Maintaining Overspeed Trip Mechanisms, Part Two

A. Maintaining Overspeed Trip Mechanisms _____

1. Adjusting Trip Pin and Plunger Clearance _____

2. Adjusting Turbine Trip Speeds _____

3. Disassembling/Assembling Trip Valves _____

4. Backseating Trip Valves _____

5. Maintaining Governor Valves _____

Session V. Maintaining Rotor Assemblies and Large Steam Turbines

A. Maintaining Rotor Assemblies _____

B. Maintaining Large Steam Turbines _____

C. Laboratory _____

Have trainees practice identifying turbine components. This laboratory corresponds to Performance Task 1.

Session VI. Review and Testing

A. Module Review _____

B. Module Examination _____

    1. Trainees must score 70% or higher to receive recognition from NCCER.

    2. Record the testing results on Craft Training Report Form 200, and submit the results to the Training Program Sponsor.

C. Performance Testing _____

    1. Trainees must perform each task to the satisfaction of the instructor to receive recognition from NCCER. If applicable, proficiency noted during laboratory exercises can be used to satisfy the Performance Testing requirements.

    2. Record the testing results on Craft Training Report Form 200, and submit the results to the Training Program Sponsor.

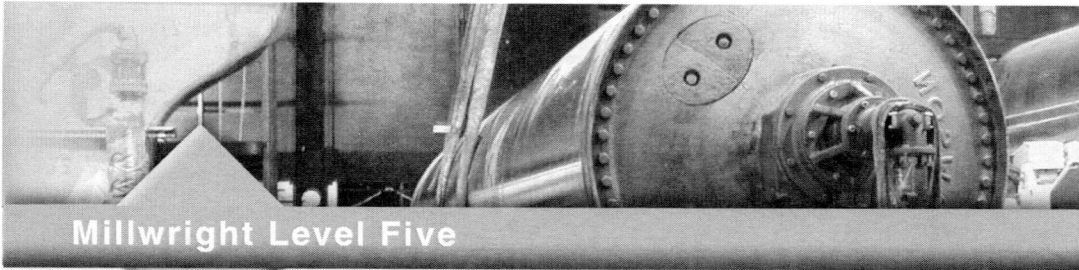

15506-09

Maintaining and Repairing Turbine Components

Assign reading of Module 15506-09.

15506-09

Maintaining and Repairing Turbine Components

Topics to be presented in this module include:

Overview

The millwright may be called upon to repair or to maintain turbine equipment. The different types of turbines share certain common maintenance requirements, such as balancing. This module presents procedures for troubleshooting and repairing or replacing turbine components. These components, including the governor, the vanes, and the blades are described, and step-by-step techniques for gaining access and replacing them are provided. The example procedures in this module are based on those for a Westinghouse Type E steam turbine.

Instructor's Notes:

_____
_____
_____
_____

Objectives

When you have completed this module, you will be able to do the following:

1. Inspect sealing glands and carbon rings.
2. Replace nozzle rings and reversing blade assemblies.
3. Inspect governor systems.
4. Replace rotor bearings.
5. Adjust overspeed trip mechanisms.
6. Inspect rotor assemblies.

Trade Term

Carbon ring

Required Trainee Materials

1. Pencil and paper
2. Appropriate personal protective equipment

Prerequisites

Before you begin this module, it is recommended that you successfully complete *Core Curriculum*; *Millwright Level One*; *Millwright Level Two*; *Millwright Level Three*; *Millwright Level Four*; and *Millwright Level Five*, Modules 15501-09 through 15505-09.

This course map shows all of the modules in the fifth level of the Millwright curriculum. The suggested training order begins at the bottom and proceeds up. Skill levels increase as you advance on the course map. The local Training Program Sponsor may adjust the training order.

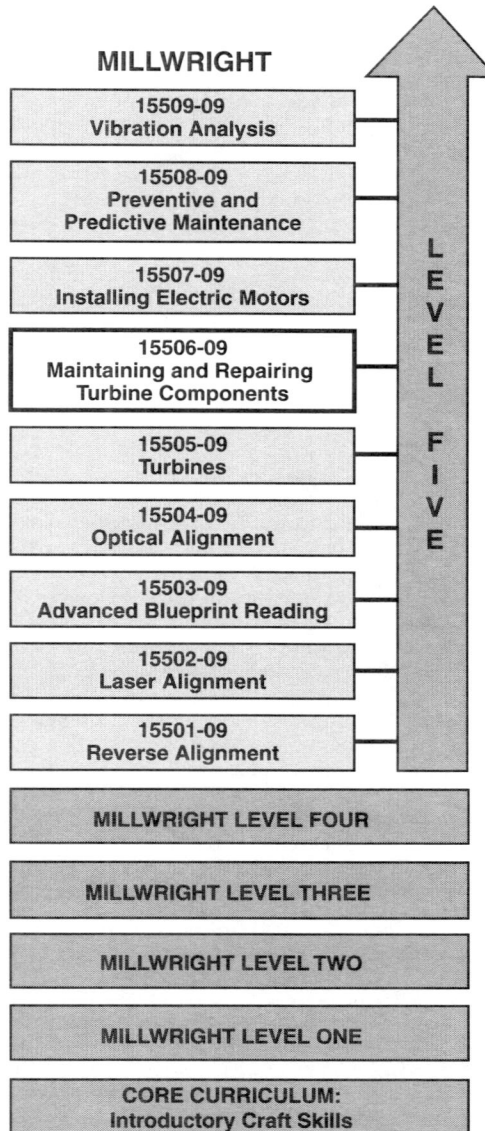

MILLWRIGHT

LEVEL FIVE
15509-09 Vibration Analysis
15508-09 Preventive and Predictive Maintenance
15507-09 Installing Electric Motors
15506-09 Maintaining and Repairing Turbine Components
15505-09 Turbines
15504-09 Optical Alignment
15503-09 Advanced Blueprint Reading
15502-09 Laser Alignment
15501-09 Reverse Alignment

MILLWRIGHT LEVEL FOUR

MILLWRIGHT LEVEL THREE

MILLWRIGHT LEVEL TWO

MILLWRIGHT LEVEL ONE

CORE CURRICULUM: Introductory Craft Skills

506CMAP.EPS

Ensure that you have everything required to teach the course. Check the Materials and Equipment List at the front of this module.

See the general Teaching Tip at the end of this module.

Explain that terms shown in bold are defined in the Glossary at the back of this module.

Show Transparency 1, Objectives, and Transparency 2, Performance Tasks. Review the goals of the module, and explain what will be expected of the trainees.

Review the modules covered in Level Five, and explain how this module fits in.

Discuss common applications of turbines and the energy conversion process involved in their operation.

Emphasize that turbine maintenance and troubleshooting procedures must be performed according to the manufacturer's recommendations.

Describe indications of turbine blade damage.

Explain that the blades can be examined using a boroscope.

Walk through the steps for maintaining and repairing turbine casings.

Show Transparency 3 (Figure 1). Point out the main parts of a turbine casing.

Stress the safety precautions required when working with compressed air.

Provide a turbine casing for trainees to examine.

1.0.0 ◆ INTRODUCTION

Turbines are used to drive equipment. Small turbines are often used to drive pumps, compressors, gearboxes, or other rotating equipment. Larger turbines are mainly used to drive power-generating equipment.

A turbine converts the kinetic energy of a moving fluid or gas into mechanical power by the impulse or reaction of the fluid or gas with a series of buckets, paddles, or blades arrayed about the circumference of a wheel or cylinder. The moving fluid may be water or steam. The moving gas may be compressed or heated air.

Steam is supplied to the turbine by a boiler. The steam provides the kinetic energy which the turbine converts to mechanical energy to turn a shaft and drive a generator, pump, or compressor.

Turbines must be properly maintained to ensure safe and efficient operation. The procedures for maintaining and repairing turbines vary from one turbine to another. This module explains some common maintenance procedures for a small Westinghouse Type E steam turbine and outlines how to overhaul a large steam turbine. When performing maintenance on a specific turbine, you must follow the procedures in the turbine manufacturer's maintenance manuals.

Some troubleshooting may apply only to gas turbines, but a few symptoms apply to any turbine. Troubleshooting must be based on the manufacturer's recommendations.

Possible indications of damage to turbine blades might be a detectable increase in fuel flow, exhaust temperature, and/or compressor discharge pressure. The blades can usually be examined with a boroscope for visible damage.

2.0.0 ◆ MAINTAINING AND REPAIRING TURBINE CASINGS

The horizontally split turbine casing surrounds the rotating element of the turbine and supports the stationary steam parts. Follow these steps to maintain and repair a turbine casing:

Step 1   Remove the cap screws from the horizontal flanges on the top half of the sealing glands.

Step 2   Remove the bolts and dowels from the horizontal casing flange.

Step 3   Lift the casing cover carefully by the eyebolt until the casing cover clears the rotor discs, using guide pins to maintain the stability of the casing.

Step 4   Move the cover to a safe location and remove sealing glands. *Figure 1* shows an opened turbine casing.

> **CAUTION**
> Care must be taken to protect the machined surfaces of the cover.

Step 5   Clean all mating sealing surfaces between the bottom half of the turbine casing, the casing cover, and the sealing glands.

Step 6   Apply plastic sealing compound according to specification to the sealing surfaces or in the sealant groove, if available.

> **CAUTION**
> Be careful not to get plastic sealant near the turbine casing bolt holes. A poor seal may result if the sealant gets in these holes.

Step 7   Lower the casing cover onto the bottom half of the casing.

Step 8   Seat the dowel pins.

Step 9   Tighten the bolts on the horizontal casing flange, starting with the bolts located closest to the sealing glands. Check the manufacturer's specifications for torque requirements.

Step 10 Clean the sealing gland flange surfaces and the mating turbine casing surfaces.

Step 11 Blow out the sealing glands, using compressed air.

> **WARNING!**
> Be sure to wear eye protection and use an OSHA-approved safety nozzle when using compressed air.

Step 12 Apply a thin coat of paste sealing compound, according to manufacturer's instructions, to the horizontal flanges and inside the bolt circles of the vertical flange faces.

Instructor's Notes:
_____
_____
_____

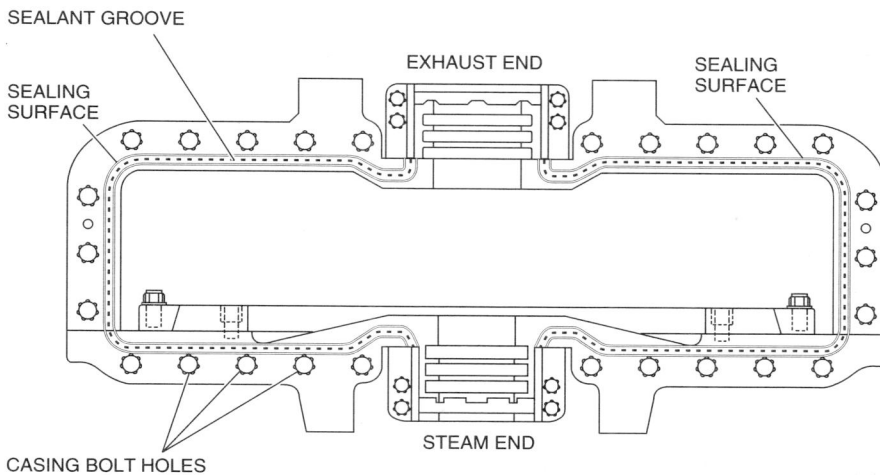

Figure 1 ◆ Turbine casing.

SEALANT GROOVE

SEALING SURFACE

EXHAUST END

SEALING SURFACE

CASING BOLT HOLES

STEAM END

506F01.EPS

Step 13 Place the top sealing glands into position.

Step 14 Install the cap screws.

Step 15 Tighten the cap screws on the vertical flange until they are snug.

Step 16 Tighten the cap screws on the horizontal flange to the torque specified by the manufacturer.

Step 17 Tighten the cap screws on the vertical flange to the specified torque.

3.0.0 ◆ MAINTAINING AND REPAIRING SEALING GLANDS AND CARBON RINGS

A turbine has either cartridge seals or two horizontally split sealing glands. One gland is bolted through a vertical flange to the steam end of the turbine casing and casing cover. The other gland is similarly attached to the exhaust end of the turbine casing and cover. In smaller turbines, the sealing glands house the **carbon rings**, which seal the casing and the rotor shaft. These rings minimize steam leakage along the shaft when the turbine operates in a noncondensing mode. Small turbines operating in a condensing mode employ carbon rings and a sealing steam arrangement to prevent air from leaking into the casing. *Figure 2* shows the sealing glands.

Explain that sealant must be applied carefully or a poor seal may result.

Show trainees how to maintain and repair turbine casings.

Discuss the various arrangements of sealing glands and carbon rings in turbines.

Describe the function of carbon rings in small turbines.

Show Transparency 4 (Figure 2). Point out the location of the sealing glands.

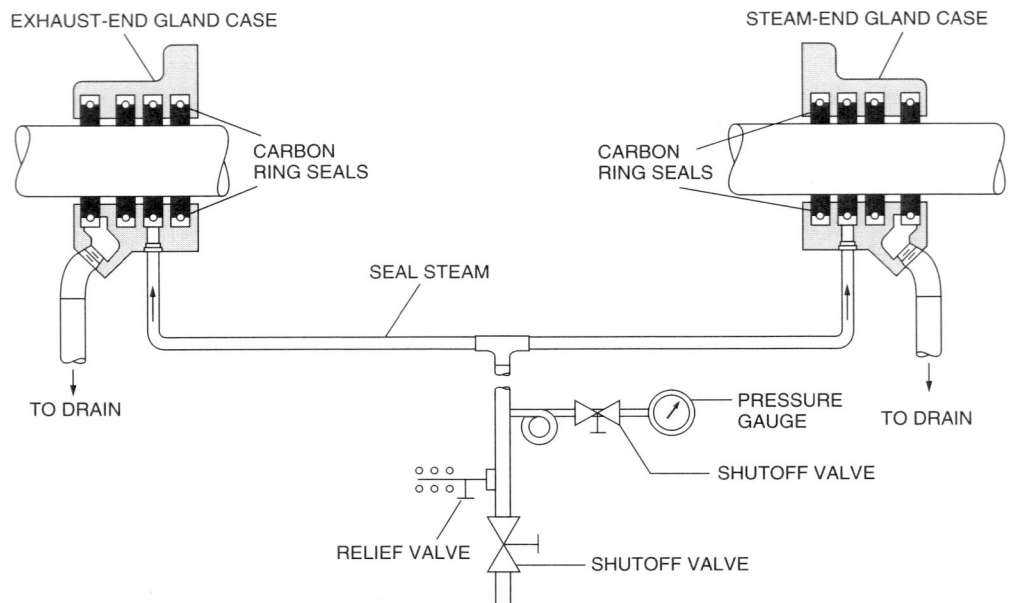

Figure 2 ◆ Sealing glands.

3.1.0 Disassembling Sealing Glands

Follow these steps to disassemble a sealing gland:

Step 1   Remove the cap screws from the horizontal and vertical flanges on the top half of the sealing glands.

Step 2   Pry the top halves of the sealing glands away from the bottom halves, using a pry bar to break the horizontal and vertical joints.

Step 3   Lift the top halves of the sealing glands straight up until the halves are clear of the carbon ring assemblies.

3.2.0 Replacing Carbon Rings

The carbon rings are not adjustable. Replace them if excessive steam leaks from the sealing glands. For best results, install new carbon rings in complete sets. Follow these steps to replace carbon rings:

Step 1   Remove the cylinder cover to provide access to the gland case (*Figure 3*), which contains the carbon rings.

Step 2   Check the radial clearance of the carbon ring using a feeler gauge (*Figure 4*).

Step 3   Lift out the upper half of the gland case (*Figure 5*). The lower half of the gland case is held by a large-headed screw that screws into the cylinder base and clamps the outer edge of the gland case.

Step 4   Unhook the carbon ring spring from around the first carbon ring.

Step 5   Slide the carbon ring stop pin off the carbon ring spring.

Step 6   Remove the top segment of the carbon ring from the lower half of the gland case.

Step 7   Rotate the carbon ring segments around the rotor shaft to remove them.

Step 8   Pull the retaining spring from the sealing gland.

Step 9   Mark each ring so that it can be returned to its original location to ensure a good seal.

Step 10   Clean the sealing gland, rotor shaft, and all sealing surfaces on the sealing gland flanges.

Describe the procedure for disassembling sealing glands.

Show trainees how to disassemble sealing glands.

Point out that carbon rings should be installed in complete sets.

Review the steps for replacing carbon rings.

Provide examples of sealing glands and carbon rings for trainees to examine.

See the two Teaching Tips for Section 3.0.0 at the end of this module.

Show Transparency 5 (Figure 3). Describe how to gain access to the rotor glands.

Show Transparency 6 (Figure 4). Explain how to check the radial clearance of the carbon ring using a feeler gauge.

Show Transparency 7 (Figure 5). Describe the components of the gland case assembly.

Instructor's Notes:

_____

_____

_____

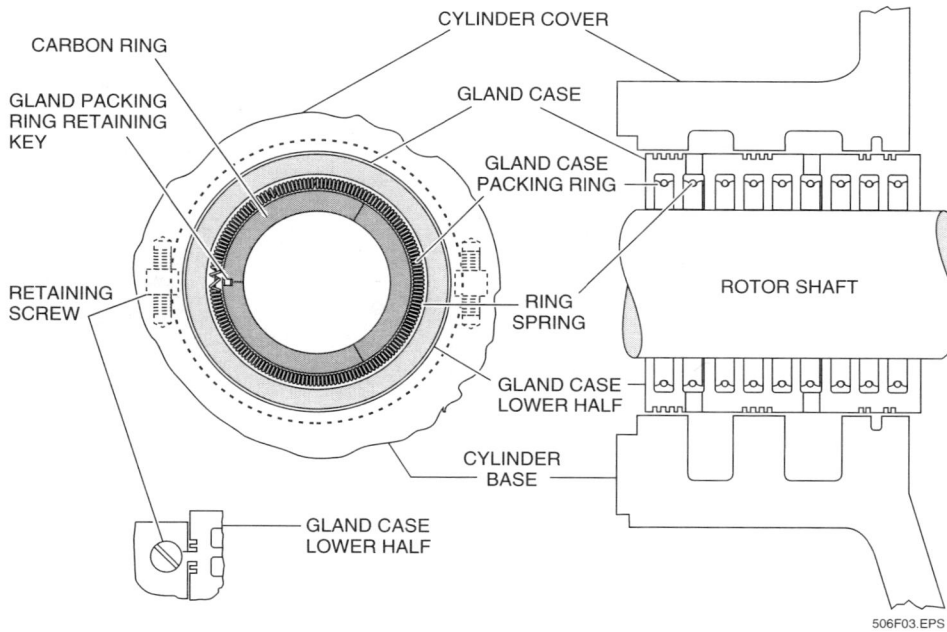

Figure 3 ◆ Rotor glands.

Labels in figure:
CARBON RING
GLAND PACKING RING RETAINING KEY
RETAINING SCREW
GLAND CASE LOWER HALF
CYLINDER COVER
GLAND CASE
GLAND CASE PACKING RING
ROTOR SHAFT
RING SPRING
GLAND CASE LOWER HALF
CYLINDER BASE
506F03.EPS

Note the importance of checking the operation of the turbine before returning it to service.

Step 11 Blow out the sealing glands, using compressed air.

Step 12 Place the carbon ring retaining springs under and part of the way around the rotor shaft.

Step 13 Roll the carbon ring segments around the shaft and into the sealing gland grooves.

Step 14 Align the match marks on the carbon ring segments to ensure proper assembly.

Step 15 Slide the carbon ring stop pin onto the retaining springs, and position the stops in the notched carbon ring segments.

Step 16 Hook the ends of each retaining spring together.

Step 17 Rotate the carbon rings so that the carbon ring stop pins are seated in the notches in the bottom half of the sealing glands.

Step 18 Install the upper half of the gland case.

Step 19 Install the turbine cylinder cover.

Step 20 Start up the turbine.

Step 21 Check the turbine operation to ensure that the glands function properly and show no signs of being too tight.

Step 22 Return the unit to service.

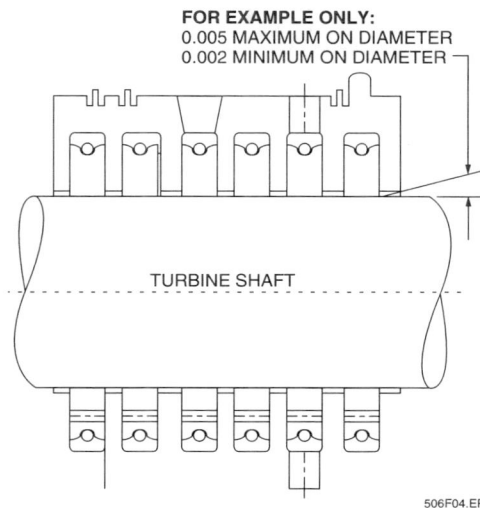

FOR EXAMPLE ONLY:
0.005 MAXIMUM ON DIAMETER
0.002 MINIMUM ON DIAMETER

TURBINE SHAFT

506F04.EPS

Figure 4 ◆ Carbon ring radial clearance.

Explain how to assemble sealing glands.

Demonstrate and/or describe how to properly apply the sealant.

Show trainees how to maintain and repair sealing glands and carbon rings.

Assign reading of Sections 4.0.0–6.0.0.

Ensure that you have everything required for teaching this session.

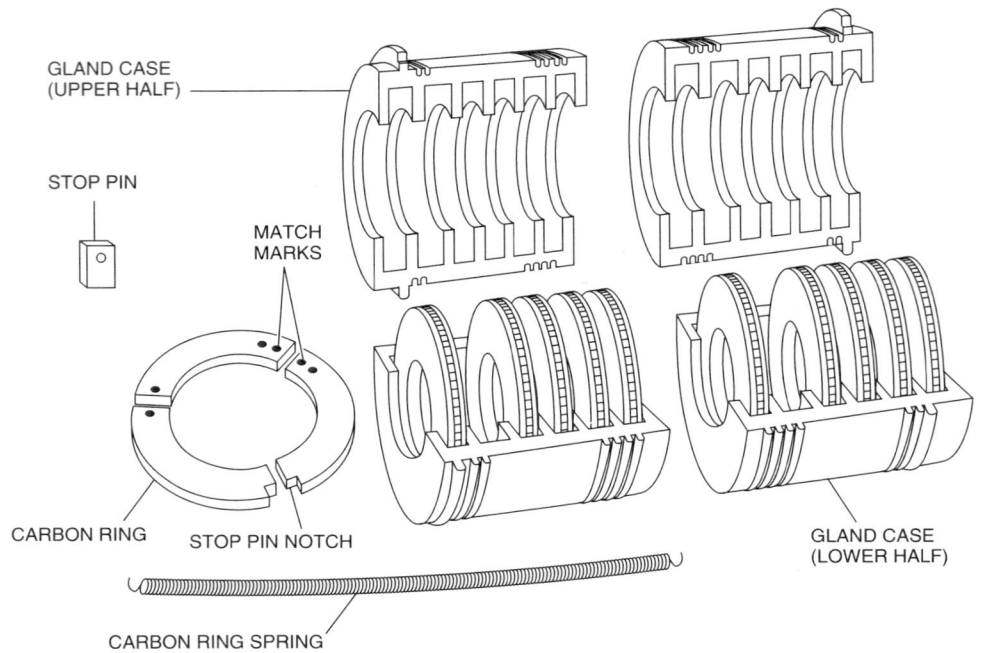

Figure 5 ◆ Gland case assembly.

3.3.0 Assembling Sealing Glands

Follow these steps to assemble sealing glands:

Step 1   Clean and hone the sealing gland surfaces.

Step 2   Clean the mating surfaces of the turbine casing.

Step 3   Blow out the sealing glands, using compressed air.

Step 4   Fill the grooves in the sealing gland vertical flange faces with a bead of plastic sealing compound $\frac{3}{16}$ inch wide.

Step 5   Cut the sealing compound, using a putty knife, to prevent it from extending beyond the horizontal flange.

Step 6   Apply the manufacturer's recommended sealing method to the horizontal flanges and inside the bolt circles of the vertical flange areas.

CAUTION

Do not use too much sealant on the sealing gland flanges because it may get in the sealing glands and adhere to the carbon rings, causing the carbon rings to seal improperly. Keep the sealant approximately $\frac{3}{16}$ inch away from the inside edges of the flanges to prevent it from squeezing into the carbon ring chambers.

Step 7   Place the top halves of the sealing glands in position.

Step 8   Install the cap screws on the horizontal and vertical flanges.

Step 9   Turn the cap screws until they are snug.

Step 10 Tighten the cap screws to the manufacturer's specified torque.

Instructor's Notes:

4.0.0 ◆ MAINTAINING GOVERNOR SYSTEMS

The following components are shared by some governor systems (*Figure 6*):

- A self-contained governor oil system, including a built-in oil pump and relief valve
- A centrifugal flyweight head and pilot valve assembly that control governor oil flow to and from a hydraulic power cylinder assembly
- A power cylinder assembly, also called a servo-motor, that operates the turbine governor valve linkage
- An internal governor compensation system for control stability
- An external, manual governor speed-setting adjustment system

Follow these guidelines to maintain the governor system:

- Check the governor oil level daily, and add oil as needed to maintain the oil level at the man-

ufacturer's specification. Do not overfill the governor reservoir because a high oil level causes foaming. A low oil level can cause sluggish governor operation and overheating, which results in damage to the turbine and the governor system.

- Inspect the governor linkage for binding, excessive play, and loose jam nuts or set screws on a daily basis.
- Clean the governor linkage.
- Lubricate the governor linkage, using a silicone grease that is resistant to high temperatures and water on an as-needed basis.

4.1.0 Removing Governor Components

Follow these steps to remove the governor components:

Step 1   Disconnect the pneumatic or electrical connections from the governor.

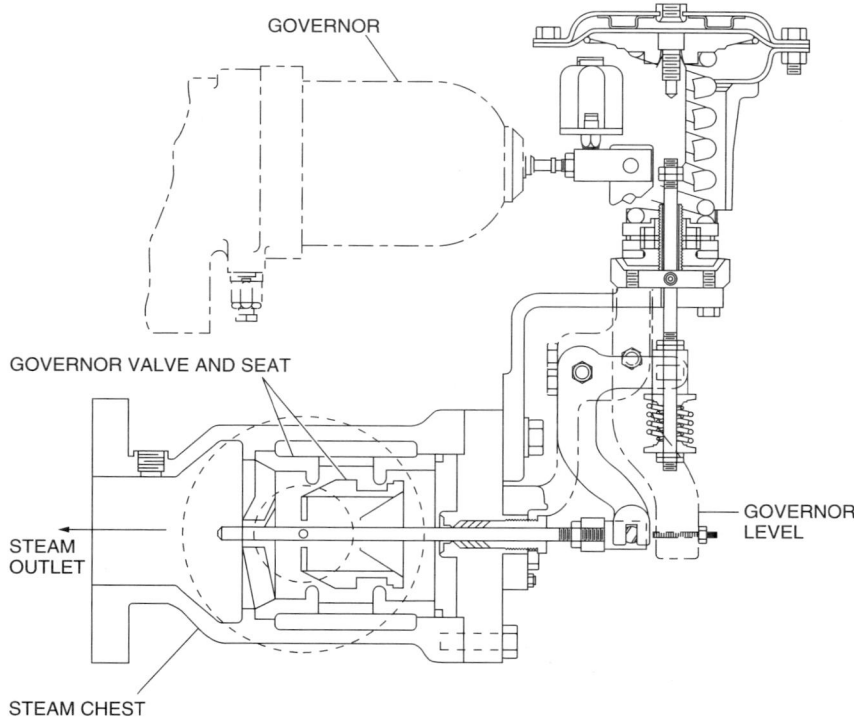

Point out the common components of governor systems.

Discuss the guidelines for maintaining a governor system.

Show Transparency 8 (Figure 6). Discuss the governor system shown in the diagram.

Provide a governor for trainees to examine.

Describe the steps for removing governor components.

GOVERNOR

GOVERNOR VALVE AND SEAT

STEAM OUTLET

STEAM CHEST

GOVERNOR LEVEL

506F06.EPS

Figure 6 ◆ Governor system.

Note that Steps 3 and 4 apply to oil-ring lubricated turbines only.

Explain how to replace the governor components.

Show trainees how to remove and replace governor components.

Describe the configuration of the nozzle rings and reversing blade assembly.

Explain the procedure for replacing the nozzle rings and reversing blade assembly.

Provide nozzle rings for trainees to examine.

Caution trainees against distorting the oil rings.

Step 2 Loosen the set screw on the lever, and remove the pin.

> **NOTE**
> Perform Steps 3 and 4 only if the turbine is oil-ring lubricated; otherwise, proceed to Step 5.

Step 3 Disconnect the tube connections from the governor drive housing.

Step 4 Remove the oil feed tube from the governor drive housing.

Step 5 Remove the dowel pins and bolts that secure the governor to the drive housing.

Step 6 Lift the governor off the drive housing.

> **CAUTION**
> Be careful to avoid damaging the worm gear and the worm wheel.

Step 7 Mark the shims before removing them to ensure that they are returned to their original position.

Step 8 Remove the shims.

4.2.0 Replacing Governor Components

Follow these steps to replace the governor components:

Step 1 Install the shims on the governor drive housing.

Step 2 Place the governor on the governor drive housing.

> **CAUTION**
> This unit must be handled carefully to prevent damage.

Step 3 Install the oil feed tube and connect the tube connections if required.

Step 4 Install the bolts and the dowels.

Step 5 Install the pin in the lever.

Step 6 Tighten the set screw.

Step 7 Connect the pneumatic or electrical connections.

5.0.0 u REPLACING NOZZLE RINGS AND REVERSING BLADE ASSEMBLIES

The nozzle ring directs the steam flow from the steam chest to the first row of rotor blades. Steam exits the first row of blades and passes through the stationary reversing blade assembly before entering the second row of rotor blades. The reversing blade assembly is positioned between the two rows of rotor blades and is bolted to the nozzle ring. The reversing blade assembly is positioned by spacers and requires no adjustment. Follow these steps to replace the nozzle rings and reversing blade assembly:

Step 1 Disconnect the coupling between the turbine and the driven machine.

Step 2 Remove the turbine casing cover.

Step 3 Remove the top halves of the sealing glands.

Step 4 Remove the carbon rings.

Step 5 Remove the cooling water piping from the bearing caps if applicable.

Step 6 Remove the dowels and bolts from the bearing cap joints.

Step 7 Pry the bearing caps away from the bearing housings, using a pry bar to break the joints.

Step 8 Raise the caps approximately 1 inch, and pry out the top liners from the bearing

> **CAUTION**
> Attempting to remove the bearing caps without prying out the top bearing liners can distort the oil rings. Distorted oil rings do not rotate to provide lubrication and result in bearing damage and failures.

caps, using a screwdriver, to release the oil rings from the caps.

Step 9 Remove the bearing caps.

Step 10 Remove the top journal bearing liners.

Step 11 Lift the rotor slightly, and roll the bottom bearing liners away from the locating lugs to remove them. When the bottom

Instructor's Notes:

_____

_____

_____

_____

Step 12 Disconnect the governor linkage.

Step 13 Remove the governor.

Step 14 Place a sling either outside or between the rotor discs.

Step 15 Lift the rotor approximately 1 inch. Lift the rotor slowly and keep it level to prevent it from binding in the casing or damaging mechanical surfaces.

Step 16 Lift the oil rings from the bearing housings.

Step 17 Place the rings to the side so that they are free of the bearing housing support casings.

Step 18 Lift the rotor assembly out of the turbine casing.

Step 19 Set the rotor in a rotor stand to prevent it from rolling.

Step 20 Wrap the rotor journals and carbon ring sealing areas with clean rags or other suitable covering to protect them from damage.

Step 21 Remove the bolts, lock washers, and spacers that secure the reversing blade assembly.

Step 22 Mark and identify each spacer so that it can be returned to its original location.

Step 23 Remove the nozzle ring bolts and lock washers.

Step 24 Remove the nozzle ring from the casing.

Step 25 Inspect the nozzle ring and the reversing blade assembly for scale or boiler compound deposits. Clean the ring and the reversing blade assembly if necessary, and replace any eroded parts.

Step 26 Clean the casing and nozzle ring sealing surfaces.

Step 27 Apply a thin coat of paste sealant to the nozzle ring sealing surface on the steam-end casing.

Step 28 Apply antigalling compound to the threads of the nozzle ring bolts.

Step 29 Bolt the nozzle ring to the turbine casing. Ensure that lock washers are used with all bolts.

Step 30 Place lock washers on the reversing blade assembly bolts.

Step 31 Apply antigalling compound to the bolt threads.

Step 32 Insert the bolts into the holes in the reversing blade assembly.

Step 33 Slip the spacers over the bolts.

Step 34 Position the reversing blade assembly in the turbine casing.

> **CAUTION**
> Ensure that the reversing blade assembly is installed in the same location from which it was removed and that the installation coincides with the rotation of the turbine to avoid causing damage to the unit.

Step 35 Bolt the reversing blade assembly to the nozzle ring.

Step 36 Lower the rotor assembly to within 1 inch of full replacement in the casing.

> **CAUTION**
> Guide the rotor assembly carefully into the casing to prevent the discs from contacting the reversing blade assembly because this will damage the blades.

Step 37 Position the oil rings so that they fall into the openings between the bearing liner supports located in the bottom of the bearing housings.

Step 38 Position the antirotation tab on the rotor locating bearing so that it engages the groove in the steam-end bearing housing.

Step 39 Lower the rotor assembly slowly into the casing.

Step 40 Install the journal bearing liners and caps.

Step 41 Install the governor.

Step 42 Connect the governor linkage.

Step 43 Install the carbon rings.

Step 44 Install the top halves of the sealing glands.

Step 45 Install the turbine casing cover.

Demonstrate and/or describe how to apply antigalling compound to the threads of the nozzle ring bolts.

Show trainees how to replace nozzle rings and reversing blade assemblies.

MODULE 15506-09 ◆ MAINTAINING AND REPAIRING TURBINE COMPONENTS   6.9

Point out the main function of the rotor bearing.

Provide examples of rotor bearings for trainees to examine.

Describe the procedure for replacing the rotor bearing.

Explain why the trip body must be heated.

Describe how to check the axial rotor float.

Show trainees how to replace the rotor bearing.

Assign reading of Sections 7.0.0–8.4.0.

Ensure that you have everything required for teaching this session.

Point out that the bearing pedestals and housings must be replaced correctly in order to maintain proper alignment of the bearings.

6.0.0 ◆ REPLACING ROTOR LOCATING BEARINGS

The rotor locating bearing maintains the correct axial position of the rotor assembly in the turbine casing. Follow these steps to replace the rotor locating bearing:

Step 1 Remove the rotor assembly.

Step 2 Remove the set screws from the over-speed trip body.

Step 3 Heat the overspeed trip body evenly.

> **CAUTION**
> Do not heat the rotor locating bearing and the rotor shaft when heating the trip body. Wrap them in an approved heat-resistant cloth to protect them from the heat.

> **NOTE**
> Apply the heat as evenly and as rapidly as possible; then pull the overspeed trip body from the rotor shaft.

Step 4 Remove the retainer ring, using ring-expanding pliers.

Step 5 Remove the rotor locating bearing, using a bearing puller.

Step 6 Mount a dial indicator perpendicular to a vertical shaft to check the axial rotor float.

Step 7 Shift the rotor as far as possible in both axial directions while observing the dial indicator.

> **NOTE**
> Refer to each manufacturer's specifications for actual measurements and tolerances of shaft float.

Step 8 Install the bearing on the shaft, using a sleeve-type bearing driver that makes contact with the bearing inner race.

Step 9 Seat the bearing solidly against the machined shoulder on the shaft so that the shielded side of the bearing faces the trip body.

Step 10 Install the retainer ring, seating it firmly in the groove on the rotor shaft, with the beveled edge of the ring positioned toward the trip body.

Step 11 Heat the trip body in hot oil or on a bearing heater.

> **CAUTION**
> To prevent warping, follow manufacturer's recommendations.

Step 12 Place the heated trip body on the rotor shaft.

Step 13 Align the set screw holes in the trip body and the shaft.

Step 14 Tighten the set screw to ensure that the trip body is properly positioned on the shaft.

Step 15 Back the set screw out one or two turns.

Step 16 Allow the trip body to cool to room temperature.

Step 17 Tighten the set screw.

Step 18 Check the trip body runout and correct it if necessary. Runout should not exceed the manufacturer's specification on the outboard end of the trip body.

Step 19 Lock the set screw.

Step 20 Check the plunger assembly to ensure that it is properly positioned in the bearing housing.

Step 21 Install the rotor in the turbine casing.

Step 22 Flush the rotor locating bearing with clean oil.

Step 23 Install the bearing cap.

7.0.0 ◆ REPLACING BEARING PEDESTALS AND HOUSINGS

The bearing pedestals and housings must be replaced accurately to ensure that the proper alignment of bearings is maintained.

7.1.0 Replacing Exhaust-End Bearing Pedestals

Follow these steps to replace the exhaust-end bearing pedestal:

Step 1 Remove the rotor assembly.

Step 2 Support the weight of the turbine exhaust-end casing with a hydraulic jack, wooden blocks, or other adequate means to prevent damaging the casing.

6.10 MILLWRIGHT ◆ LEVEL FIVE

Instructor's Notes:

_____

_____

_____

_____

Step 3 Remove the hold-down bolts and the dowel pins from the pedestal support feet.

Step 4 Remove the tapered pins from the combining studs.

Step 5 Loosen the four cap screws three or four turns.

Step 6 Pry the pedestal away from the casing, using a pry bar, until the spacers are free to move.

Step 7 Remove the cap screws and the spacers, and mark each spacer as it is removed so that it can be returned to the location from which it was removed. If the spacers are not returned to their original locations, bearing misalignment may occur and cause uneven bearing wear or possible failure.

Step 8 Slide the pedestal off the combining studs and dowel pins.

NOTE

Bearing liner locating grooves are located at the horizontal split on replacement bearing pedestals.

Step 9 Slide the original pedestal onto the combining studs and dowel pins.

Step 10 Install the spacers and cap screws, ensuring that the spacers are returned to their original positions.

Step 11 Tighten the cap screws.

Step 12 Insert the taper pins into the pedestal and the combining studs.

Step 13 Install the rotor assembly.

Step 14 Install the bottom half of the journal bearing liners.

Step 15 Check the bearing alignment and adjust if necessary.

7.2.0 Replacing Steam-End Bearing Housings

Follow these steps to replace the steam-end bearing housing:

Step 1 Remove the rotor assembly.

Step 2 Place a hydraulic jack, wooden blocks, or other adequate support under the steam end of the turbine casing and steam chest to prevent damaging them.

Step 3 Remove the hold-down bolts and dowel pins from the steam-end bearing support.

Step 4 Remove the bolts that secure the support to the bearing housing.

Step 5 Loosen the socket head cap screws.

Step 6 Pry the bearing housing away from the turbine casing, using a pry bar, until the spacers are free to move.

Step 7 Remove the cap screws and spacers, and mark the spacers as they are removed so that they can be returned to the original positions.

Step 8 Pull the bearing housing off the dowel pins.

Step 9 Push the bearing housing onto the dowel pins in the steam-end turbine casing.

Step 10 Install the cap screws and spacers, ensuring that the spacers are returned to their original positions.

Step 11 Bolt the support to the bearing housing.

Step 12 Install the rotor assembly.

Step 13 Install the bottom halves of the journal bearing liners.

Step 14 Check the bearing alignment and adjust if necessary.

7.3.0 Aligning Exhaust-End Bearing Pedestals and Steam-End Bearing Housings

Follow these steps to align the exhaust-end bearing pedestal and the steam-end bearing housing:

Step 1 Remove the rotor assembly.

Step 2 Clean the shaft journals.

Step 3 Apply a light coat of Prussian blue to both shaft journals.

Step 4 Install the bottom-half journal bearing liners in the bearing pedestal and steam-end bearing housing.

CAUTION

Ensure that the liners are properly seated to prevent damage.

Provide examples of bearing pedestals and housings for trainees to examine.

Describe the steps for replacing the exhaust-end bearing pedestals and steam-end bearing housing.

Show trainees how to replace the exhaust-end bearing pedestal and the steam-end bearing housing.

Describe how to align the exhaust-end bearing pedestal and the steam-end bearing housing.

Explain how to seat the rotor shaft to avoid damaging the bearings.

Explain how shims are used to correct misalignment.

Show trainees how to align the exhaust-end bearing pedestal and the steam-end bearing housing.

Discuss the function of the overspeed trip mechanism.

Show Transparency 9 (Figure 7).

Provide an example of an overspeed trip mechanism for trainees to examine.

Explain the procedure for disassembling overspeed trip mechanisms.

Emphasize that all stored energy and energy systems must be blocked and locked out before initiating any maintenance.

Step 5 Lower the rotor assembly until the full weight of the rotor is supported by the journal bearing liners.

Step 6 Rotate the rotor assembly one-quarter turn in each direction.

> **CAUTION**
> To prevent damage to the bearings, ensure that the rotor shaft is seated on the bottom of the bearing liners and does not move sideways or upward while being rotated.

Step 7 Remove the rotor assembly from the turbine casing, and check the bearing contact.

> **NOTE**
> The exhaust-end bearing pedestal and the steam-end bearing housing are considered to be aligned when the bearing contact with the shaft journals is no less than 85 percent along the bottom of the bearing liners and when the contact along the sides of the liners is parallel with the bearing bore and equal on each side.

Step 8 Place shim stock in increments of 0.002 inch behind the spacers that need shims to correct any misalignment.

Step 9 Recheck the bearing contact and continue to add shims until the bearings are properly aligned.

Step 10 Remove the shims from each spacer, and alter the thickness of the opposite spacer accordingly. Surface grinding is the preferred method for adjusting the thickness of the spacers.

Step 11 Recheck the bearing contact.

8.0.0 ◆ MAINTAINING OVERSPEED TRIP MECHANISMS

The overspeed trip device is mounted in a housing carried on the governor end of the rotor shaft. It is actuated by centrifugal force when the turbine reaches a predetermined limiting speed, which is usually 10 percent above normal full-load speed. The linkage is arranged to control not only the governor valve but also the emergency quick-closing valve, located ahead of the governor valve. The overspeed trip mechanism (*Figure 7*) should be checked on a regular PM basis to ensure correction operation.

Two factors should be considered when maintaining the overspeed trip system: the sticking of governor and valve stems due to deposits of scale or carryover and the excessive clearance between the turbine shaft and the trip lever due to wear or incorrect adjustment.

8.1.0 Disassembling Overspeed Trip Mechanisms

Frequently, overspeed mechanisms may only be worked on by certified specialists. Follow these steps to disassemble the overspeed trip mechanism:

> **WARNING!**
> Ensure that all stored energy and energy systems are blocked and locked out before initiating any maintenance.

Step 1 Remove the steam-end bearing cap.

Step 2 Pry the U-lock staple surrounding the adjustment nut out of the trip body, using a screwdriver.

Step 3 Remove the adjusting nut, trip spring, and washers. Record the number of turns required to remove the adjusting nut so that it can be returned to the original setting during assembly.

Step 4 Rotate the rotor shaft 180 degrees, and remove the U-lock staple surrounding the weighted end of the trip pin.

Step 5 Remove the trip pin from the trip body.

Step 6 Remove the auxiliary weight, if furnished.

Step 7 Remove the oil pump.

Step 8 Remove the rotor assembly from the turbine casing.

Step 9 Remove the set screw from the trip body.

Step 10 Heat the trip body evenly, using a torch.

> **CAUTION**
> Do not heat the rotor locating bearing and the rotor shaft because damage could result.

Step 11 Apply the heat as rapidly as possible.

Step 12 Pull the trip body from the rotor shaft.

Instructor's Notes:

_____

_____

_____

_____

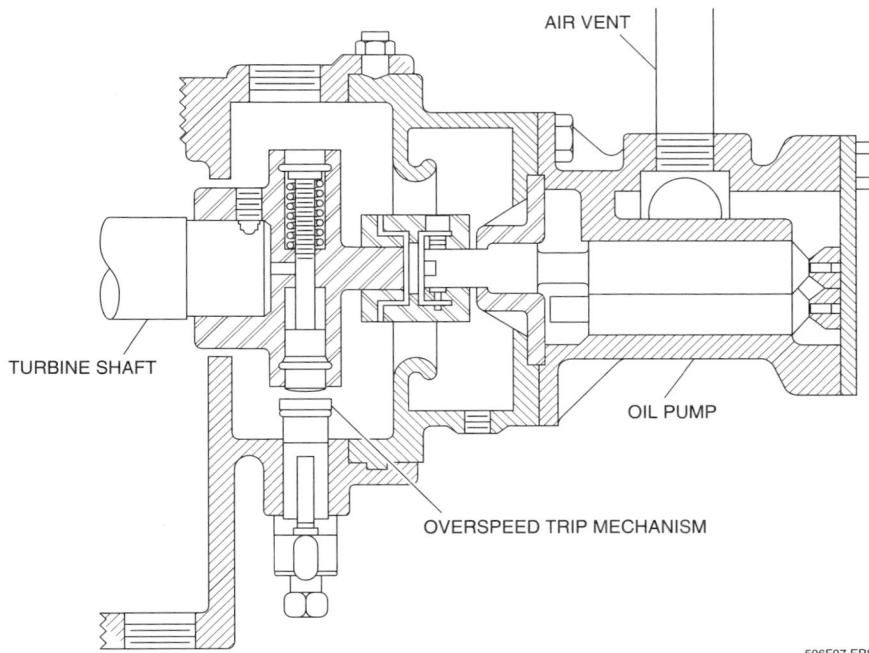

AIR VENT

TURBINE SHAFT

OIL PUMP

OVERSPEED TRIP MECHANISM

506F07.EPS

Figure 7 ◆ Overspeed trip mechanism.

8.2.0 Replacing Plunger Assemblies

Follow these steps to replace the plunger assembly:

> **WARNING!**
> Ensure that all stored energy and energy systems are blocked and locked out before initiating any maintenance.

Step 1 Remove the steam-end bearing.

Step 2 Remove the governor and adapter piece from the steam-end bearing housing.

Step 3 Loosen the jam nut.

Step 4 Remove the set screw from the side of the bearing housing.

Step 5 Remove the set screw from the plunger assembly.

Step 6 Separate the two halves of the plunger.

Step 7 Remove the two plunger halves from the bearing housing.

Step 8 Install the two new plunger halves into the bearing housing, making sure that they fit together tightly.

Step 9 Screw the set screw into the plunger assembly, using a screwdriver, to ensure proper alignment.

Step 10 Screw the set screw into the side of the bearing housing, using a screwdriver.

Step 11 Tighten the jam nut, using an open-end wrench.

Step 12 Install the governor and adapter piece into the steam-end bearing housing.

Step 13 Install the steam-end bearing, making sure not to damage it.

8.3.0 Replacing Trip Bodies

Follow these steps to replace the trip body:

> **WARNING!**
> Ensure that all stored energy and energy systems are blocked and locked out before initiating any maintenance.

Step 1 Heat the trip body in hot oil or in an oven.

Review the steps for assembling the overspeed trip mechanisms.

Assign reading of Sections 8.5.0–8.9.0.

Ensure that you have everything required for teaching this session.

Explain the procedure for adjusting trip pin and plunger clearance.

CAUTION

Do not exceed 500°F (260°C) because excessive heat may damage the trip body.

Step 2   Place the heated trip body on the rotor shaft.

Step 3   Align the set screw holes in the trip body and the shaft.

Step 4   Tighten the set screw to ensure that the trip body is properly positioned on the shaft.

Step 5   Back the set screw out of the body one or two turns.

Step 6   Allow the trip body to cool to room temperature.

Step 7   Tighten the set screw.

Step 8   Check the trip body runout and correct if necessary. The runout should not exceed the manufacturer's specification on the outboard end of the trip body.

Step 9   Lock the set screw.

Step 10 Ensure that the plunger assembly is properly positioned in the bearing housing.

Step 11 Return the rotor to the turbine casing.

8.4.0 Assembling Overspeed Trip Mechanisms

Follow these steps to assemble the overspeed trip mechanism:

WARNING!

Ensure that all stored energy and energy systems are blocked and locked out before initiating any maintenance.

Step 1   Place the auxiliary weight on the trip pin.

Step 2   Insert the trip pin into the trip body.

Step 3   Position the weighted end of the pin on the opposite side of the trip body set screw.

Step 4   Press the U-lock staple into the trip body to secure the weighted end of the trip pin. The staple must be fully seated in the circular groove in the trip body.

Step 5   Place the trip spring in the trip body.

Step 6   Install the washers.

Step 7   Return the adjusting nut to the original setting. Tighten the nut the same number of turns that were made during disassembly.

Step 8   Press the U-lock staple into the trip body to lock the adjusting nut. Ensure that the staple is fully seated in the circular groove in the trip body.

8.5.0 Adjusting Trip Pin and Plunger Clearance

Follow these steps to adjust the trip pin and plunger clearance:

WARNING!

Ensure that all stored energy and energy systems are blocked and locked out before initiating any maintenance.

Step 1   Remove the inspection plug from the steam-end bearing cap.

Step 2   Rotate the rotor shaft by hand until the adjusting nut can be seen through the inspection hole. This positions the weighted end of the trip pin directly above the plunger assembly.

Step 3   Latch the resetting lever.

Step 4   Loosen the jam nut on the trip lever jackscrew.

Step 5   Push the plunger assembly upward and into the bearing housing. Ensure that the plunger assembly is in solid contact with the trip pin.

Step 6   Adjust the jackscrew to obtain ¹⁄₁₆ inch of clearance between the plunger and the jackscrew.

Step 7   Tighten the jam nut and lock it in place

CAUTION

The jam nut must be locked at all times to prevent the jackscrew from vibrating loose during operation. A loose jackscrew can render the trip system inoperative.

Step 8   Recheck the clearance of the trip pin and plunger.

Instructor's Notes:

_____
_____
_____
_____

8.6.0 Adjusting Turbine Trip Speeds

Follow these steps to adjust the turbine trip speed

> **WARNING!**
> Ensure that all stored energy and energy systems are blocked and locked out before initiating any maintenance.

Step 1 Remove the inspection plug from the steam-end bearing cap.

Step 2 Rotate the rotor shaft by hand until the adjusting nut can be viewed through the inspection hole.

Step 3 Latch the resetting lever.

Step 4 Place a nonferrous drift pin on the adjusting nut.

Step 5 Strike the drift pin sharply to ensure that the trip pin, trip valve, and trip linkage function properly.

Step 6 Latch the resetting lever.

Step 7 Start the turbine, and monitor the turbine speed.

Step 8 Override the governor and gradually overspeed the turbine.

Step 9 Close the steam inlet shutoff valve.

Step 10 Turn the shaft by hand after the rotor shaft starts rotating until the adjusting nut is visible through the bearing cap inspection hole.

Step 11 Pry the U-lock staple partially away from the trip body, using a screwdriver, until the adjusting nut is free to turn.

Step 12 Turn the adjusting nut to change the trip speed. To decrease the trip speed, turn the nut counterclockwise. To increase the trip speed, turn the nut clockwise.

Step 13 Push the U-lock staple into the trip body, making sure that the staple is firmly seated.

Step 14 Check that the trip pin moves freely.

Step 15 Start the turbine.

Step 16 Check the trip speed.

Step 17 Make trip speed adjustments until the turbine trips at the rated trip speed.

8.7.0 Disassembling and Assembling Trip Valves

Follow these steps to disassemble and assemble the trip valve:

> **WARNING!**
> Ensure that all stored energy and energy systems are blocked and locked out before initiating any maintenance.

Step 1 Place the trip valve in the tripped position.

Step 2 Disconnect the closing spring from the resetting lever.

Step 3 Remove the cap screws from the valve cover.

Step 4 Lift the trip valve assembly and cover from the steam chest body.

Step 5 Remove the nut, the spring, the bushing, and the spring seats from the valve stem.

Step 6 Remove the valve assembly from the cover.

Step 7 Drive the bushings out of the valve cover, using a nonferrous drift pin.

Step 8 Clean the valve cover thoroughly.

Step 9 Press new bushings into the valve cover.

Step 10 Lock the bushings in place.

Step 11 Clean the sealing surfaces on the valve cover flange and the steam chest body.

Step 12 Insert the valve stem into the lower guide bushing.

Step 13 Push the valve stem through the valve cover.

Step 14 Turn the valve stem into and through the connection.

Step 15 Install the spring seats, bushings, spring, and locknut.

Step 16 Apply a combination of plastic sealant and paste sealant to the sealing surfaces of the steam chest valve cover flange.

Step 17 Return the valve assembly and cover to the steam chest body.

Step 18 Tighten the cap screws.

Describe the proper procedure for adjusting turbine trip speeds.

Explain how the adjusting nut is turned to increase or decrease the trip speed.

Review the steps for disassembling and assembling trip valves.

Describe the steps for backseating the trip valve.

Show Transparency 10 (Figure 8). Describe the trip valve components and the backseating procedure as shown in the diagram.

Explain that turning the valve stem clockwise decreases the dimension, and turning it counterclockwise increases the dimension.

See the Teaching Tip for Section 8.0.0 at the end of this module.

Review the procedure for maintaining the governor valve.

Step 19 Backseat the trip valve.

> **NOTE**
> Follow the procedure in the next section of this module to backseat the trip valve.

Step 20 Connect the closing spring to the resetting lever.

8.8.0 Backseating Trip Valves

Follow these steps to backseat the trip valve:

> **WARNING!**
> Ensure that all stored energy and energy systems are blocked and locked out before initiating any maintenance.

Step 1 Disassemble the trip valve and linkage, and ensure that all parts are clean.

Step 2 Replace the worn linkage pins, guide bushings, valve stem, knife edge, and latch.

Step 3 Reassemble the trip valve and linkage.

Step 4 Disconnect the closing spring from the resetting lever.

Step 5 Remove the locknut from the trip valve stem.

> **WARNING!**
> While removing the locknut, firmly grasp the spring to prevent it from hitting someone due to rapid decompression.

Step 6 Pry against the bottom of the connection and the valve cover, using a long screwdriver, to raise the connection, backseating the valve against the lower guide bushing. *Figure 8* shows backseating the trip valve.

Step 7 Release pressure on the screwdriver slightly, and turn the valve stem until the bottom of the resetting liner knife edge is 0.12 inch below the top edge of the hand trip lever latch. Turning the valve stem clockwise decreases the dimension; turning the valve stem counterclockwise increases the dimension.

Step 8 Replace and fully tighten the locknut until the upper spring seal is seated against the bushing.

Step 9 Place a wrench on the valve stem flats below the connection to prevent the valve stem from turning.

Step 10 Raise the resetting liner until the valve backseats against the bushing.

Step 11 Check to ensure that the bottom of the resetting lever knife edge is still 0.12 inch below the top of the hand trip lever latch.

Step 12 Latch the resetting lever, and verify that the spring compresses. If the spring does not compress, readjustments are required.

Step 13 Connect the closing spring.

Step 14 Check the trip valve operation.

8.9.0 Maintaining Governor Valves

Follow these steps to maintain the governor valve:

> **WARNING!**
> Ensure that all stored energy and energy systems are blocked and locked out before initiating any maintenance.

Step 1 Remove the linkage that connects the governor valve to the governor.

Step 2 Remove the bolts from the valve cover.

Step 3 Pull the cover and the valve away from the steam chest body.

Step 4 Remove the valve stem connection and the jam nut from the valve stem.

Step 5 Remove the valve stem from the cover assembly.

Step 6 Chip or grind the welded blocks, using a chipping hammer or a hand grinder, to remove the blocks from the steam chest.

Step 7 Pack the valve seat with dry ice to chill it, and pull the seat from the steam chest, using a puller.

Step 8 Drive the bushing from the valve seat, using a nonferrous rod, to remove the bushing.

Step 9 Press the bushing into the valve seat to replace the bushing.

Step 10 Stake the valve seat to lock the bushing in place.

Instructor's Notes:

_____

_____

_____

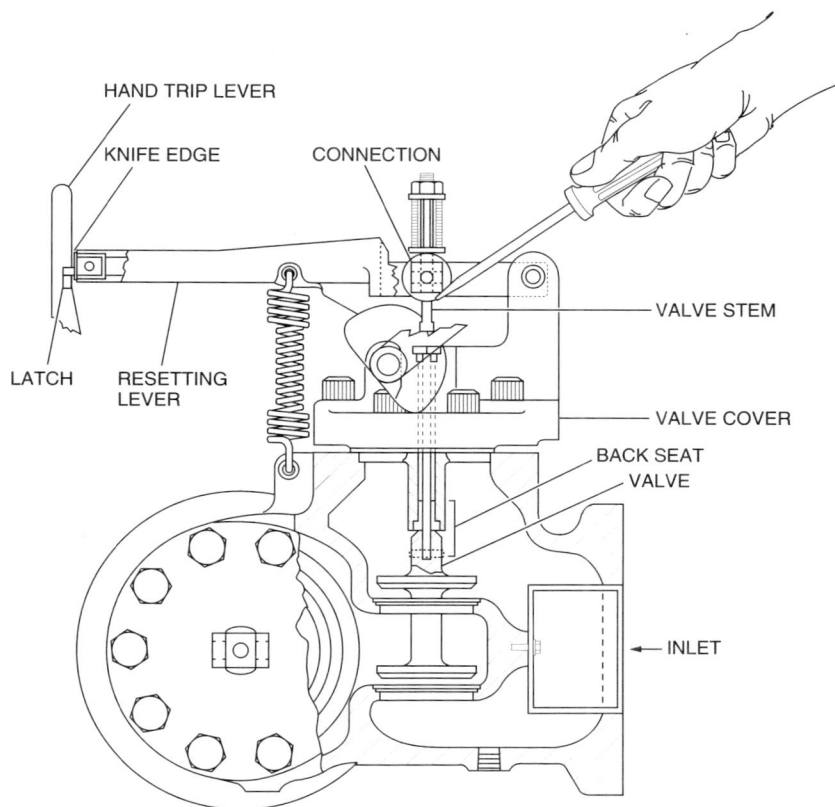

HAND TRIP LEVER
KNIFE EDGE
CONNECTION
VALVE STEM
LATCH
RESETTING LEVER
VALVE COVER
BACK SEAT VALVE
INLET

506F08.EPS

Figure 8 ◆ Backseating trip valve.

Step 11 Turn the valve stem from the connection until the valve is fully seated.

Step 12 Adjust the jam nut according to the equipment specifications.

Step 13 Screw the valve stem into the connection until the jam nut contacts the face of the connection.

Step 14 Tighten the jam nut against the connection to lock the jam nut.

Step 15 Install the governor linkage pins and bushings.

Step 16 Inspect the governor valve stem and guide bushings. Replace any worn valve stem and bushings if necessary.

Step 17 Inspect the valve for excessive steam leakage.

Step 18 Remove the packing follower, and replace the valve stem packing.

CAUTION

Do not overtighten the packing follower because this causes the governor valve stem to bind in the valve cover and results in erratic speed control.

Step 19 Lubricate the governor linkage pins, using a silicone grease that can withstand high temperatures and is water-resistant.

Step 20 Chill the valve seat, using dry ice.

Step 21 Press the valve seat into the steam-chest body.

Step 22 Weld blocks to the steam chest to secure the valve seat. Make sure the blocks are 180 degrees apart.

MODULE 15506-09 ◆ MAINTAINING AND REPAIRING TURBINE COMPONENTS   6.17

Assign reading of
Sections 9.0.0–10.0.0.

Ensure that you have
everything required for
teaching this session.

Review the steps for
maintaining the rotor
assembly.

Provide an example of
a rotor assembly for
trainees to examine.

Discuss some of the
common applications
of large steam tur-
bines.

Discribe how steam
turbines use steam to
generate power.

Step 23 Place the governor valve stem in the valve cover.

Step 24 Install the connection and the jam nut on the valve stem.

Step 25 Clean the joint between the valve cover and the steam-chest body.

Step 26 Apply a compound of paste sealant and plastic sealant on the sealing surfaces.

Step 27 Replace the cover.

Step 28 Tighten the cover bolts.

Step 29 Connect the governor valve linkage.

Step 30 Adjust the valve travel.

Step 31 Return the valve to service.

9.0.0 ◆ MAINTAINING ROTOR ASSEMBLIES

Follow these steps to maintain the rotor assembly:

> **WARNING!**
> Ensure that all stored energy and energy systems are blocked and locked out before initiating any maintenance.

Step 1 Disconnect the coupling between the turbine and the driven machine.

Step 2 Remove the turbine casing cover.

Step 3 Remove the top half of the sealing glands and carbon rings.

Step 4 Remove the journal bearing liners.

Step 5 Disconnect the governor linkage.

Step 6 Remove the governor.

Step 7 Lift the rotor approximately 1 inch using a sling. When lifting the rotor, keep it level to prevent it from binding in the casing or damaging machined surfaces.

Step 8 Lift the oil rings from the bearing housings, and set them aside.

Step 9 Ensure that the rings are free of the bearing housing support casings.

Step 10 Lift the rotor out of the turbine casing.

> **CAUTION**
> Chock the rotor assembly with wooden blocks to prevent the rotor from rolling when it is removed from the casing. Protect the rotor journals and carbon ring sealing areas by wrapping them with a suitable covering.

Step 11 Lower the rotor assembly to within 1 inch of full replacement in the casing.

> **CAUTION**
> Guide the rotor carefully into the casing to prevent the discs from contacting the reversing blade assembly and damaging the rotor.

Step 12 Place the oil rings into the openings between the bearing liner supports in the bottom of the bearing housing.

Step 13 Position the antirotation tab on the rotor locating bearing to engage the groove in the steam-end bearing housing.

Step 14 Lower the rotor slowly into the casing. Be sure to guide the rotor correctly.

Step 15 Replace the journal bearing liners and caps.

Step 16 Replace the governor.

Step 17 Connect the governor linkage.

Step 18 Replace the carbon rings and the top half of the sealing glands.

Step 19 Replace and secure the casing cover.

10.0.0 ◆ MAINTAINING LARGE STEAM TURBINES

Steam turbines are used primarily in fossil-fuel and nuclear power plants to drive electric generators for producing electric power for consumers. Steam turbines are also used to drive electric generators in large industrial plants to provide electricity to the plant. The steam needed to operate the turbine is produced in a boiler and piped under pressure to the steam turbine. Because of the velocity of the steam, the rotational speed of the turbine can exceed 10,000 revolutions per minute.

Instructor's Notes:

_____

_____

_____

Modern industrial turbines can have 50 or more stages set on a horizontal axle. Each stage consists of a turbine wheel of rotating blades and a set of stationary blades. The curved wheel blades and the stationary blades are shaped so that the spaces between them act as nozzles that aim the steam and increase the speed of the steam before it enters the next stage. As the steam travels through the turbine, it expands because of pressure and temperature drops. Therefore, to maintain useful energy in the turbine, each stage is larger than the one that precedes it.

Steam turbines may be either condensing or noncondensing, depending on how the steam exiting the turbine is used. In the condensing turbine, the steam exiting the turbine enters a condenser to convert the steam back to water. This creates a vacuum which helps to force the steam through the turbine. The water is then pumped back to the boiler to be converted into steam again.

The steam coming from a noncondensing turbine is not converted into water but is sent to another unit or system in the plant operating at a lower pressure. In a noncondensing turbine, steam can be extracted from the turbine at extraction points to obtain desired pressure rates. Controlling valves at each extraction point regulate the amount of steam extracted from the turbine.

Because of the many types, sizes, and configurations of large turbines, the following checklist provides a summary of the maintenance activities required on most large turbines:

- Check blind piping as necessary.
- Remove and replace instrument probes as necessary.
- Disconnect and drain pressure oil piping.
- Remove and replace the governor.
- Remove and replace the amplifier servo/control linkage.

- Remove and replace the turning gear and motor.
- Remove and replace the insulation.
- Remove and replace the coupling guard and spool.
- Check cold alignment.
- Remove and replace the coupling halves.
- Remove and replace the horizontal case bolts.
- Remove the TTVs and extraction valves.
- Remove and replace the control cylinder and linkage.
- Remove and replace the external valve rack.
- Split the case.
- Remove the top case half and set it at a lower level.
- Remove, recondition, and replace the hydraulic relay valve.
- Overhaul all valves.
- Check and record the axial float and bucket clearances.
- Remove and replace the upper-case diaphragms.
- Remove and replace the first-stage inner housing.
- Check internal clearances.
- Remove the rotor and set it at a lower level.
- Inspect the rotor.
- Remove and replace the lower-case diaphragms.
- Fabricate guides and holders for crosshair targets.
- Remove and replace the internal valve rack.
- Calibrate the rotor.
- Remove, repair, and replace linkage bearings and valves.
- Disassemble and recondition the servo/control cylinder.
- Buff and hone the lower turbine case.

MODULE 15506-09 ◆ MAINTAINING AND REPAIRING TURBINE COMPONENTS 6.19

Have trainees complete the Review Questions, and go over the answers prior to administering the Module Examination.

1. The most common use for large turbines is _____.
   a. cogeneration
   b. aircraft engines
   c. power generation
   d. pump drivers

2. A turbine converts the _____ of a moving fluid or gas into mechanical power.
   a. potential energy
   b. kinetic energy
   c. power
   d. motion

3. The most common type of turbine is the _____ turbine.
   a. steam-driven
   b. horizontal
   c. gas-fired
   d. split

4. A possible indication of damage to turbine blades might be a(n) _____.
   a. decrease in fuel flow
   b. increase in power output
   c. increase in fuel flow
   d. decrease in exhaust temperature

5. The casing cover must be stabilized during removal until it clears the rotor disks by using the _____.
   a. guide pins
   b. hold-down bolts
   c. cap screws
   d. dowels

6. The sealing glands house the _____, which seal the casing and the rotor shaft.
   a. reversing blade
   b. nozzle rings
   c. stationary blade
   d. carbon rings

7. For best results, install carbon rings in complete sets.
   a. True
   b. False

8. Most governor systems are equipped with a self-contained _____.
   a. pneumatic system
   b. nozzle ring
   c. oil system
   d. sleeve seal

9. Check the governor oil level _____.
   a. annually
   b. daily
   c. weekly
   d. monthly

10. Overfilling the governor reservoir can cause _____.
    a. foaming
    b. vibration
    c. sparking
    d. overheating

11. A low oil level in the governor reservoir can cause _____.
    a. excessive cooling
    b. overspeed
    c. overheating
    d. stoppage

12. The steam flow from the steam chest to the first row of rotor blades is directed by the _____.
    a. nozzle ring
    b. nozzle chamber
    c. carbon rings
    d. governor linkage

13. The reversing blade assembly is bolted to the rotors.
    a. True
    b. False

14. When the bottom bearing liners are removed, the rotor shaft rests on the _____.
    a. shaft sleeve seals
    b. oil rings
    c. casing gaskets
    d. governor linkage

Instructor's Notes:
_____
_____
_____
_____

15. The rotor locating bearing maintains the correct _____ of the rotor assembly in the turbine casing.
    a. tension
    b. axial position
    c. timing
    d. location

16. When heating the trip body, do *not* heat the rotor locating bearing and the _____.
    a. valve linkage
    b. retainer ring
    c. servomotor
    d. rotor shaft

17. Bearing liner locating grooves for this unit are located at the _____ on replacement bearing pedestals.
    a. vertical split
    b. horizontal split
    c. longitudinal split
    d. lower pedestals

18. The overspeed trip device is actuated by _____.
    a. high temperature
    b. centrifugal force
    c. high pressure
    d. low flow rates

19. As the steam travels through the turbine, it expands because of _____.
    a. pressure increases
    b. temperature increases
    c. steam flow increases
    d. pressure and temperature drop

20. In the condensing steam turbine, the steam is converted back to water, creating a _____.
    a. pressure increase
    b. temperature increase
    c. vacuum
    d. flow increase

_____
_____
_____
_____

Summarize the major concepts presented in the module.

Administer the Module Examination. Be sure to record the results of the Examination on Craft Training Report Form 200, and submit the results to the Training Program Sponsor.

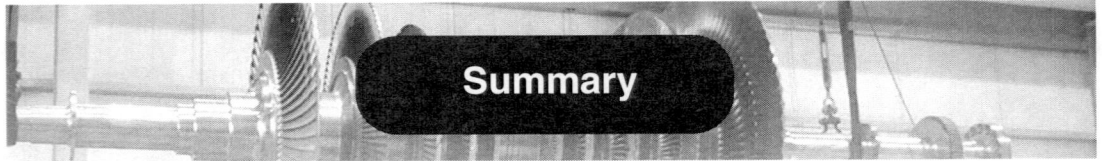

Administer the Performance Tests, and fill out Performance Profile Sheets for each trainee. If desired, trainee proficiency noted during laboratory sessions may be used to complete the Performance Test. Record the results on Craft Training Report Form 200, and submit the results to the Training Program Sponsor.

Summary

The principal function of a turbine is to rotate a shaft by moving steam, water, air, or gas through a series of blades attached to the shaft to drive a generator or other piece of equipment. As a millwright, you will be expected to identify the different types of turbine problems, know the basic turbine components, and understand the steps involved in repairing different types of turbines. Always follow all safety guidelines that pertain to rotating equipment when working in the vicinity of operating turbines.

Notes

Instructor's Notes:

_____

_____

_____

Carbon ring: A gland shaft packing made of carbon
to seal the steam chest.

Resources & Acknowledgments

Additional Resources

This module is intended to present thorough resources for task training. The following reference works are suggested for further study. These are optional materials for continued education rather than for task training.

http://mysite.du.edu/~jcalvert/tech/fluids/turbine.htm

http://www.epa.gov/CHP/documents/tech_turbines.pdf

http://www.powergeneration.siemens.com/home

http://www.gepower.com/home/index.htm

Figure Credits

Bill Wall, 506F01 (photos)

Instructor's Notes:

_____

_____

_____

MODULE 15506-09 — TEACHING TIPS

The following are suggested activities or instructional methods to help you teach the material in this module.

General

When you call on someone to answer a question, the rest of the class relaxes or even tunes out because they expect that the question and answer will take place only between you and the trainee you called on. Instead, use this technique to involve more trainees in answering questions and to keep them on their toes.

1. Ask trainees to define a term or explain a concept.
2. After one trainee has answered, ask a trainee seated nearby if the answer is right. Then ask whether a trainee in the back of the room agrees.
3. Ask trainees to explain why they think an answer is right or wrong.
4. Use the session to clear up incorrect ideas and encourage trainees to learn from their mistakes.

Sections 1.0.0–10.0.0

Quick Quiz

This Quick Quiz will familiarize trainees with terms commonly used in relation to maintaining and repairing turbines and their components. You will need photocopies of the quiz provided on the following page. Trainees will need pencils. If you allow trainees to use the Trainee Guide, decrease the amount of time you give them to complete the quiz, and remind them to bring their books to class.

1. Make a photocopy of the quiz for each trainee.
2. Give trainees between 5 and 10 minutes to complete the quiz.
3. Go over the answers to the quiz.
4. Ask trainees if they have questions.

Answers to Quick Quiz

1. f
2. b
3. a
4. d
5. j
6. e
7. m
8. l
9. k
10. g

Quick Quiz   *Maintaining and Repairing Turbine Components*

For each description listed, identify the term that the text best describes. Write the corresponding letter in the blank provided.

_____ 1. A turbine converts the kinetic energy of a moving fluid or gas into _____ power.

_____ 2. The horizontally split turbine casing surrounds the _____ of the turbine and supports the stationary steam parts.

_____ 3. A turbine has either cartridge seals or two horizontally split _____.

_____ 4. The steam flow from the steam chest to the first row of rotor blades is directed by the _____.

_____ 5. The _____ assembly is positioned between the two rows of rotor blades and is bolted to the nozzle ring.

_____ 6. In smaller turbines, the sealing glands house the _____, which seal the casing and the rotor shaft.

_____ 7. Turbine blades can usually be checked for damage by examination with a _____.

_____ 8. The device _____ is arranged to control not only the governor valve but also the emergency quick-closing valve.

_____ 9. To ensure that the proper alignment of bearings is maintained, the bearing _____ must be replaced accurately.

_____ 10. The correct axial position of the rotor assembly is maintained in the turbine casing by the _____.

a. sealing glands
b. rotating element
d. nozzle ring
e. carbon rings
f. mechanical
g. rotor locating bearing
j. reversing blade
k. pedestals and housings
l. trip linkage
m. borescope

Section 3.0.0   *Identifying Sealing Gland/Sealing Steam Arrangement Components*

This Quick Quiz will familiarize trainees with sealing glands and the sealing steam arrangement components. You will need photocopies of the quiz provided on the following page. Trainees will need pencils. If you allow trainees to use the Trainee Guide, decrease the amount of time you give them to complete the quiz, and remind them to bring their books to class.

1. Make a photocopy of the quiz for each trainee.
2. Give trainees between 5 and 10 minutes to complete the quiz.
3. Go over the answers to the quiz.
4. Ask trainees if they have questions.

Answers to Quick Quiz

1. F
2. C
3. A
4. D
5. B, E
6. G
7. H, I

Identifying Sealing Gland/Sealing Steam Arrangement Components

Identify the sealing gland and sealing steam arrangement components. In the blank provided, write the letter corresponding to the component in *Figure 1*. Note that some components appear twice in the diagram. In that case, write both corresponding letters in the blank provided.

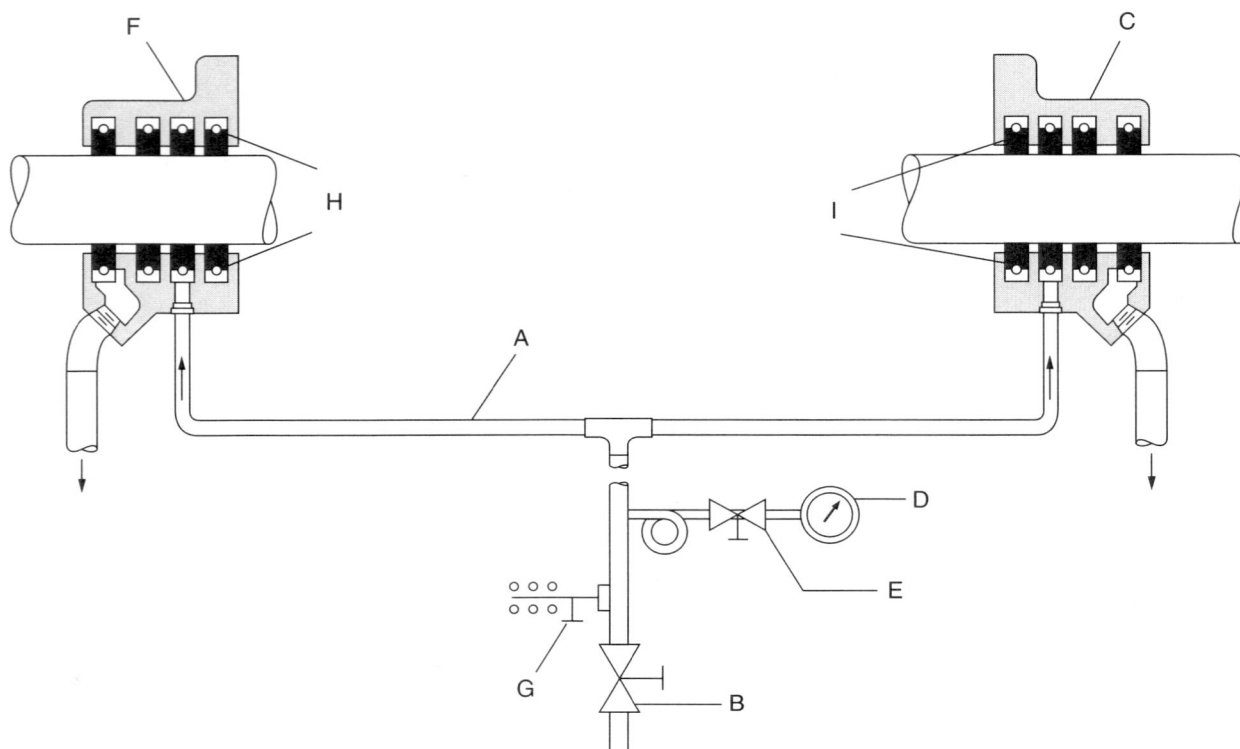

506IG01.EPS

Figure 1

_____	1.	Exhaust-end gland case
_____	2.	Steam-end gland case
_____	3.	Seal steam
_____	4.	Pressure gauge
_____	5.	Shutoff valve
_____	6.	Relief valve
_____	7.	Carbon ring seals

Section 3.0.0 *Identifying Rotor Gland Components*

This Quick Quiz will familiarize trainees with the components of the rotor glands, which contain the carbon rings. You will need photocopies of the quiz provided on the following page. Trainees will need pencils. If you allow trainees to use the Trainee Guide, decrease the amount of time you give them to complete the quiz, and remind them to bring their books to class.

1. Make a photocopy of the quiz for each trainee.
2. Give trainees between 5 and 10 minutes to complete the quiz.
3. Go over the answers to the quiz.
4. Ask trainees if they have questions.

Answers to Quick Quiz

1. F
2. J
3. C
4. B
5. K
6. I
7. D
8. A
9. G, H
10. E

Quick Quiz

Identifying Rotor Gland Components

Identify the components of the rotor glands. In the blank provided, write the letter corresponding to the component in *Figure 2*. Note that some components appear twice in the diagram. In that case, write both corresponding letters in the blank provided.

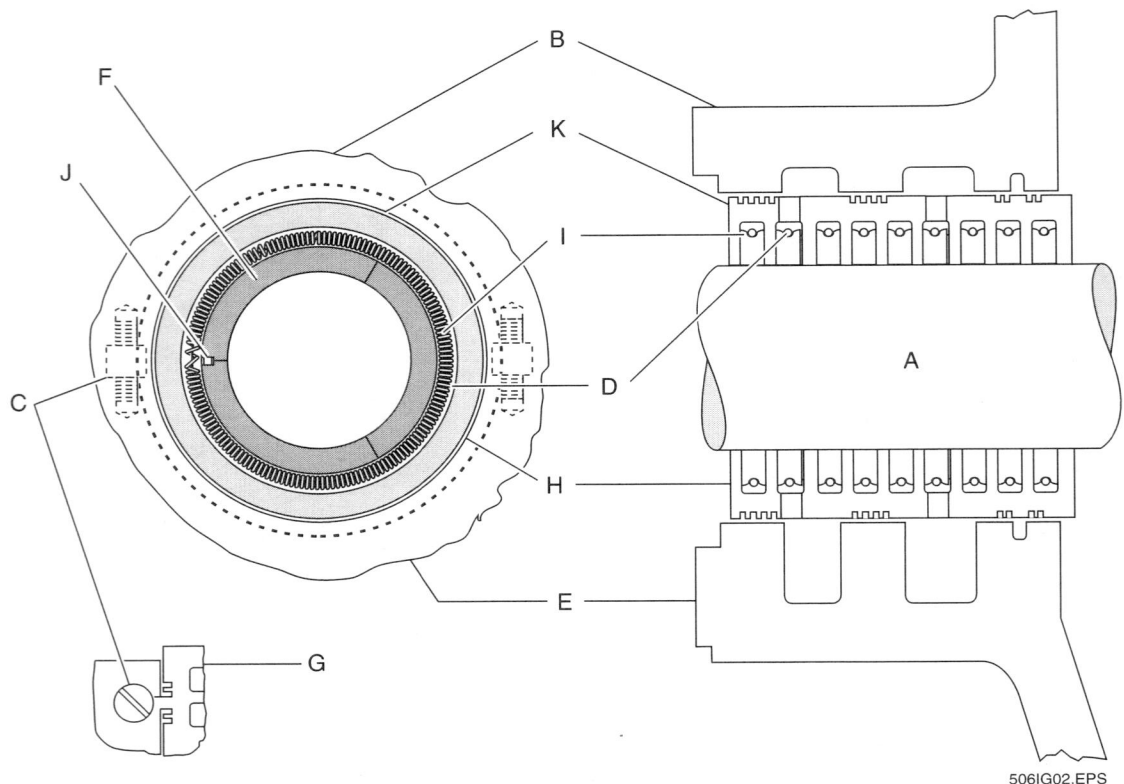

506IG02.EPS

Figure 2

_____	1.	Carbon ring
_____	2.	Gland packing ring retaining key
_____	3.	Retaining screw
_____	4.	Cylinder cover
_____	5.	Gland case
_____	6.	Gland case packing ring
_____	7.	Ring spring
_____	8.	Rotor shaft
_____	9.	Gland case lower half
_____	10.	Cylinder base

Sections
3.0.0–10.0.0 *Identifying Turbine Components*

This exercise will provide trainees with practice in identifying turbine components. Trainees will need appropriate personal protective equipment, pencils, and paper. You will need to arrange to obtain different types of turbine components and have them set up at several workstations. Allow 30–45 minutes for this exercise. This exercise corresponds to Performance Task 1.

Alternatively, obtain manufacturers' literature, photographs, or illustrations of turbine components. Discuss each component with trainees. Ask trainees to identify each type. If desired, have trainees identify the function of each component.

1. Show trainees various turbine components, and discuss how they function in the turbine.
2. Set up one or two of the components at each workstation, depending on the space available.
3. Have trainees circulate to all of the stations and identify each component. If desired, have trainees identify the function of the component.
4. Answer any questions trainees might have.

Section 8.0.0 *Identifying Trip Valve Components*

This Quick Quiz will familiarize trainees with trip valve components encountered when backseating the trip valve. You will need photocopies of the quiz provided on the following page. Trainees will need pencils. If you allow trainees to use the Trainee Guide, decrease the amount of time you give them to complete the quiz, and remind them to bring their books to class.

1. Make a photocopy of the quiz for each trainee.
2. Give trainees between 5 and 10 minutes to complete the quiz.
3. Go over the answers to the quiz.
4. Ask trainees if they have questions.

Answers to Quick Quiz

1. I
2. D
3. G
4. C
5. A
6. E
7. J
8. F
9. H
10. B

Quick Quiz   *Identifying Trip Valve Components*

Identify the trip valve components. In the blank provided, write the letter corresponding to the component in *Figure 3*.

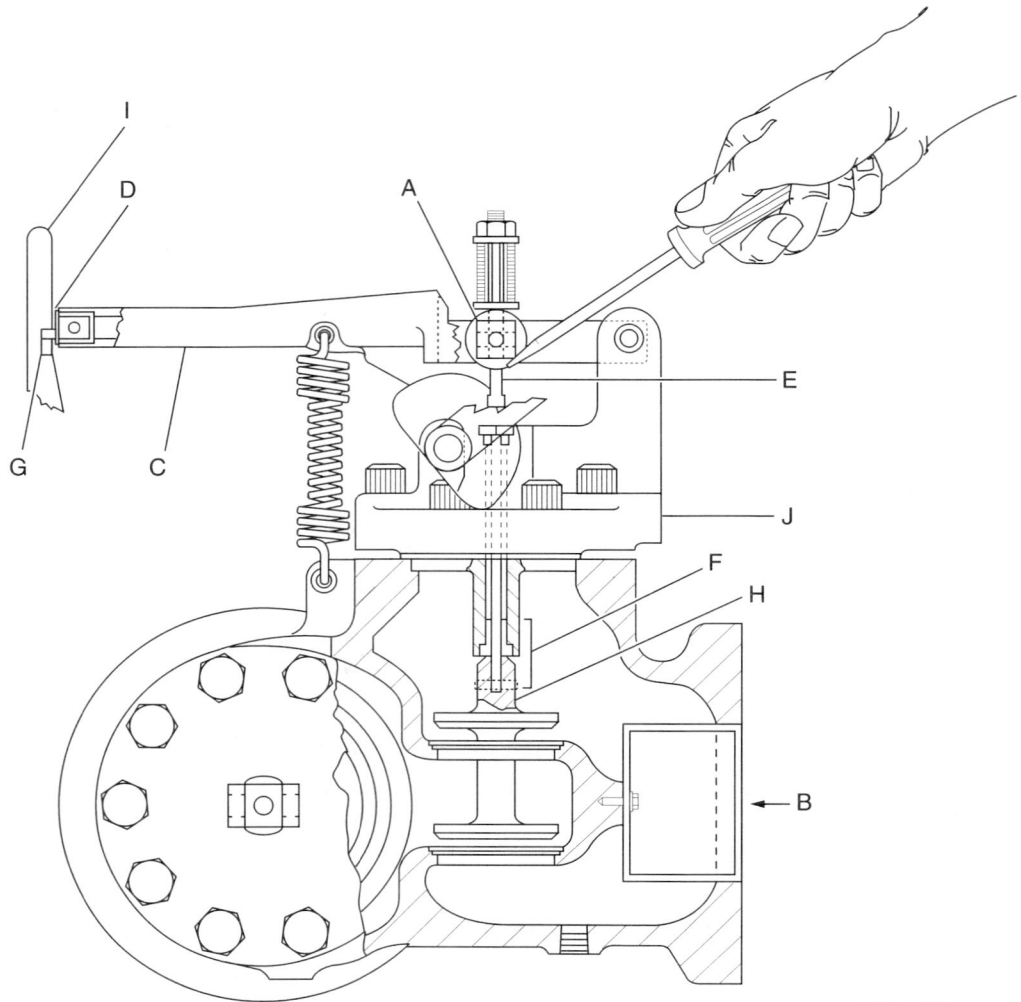

506IG03.EPS

Figure 3

_____ 1. Hand trip lever

_____ 2. Knife edge

_____ 3. Latch

_____ 4. Resetting lever

_____ 5. Connection

_____ 6. Valve stem

_____ 7. Valve cover

_____ 8. Back seat

_____ 9. Valve

_____ 10. Inlet

MODULE 15506-09 — ANSWERS TO REVIEW QUESTIONS

Answer		Section
1.	c	1.0.0
2.	b	1.0.0
3.	c	1.0.0
4.	c	1.0.0
5.	a	2.0.0
6.	d	3.0.0
7.	a	3.2.0
8.	c	4.0.0
9.	b	4.0.0
10.	a	4.0.0
11.	c	4.0.0
12.	a	5.0.0
13.	b	5.0.0
14.	a	5.0.0
15.	b	6.0.0
16.	d	6.0.0
17.	b	7.1.0
18.	b	8.0.0
19.	d	10.0.0
20.	c	10.0.0

Craft _____ Module Name _____

Copyright Date _____ Module Number _____ Page Number(s) _____

Description _____

_____

_____

_____

_____

(Optional) Correction _____

_____

_____

(Optional) Your Name and Address _____

_____

_____

_____

Installing Electric Motors

NCCER STANDARDIZED CRAFT TRAINING PROGRAM

The National Center for Construction Education and Research (NCCER) provides a standardized national program of accredited craft training. Key features of the program include instructor certification, competency-based training, and performance testing. The program provides trainees, instructors, and companies with a standard form of recognition through a National Craft Training Registry. The program is described in full in the *Guidelines for Accreditation*, published by NCCER. For more information on standardized craft training, contact the NCCER by writing us at 3600 NW 43rd St., Bldg. G, Gainesville, FL 32606; calling 352-334-0911; or emailing info@nccer.org. More information may be found at our website, www.nccer.org.

HOW TO USE THIS ANNOTATED INSTRUCTOR'S GUIDE

Each page presents two sections of information. The larger section displays each page exactly as it appears in the Trainee Module. The narrow column ties suggested trainee and instructor actions to each page and provides icons (detailed below) to call your attention to material, safety, audiovisual, or testing requirements. The bottom of each page includes space for your notes.

The **Audiovisual** icon indicates an appropriate time to show a transparency or other audiovisual aid.

The **Classroom** icon prompts you to define a term, stress a point, ask trainees to explain a concept, or give examples.

The **Demonstration** icon directs you to show trainees how to perform tasks.

The **Examination** icon tells you to administer the written module examination.

The **Homework** icon is placed where you may wish to assign reading for the next class, assign a project, or advise trainees to prepare for an examination.

The **Laboratory** icon is used when trainees are to practice performing tasks.

The **Materials** icon is a reminder for you to gather materials needed for classes, labs, and testing.

The **Performance Testing** icon tells you to administer a performance test or a portion thereof.

The **Safety** icon is used to emphasize safety issues. It is often keyed to *Caution* and *Warning!* statements in the Trainee Module.

The **Teaching Tip** icon indicates additional guidance is available, such as how to conduct an exercise, get the most educational value from a field trip, or encourage class participation. Teaching Tips may expand on a feature (*Think About It, Did You Know?*) or provide *Quick Quizzes* or similar exercises. You will be referred to the Teaching Tips section at the back of the module if there is additional material.

The **Combination** icon indicates that the laboratory listed corresponds with a performance task. If desired, you can note the proficiency of the trainees during the laboratory, and use it to satisfy performance testing requirements.

PREPARATION

Before teaching this module, you should review the Objectives, Performance Tasks, Materials and Equipment List, and Module Outline. Be sure to allow ample time to prepare your own training or lesson plan and gather all required materials and equipment.

MODULE OVERVIEW

In this module, the trainee will learn to rig, move, and store motors properly. The trainee will also learn how to properly install the motor, and will gain a basic understanding of maintenance procedures. Because installation requires basic alignment to a driven machine, information on couplings and shaft alignment is also included.

PREREQUISITES

Prior to training with this module, it is recommended that the trainee shall have successfully completed *Core Curriculum; Millwright Level One; Millwright Level Two; Millwright Level Three, Millwright Level Four;* and *Millwright Level Five,* Modules 15501-09 through 15506-09.

OBJECTIVES

Upon completion of this module, the trainee will be able to do the following:

1. Explain the proper methods for motor storage.
2. Explain the proper rigging and handling of motors.
3. Determine if a motor has a thrust bearing or relies on electromagnetic force to determine rotor location.
4. Properly align the motor to the specified equipment.
5. Verify rotation and coupling gap.

PERFORMANCE TASKS

Under the supervision of the instructor, the trainee should be able to do the following:

1. Demonstrate proper storage methods for a motor.
2. Properly install a motor.

MATERIALS AND EQUIPMENT LIST

Overhead projector and screen

Transparencies

Blank acetate sheets

Transparency pens

Whiteboard/chalkboard

Markers/chalk

Pencils and scratch paper

Appropriate personal protective equipment

Wrenches

Allen wrenches

Bluing

Dial indicators

Assortment of different types of bearings

Tachometer

Ammeter

Feeler gauge

Thickness gauge

Straightedge

Drift punch

Oil, grease, and lubrication devices

Rags

Manufacturers' literature for various types of motors

Safety video or DVD, and appropriate devices for viewing, or online safety training

Photographs/illustrations of electric motors

Samples of AC and DC motors, including motors with damaged bearings, if possible

Appropriate rigging equipment for lifting motors

Copies of the Quick Quiz*

Module Examinations**

Performance Profile Sheets**

* Located at the back of this module.

** Located in the Test Booklet.

SAFETY CONSIDERATIONS

Ensure that the trainees are equipped with appropriate personal protective equipment and know how to use it properly. This module requires trainees to rig and lift motors for storage, and safely install a motor. Be sure trainees are briefed on site safety procedures.

ADDITIONAL RESOURCES

This module is intended to present thorough resources for task training. The following reference works are suggested for both instructors and motivated trainees interested in further study. These are optional materials for continued education rather than for task training.

R + W America L.P.
   www.rw-america.com

Coupling Corporation of America
   www.couplingcorp.com

TEACHING TIME FOR THIS MODULE

An outline for use in developing your lesson plan is presented below. Note that each Roman numeral in the outline equates to one session of instruction. Each session has a suggested time period of 2½ hours. This includes 10 minutes at the beginning of each session for administrative tasks and one 10-minute break during the session. Approximately 10 hours are suggested to cover *Installing Electric Motors*. You will need to adjust the time required for hands-on activity and testing based on your class size and resources. Because laboratories often correspond to Performance Tasks, the proficiency of the trainees may be noted during these exercises for Performance Testing purposes.

Topic	Planned Time
Session I. Introduction; Inspecting Equipment; Setting the Motor	
A. Introduction	_____
B. Inspecting Equipment	_____
C. Setting the Motor	_____
D. Laboratory	_____
Have trainees practice installing a motor. This laboratory corresponds to Performance Task 2.	
Session II. Motor Maintenance	
A. Motor Maintenance	_____
B. Practical Maintenance Techniques	_____
C. Motor Bearing Maintenance	_____
Session III. Lubrication; Troubleshooting; Storage; Recordkeeping	
A. Lubrication	_____
B. Troubleshooting	_____
C. Storing Motors	_____
D. Recordkeeping	_____
E. Laboratory	_____
Have trainees practice rigging and storing a motor. This laboratory corresponds to Performance Task 1.	

Session IV. Review and Testing

    A. Review   _____

    B. Module Examination   _____

        1. Trainees must score 70% or higher to receive recognition from NCCER.

        2. Record the testing results on Craft Training Report Form 200, and submit the results to the Training Program Sponsor.

    C. Performance Testing   _____

        1. Trainees must perform each task to the satisfaction of the instructor to receive recognition from NCCER. If applicable, proficiency noted during laboratory exercises can be used to satisfy the Performance Testing requirements.

        2. Record the testing results on Craft Training Report Form 200, and submit the results to the Training Program Sponsor.

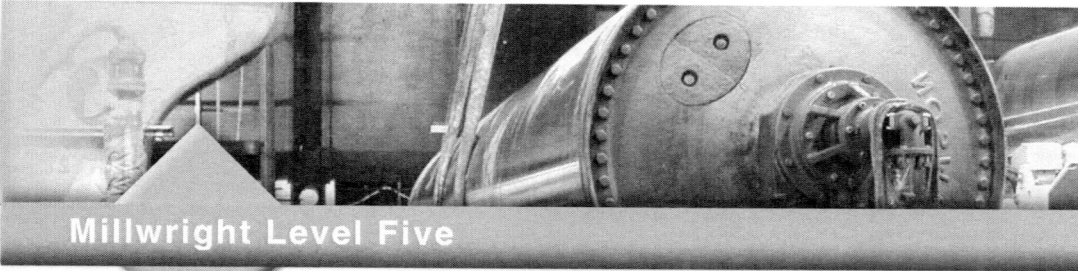

Millwright Level Five

15507-09

Installing Electric Motors

Assign reading of Module 15507-09.

15507-09
Installing Electric Motors

Topics to be presented in this module include:

Overview

Electric motors are commonly used to drive pumps, power tools, and heavy machinery. In this module, trainees will learn how to properly inspect, maintain, and install a motor. Because installation requires alignment to a driven machine, information on couplings and shaft alignment is also provided. Trainees will also learn how to properly rig, move, and store motors.

Instructor's Notes:

_____

_____

_____

_____

Objectives

When you have completed this module, you will be able to do the following:

1. Explain proper methods for motor storage.
2. Explain proper rigging and handling of motors.
3. Determine if a motor has a thrust bearing or relies on electromagnetic force to determine rotor location.
4. Properly align the motor to the specified equipment.
5. Verify rotation and coupling gap.

Trade Terms

Axial movement
Brush
Commutator

Required Trainee Materials

1. Pencil and paper
2. Appropriate personal protective equipment

Prerequisites

Before you begin this module, it is recommended that you successfully complete *Core Curriculum*; *Millwright Level One*; *Millwright Level Two*; *Millwright Level Three*; *Millwright Level Four*; and *Millwright Level Five*, Modules 15501-09 through 15506-09.

This course map shows all of the modules in the fifth level of the Millwright curriculum. The suggested training order begins at the bottom and proceeds up. Skill levels increase as you advance on the course map. The local Training Program Sponsor may adjust the training order.

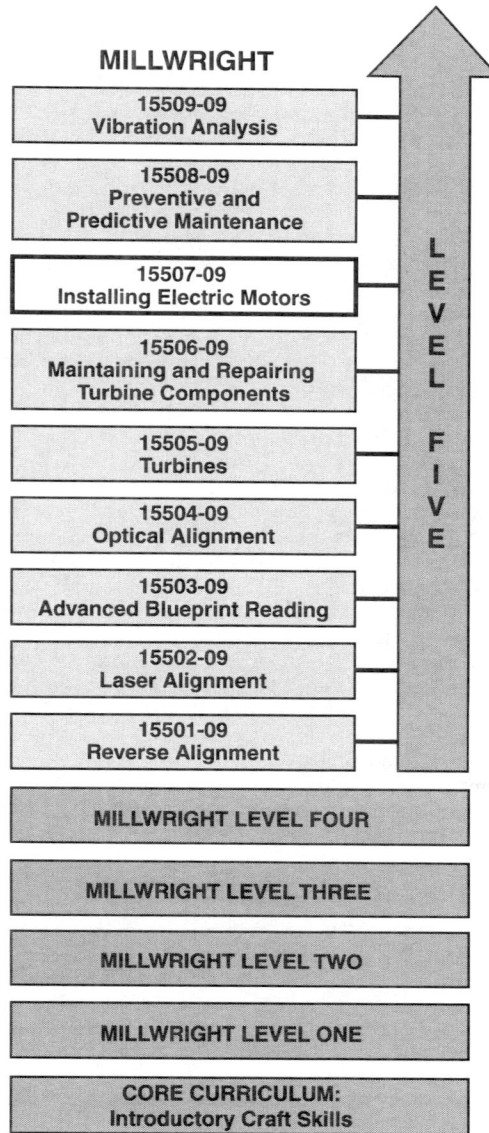

MILLWRIGHT

15509-09 Vibration Analysis
15508-09 Preventive and Predictive Maintenance
15507-09 Installing Electric Motors
15506-09 Maintaining and Repairing Turbine Components
15505-09 Turbines
15504-09 Optical Alignment
15503-09 Advanced Blueprint Reading
15502-09 Laser Alignment
15501-09 Reverse Alignment

LEVEL FIVE

MILLWRIGHT LEVEL FOUR
MILLWRIGHT LEVEL THREE
MILLWRIGHT LEVEL TWO
MILLWRIGHT LEVEL ONE
CORE CURRICULUM: Introductory Craft Skills

507CMAP.EPS

Ensure that you have everything required to teach the course. Check the Materials and Equipment List at the front of this module.

See the general Teaching Tip at the end of this module.

Explain that terms shown in bold are defined in the Glossary at the back of this module.

Show Transparency 1, Objectives, and Transparency 2, Performance Tasks. Review the goals of the module, and explain what will be expected of the trainees.

Review the modules covered in Level Five and explain how this module fits in.

1.0.0 ◆ INTRODUCTION

There are two basic types of electric motors, AC and DC. AC motors include the ordinary induction motor used in washing machines and table saws, as well as the three-phase motors that drive large machinery. The AC induction motor is the simplest and most robust motor commercially available.

DC motors range in size from very small to very large. Small servomotors are often brushless DC motors, because this type does not produce carbon dust from the friction of the spinning **commutator** and the **brush**. These types of motors are used in CD and DVD players, and in some computer applications.

Industrial applications use DC motors because of the ability to control torque and speed to almost any specification. The DC motor can control speed smoothly at any required rate, and can also be reversed or used for braking. However, the maintenance requirements of the brushes and commutator make the DC motor more demanding of maintenance than the AC induction motor.

Variable frequency drives (VFDs) are now more commonly used than DC drives. The VFD controls the rotational speed of an AC motor by controlling the frequency of the current supplied to the motor. This means that less energy is used when there is less energy demand, saving both cost and energy. It retains the low maintenance characteristics of the AC drive, while allowing the motor output to be controlled. The VFD unit itself is a control box driving an AC motor.

2.0.0 ◆ INSPECTING EQUIPMENT

Prealignment is performed when the equipment is originally set in place. Before installing and prealigning a piece of equipment, the equipment must be inspected to ensure that it is suitable for service. Follow these steps to inspect a STAT and motor before installation:

Step 1 Inspect the STAT and motor for physical damage.

Step 2 Rotate the STAT and motor by hand to ensure that they are not locked up or binding, that there are no flat places in the bearings, and that the shafts are turning true.

NOTE

The shafts of the machines should turn smoothly. If there are flat places in the bearings, you can feel a roughness as you rotate the shaft. Rough places in the bearing can be caused by many things, including improperly attached welding grounds, or improperly installed metal bearings.

Step 3 Inspect the equipment baseplate to ensure that it is not warped or damaged.

A motor can be damaged by condensation on the windings. If the motor has been brought from a cold location to a warm one, do not unpack the motor until the motor reaches room temperature to prevent condensation on the windings. When the motor reaches room temperature, remove all of the protective wrappings before installation. Many motors are now supplied with built-in stator heaters, which can be plugged in during storage or while offline.

3.0.0 ◆ SETTING THE MOTOR

Once the STAT is set and bolted, the motor is ready to be set. The motor must be set and roughly aligned with the STAT before drilling and tapping the motor bolt holes into the baseplate. Follow these steps to set the motor:

Step 1 Place the motor on the baseplate, aligning it with the lengthwise center line.

Step 2 Move the motor toward the STAT until the couplings are close enough together to set the gap between the couplings. If there is a key on the shaft, it must fill half the keyway outside the coupling.

Setting a motor up for operation requires a series of checks to be sure that the motor is not damaged or improperly installed. These include the following:

• Make sure all packaging and bracing has been removed from the motor and shaft.

• Rotate the shaft by hand, to be sure it turns freely.

• Replace any panels or covers that may have been removed during installation. Follow all standard procedures for starting up powered equipment.

Instructor's Notes:

_____

_____

_____

- Turn the machine on momentarily to determine whether the direction of rotation is correct.
- If the direction of rotation is incorrect, the electrician will be required to reverse the connections, and the direction of rotation should be rechecked.
- Start the motor and ensure that it operates smoothly, without noise or vibration. If the motor runs smoothly and quietly, operate the motor for approximately an hour with no load.

Prior to having the electrician hook up the wiring, use the following steps as preparation to couple and align the machines:

Step 1 Set the gap between the couplings using a thickness gauge.

Step 2 Align the couplings as closely as possible by eyesight. The gap setting may vary depending on the type of coupling used. Refer to the coupling manufacturer's specifications for the proper gap setting.

Step 3 Place a straightedge on the side of the couplings.

Step 4 Measure the offset misalignment using a thickness gauge (see *Figure 1*).

Step 5 Move the motor to correct the offset misalignment.

Step 6 Attach a dial indicator to the motor coupling.

Step 7 Set the dial indicator on the face of the driver coupling at the 3-o'clock position and zero it.

Step 8 Turn the coupling until the indicator is on the 9-o'clock face of the driven coupling. Adjust the motor until the indicator reads zero in both positions.

Step 9 Set the dial indicator at the 3-o'clock position on the rim of the driver coupling.

Step 10 Set the dial indicator to read zero.

Step 11 Rotate the couplings until the dial indicator is at the 9-o'clock position on the driver half of the coupling.

Step 12 Adjust the motor and take readings at the 3-o'clock and the 9-o'clock positions until zero readings on both sides of the coupling are achieved.

Step 13 Recheck the angularity.

Step 14 Use a transfer punch to transfer the bolt holes.

NOTE

The offset misalignment of the couplings must be minimal before the motor hold-down bolt holes are drilled and tapped, because once the bolt holes are drilled, there is very little side-to-side movement of the motor. After the offset has been corrected as much as possible with a straightedge and thickness gauge, use a dial indicator to further align the coupling from side to side.

Explain how to mark the EMC. Emphasize that the shaft must be set to the scribed line before doing the alignment.

Describe how to perform prealignment with a straightedge and thickness gauge.

Show Transparency 3 (Figure 1).

Show trainees how to prepare to couple and align equipment. Demonstrate the procedure for prealigning and setting a motor.

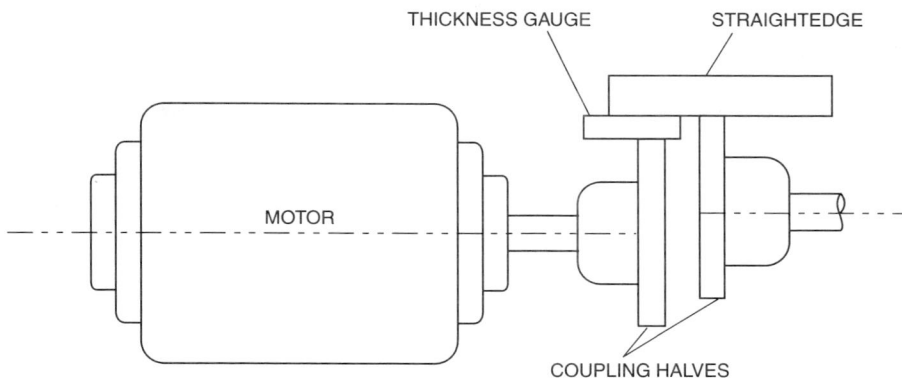

THICKNESS GAUGE   STRAIGHTEDGE

MOTOR

COUPLING HALVES

507F01.EPS

Figure 1 ◆ Measuring horizontal offset.

NOTE

If you are able to use a transfer punch, you can skip Steps 17 and 18.

Step 15 If transfer punches cannot be used, scribe the outline of the holes carefully.

Step 16 Remove the motor.

Step 17 Find the hole center.

Step 18 Punch mark the hole centers.

CAUTION

Ensure that the baseplate holes are clearly marked before moving the motor, or the entire alignment procedure will have to be repeated.

Step 19 Drill and tap the baseplate holes.

Step 20 Place the motor back onto the baseplate.

Step 21 Install the motor mounting bolts.

4.0.0 ◆ MOTOR MAINTENANCE

AC motor failure accounts for a high percentage of electrical repair work. The care given to an electric motor while it is being stored and operated affects the life and usefulness of the motor. A motor that receives good maintenance will outlast a poorly maintained motor many times over. A motor that is installed correctly and has been properly selected for the job will require very little maintenance, provided it is cleaned and lubricated at regular intervals.

The frequency for cleaning an AC motor depends on the type of environment in which it is used. In general, keep both the interior and exterior of the motor free from dirt, water, oil, and grease. Motors operating in dirty areas should be periodically disassembled and thoroughly cleaned.

For motors that are totally enclosed (fan-cooled or nonventilated) that are equipped with automatic drain plugs (*Figure 2*), care must be taken to keep the plugs free of oil, grease, paint, grit, and dirt so they do not become clogged.

Most motors are properly lubricated at the time of manufacture, and it is not necessary to lubricate them again on installation. However, if a motor has been in storage for a period of six months or longer, it should be relubricated before it is started.

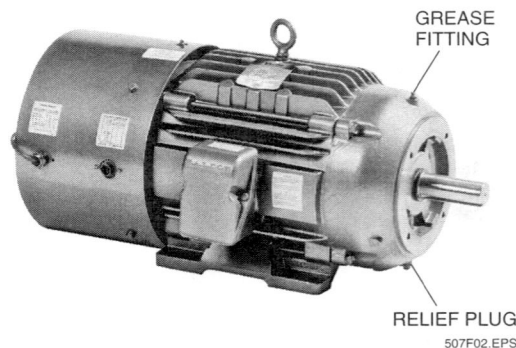

Figure 2 ◆ Totally enclosed fan-cooled motor.

Follow these steps to lubricate conventional motors with ball bearings:

Step 1 Stop the motor.

Step 2 Wipe clean all grease fittings (filler and drain).

Step 3 Remove the filler and drain plugs (*Figure 3*).

Step 4 Free the drain hole of any hard grease, using a piece of wire if necessary.

Step 5 Add grease using a low-pressure grease gun.

Step 6 Start the motor and let it run for approximately 30-40 minutes.

Step 7 Stop the motor, wipe off any drained grease, and replace the filler and drain plugs.

Step 8 The motor is now ready for operation.

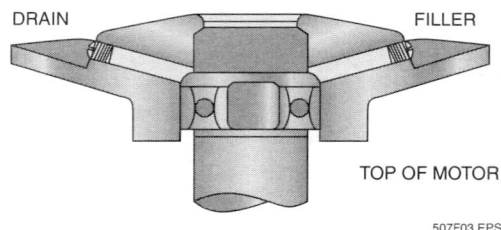

Figure 3 ◆ Location of motor filler and drain plugs.

Instructor's Notes:

_____

_____

_____

Every four years (or every year, in the case of severe duty), motors with open bearings should be thoroughly cleaned, washed, and repacked with grease. *Table 1* shows the relubrication periods for various motors. Standard conditions mean operation of 8 hours per day, normal or light loading, at 100°F maximum ambient temperature. Severe conditions represent operation of 24 hours per day, shock loadings, vibration, dirty or dusty areas, or areas of 100°F to 150°F ambient temperature. Extreme conditions are defined as heavy shock or vibration, large amounts of dirt or dust, or high ambient temperatures (above 150°F).

Bearings provide minimum resistance and align the motor rotor during operation. Therefore, the quantity of grease is important for proper bearing operation. The grease cavity should be filled one-third to one-half full. Always remember that too much grease is as detrimental as insufficient grease. *Table 2* shows the amount of grease required and *Table 3* gives the recommended grease for a Class B or F motor. However, always check with the motor manufacturer or the customer for their recommendation and specifications for greasing motors.

> **CAUTION**
> The amount of grease added to motor bearings is very important. Only enough grease should be added to replace the grease used by the bearings. Too much grease can be as harmful as insufficient grease.

After the coupling is bolted up and is ready to run, the motor should be run with no load on the driven machine for approximately one hour to test for vibration or other symptoms of problems.

5.0.0 ◆ PRACTICAL MAINTENANCE TECHNIQUES

Once the motor has been sized and installed properly, the key to long, trouble-free motor life is proper maintenance. Maintaining a motor in good operating condition requires periodic inspection to determine if any faults exist, and then promptly correcting these faults. The frequency and thoroughness of these inspections depend on factors such as:

- Number of hours and days the motor operates
- Importance of the motor in the production scheme
- Nature of service
- Environmental conditions

Each week, every motor in operation should be inspected to see if the windings are exposed to any dripping water, acid, or alcohol fumes. Also look for excessive dust, chips, or lint on or near the motor. Make certain that objects that will cause problems with the motor's ventilation system are not placed too close to the motor and that nothing is making direct contact with the motor's moving parts.

In sleeve-bearing motors, check the oil level frequently (at least once a week) and fill the oil cups

Show Transparency 5 (Table 1). Define standard, severe, and extreme operating conditions.

Explain how to grease bearings. Note that too much grease is as harmful as insufficient grease.

Show Transparencies 6 and 7 (Tables 2 and 3).

Discuss the importance of proper maintenance and regular periodic inspections.

Describe motor inspections that should be performed weekly.

Frame Size	Relubrication Period		
900 rpm, 1,200 rpm, and Variable Speed	**Standard Conditions**	**Severe Conditions**	**Extreme Conditions**
140–180	4.5 years	18 months	9 months
210–280	4 years	16 months	8 months
320–400	3.5 years	14 months	7 months
440–508	3 years	12 months	6 months
510	2.5 years	11.5 months	6 months
1,800 rpm	**Standard Conditions**	**Severe Conditions**	**Extreme Conditions**
140–180	3 years	1 year	6 months
210–280	2.5 years	10.5 months	5.5 months
320–400	2 years	9 months	4.5 months
440–508	1.5 years	8 months	4 months
510	1 year	6 months	3.5 months
Over 1,800 rpm	6 months	3 months	3 months

Table 2 Typical Amount of Grease Required When Regreasing Electric Motors

Bearing Number	Amount in Cubic Inches	Approximate Equivalent Teaspoons
203	0.15	0.5
205	0.27	0.9
206	0.34	1.1
207	0.43	1.4
208	0.52	1.7
209	0.61	2.0
210	0.72	2.4
212	0.95	3.1
213	1.07	3.6
216	1.49	4.9
219	2.8	7.2
222	3.0	10.0
307	0.53	1.8
308	0.66	2.2
309	0.81	2.7
310	0.97	3.2
311	1.14	3.8
312	1.33	4.4
313	1.54	5.1
314	1.76	5.9
316	2.24	7.4
318	2.78	9.2

to the specified line with the recommended lubricant. If the journal or motor shaft diameter is less than 2 inches, always stop the motor before checking the oil level. For special lubricating systems such as forced, flood-and-disc, and wool-packed lubrication, follow the manufacturer's recommendations. Oil should be added to the bearing housing only when the motor is stopped; make sure that no oil creeps along the shaft toward the windings, where it may harm the insulation.

Always be alert to any unusual noise, which may be caused by metal-to-metal contact such as bad bearings. Learn to detect any abnormal odor, which might indicate scorching insulation varnish. Feel the bearing housing each week for evidence of excess heat and vibration.

WARNING!
Before performing any motor maintenance procedures other than external visual inspection or noise monitoring, always lock out and tag the equipment according to approved procedures.

The air gap on sleeve-bearing motors should be checked frequently, especially if the motor has recently been rewound or otherwise repaired. After new bearings have been installed, for example, make sure that the average reading is within specified tolerances. Check the air passages through punchings and make sure they are free of all foreign matter.

WARNING!
When using compressed air, exercise caution by wearing appropriate personal protective equipment and using an airflow tip or pulse nozzle. Excess air pressure can result in personnel injury and cause damage to the motor windings.

Always disconnect the power, and lock out/tag the equipment according to site safety procedures prior to performing any motor maintenance tasks. Check other motor parts and accessories such as belts, gears, couplings, chains, and sprockets for excessive wear or misalignment. Inspect the motor starter and verify that the motor reaches the proper speed each time it is started. To check the rotating speed of motors, use a tachometer.

Where motors having ball or roller bearings are exposed to extreme conditions or constant usage, bearings should be serviced at least monthly by purging out the old grease through the drain hole and applying new grease. After changing the grease in these bearings, inspect the bearing housing for leaking grease. If present, correct the leakage prior to starting the motor.

Table 3 Recommended Grease for Motor Lubrication

Insulation Class Shown on Nameplate	Grease Designation	Grease Supplier
B or F	Chevron SRI-2	Standard Oil of California or equivalent

7.6 MILLWRIGHT ◆ LEVEL FIVE

Instructor's Notes:

_____

_____

_____

On motors with sleeve bearings, check the sleeve bearings for wear at least once every two months. Clean out the oil wells if there is evidence of dirt or sludge. Flush with lighter oil before refilling with the specified bearing lubricant.

On motors with gears, open the drain plug and check the oil for the presence of metal scale, sand, grit, or water. If any of these conditions are present, drain, flush, and refill as recommended by the motor manufacturer. Check the rotor for slack or backlash by carefully rocking the rotor.

Loads being driven by motors have a tendency to change from time to time due to wear on the machine or the product being processed through the machine. Therefore, all loads should be checked periodically for any changed conditions, bad adjustment, and poor handling or control.

During the monthly inspection, check the tightness of all belts and adjust them, as necessary. Check the belts to verify that they are running steadily and near the inside edge of the pulley. On chain-driven machines, check the chain for evidence of wear and stretch, and clean the chain thoroughly. Check the chain lubricating system and note the incline of the slanting base to make sure it does not cause oil rings to rub on the housing.

Motor loads should be reevaluated from time to time, as they can vary for several reasons. Have a qualified person use an ammeter to take an ampere reading on the motor, first with no load, then at full load, and finally through a full cycle of no load to full load. This test should provide information concerning the mechanical condition of the driven machine (load).

Without proper maintenance, no motor can be expected to perform well for any length of time or remain in service as long as it should. Although motor maintenance is costly, it is less expensive than continually replacing or overhauling motors.

6.0.0 ◆ MOTOR BEARING MAINTENANCE

AC motors account for a large percentage of industrial maintenance and repair, with many motor failures caused by faulty bearings. Consequently, most industrial facilities place great emphasis on the proper care of motor bearings. Electric motors last much longer and perform better when a carefully planned motor lubrication schedule is followed.

If an AC motor failure does occur, the first step is to find out why the motor failed. There are various causes of motor failures, including excessive load, binding or misalignment of motor drives, wet or dirty environments, and bearing failures.

Bearing failures can occur in newer motors with high-quality bearings as often as in older motors with less reliable bearings if the bearings are not maintained properly. A notable exception in bearing failures may be found in motors equipped with sealed bearings, which are much less prone to failure.

6.1.0 Types of Bearings

Ball bearings are the most common type of bearings used in the construction of electric motors. This type of bearing is found on various sizes of motors, with bearing design types including:

- Open
- Single-shielded
- Double-shielded
- Sealed
- Double-row and other special types

Open bearings, as the name implies, are open construction and must be installed in a sealed housing. These bearings are less apt to cause churning of grease and are therefore used mostly on large motors.

The single-shield bearing has a shield on one side to keep grease from the motor windings. Double-shielded bearings have a shield on both sides of the bearing. This type of bearing is less susceptible to contamination and, because of its design, reduces the possibility of over-greasing. Sealed bearings have a double shield on each side of the bearing, which forms an excellent seal. This bearing requires no maintenance, affords protection from contamination at all times, and does not require regreasing. It is normally used on small or medium motor sizes.

Many large motors are furnished with oil-ring sleeve bearings, while some of the smaller fractional-horsepower motors are equipped with simple sleeve bearings without oil rings.

Each bearing type has characteristics that are suited for a particular application. Replacement of bearings should be made using the same type of bearing originally installed in the motor or equipment. *Figure 4* shows several types of bearings used in electric motors. The following list serves as a basic overview of bearing applications and a guide to analyzing bearing failures:

- *Self-aligning ball bearing* – The self-aligning ball bearing has two rows of balls rolling on the spherical surface of the outer ring, and this design compensates for angular misalignment due to errors in mounting, shaft deflection, or distortion of the foundation. This design also

Discuss the maintenance requirements for sleeve bearings and gears.

Describe the required maintenance for belts and chains.

Demonstrate and/or describe how to take an ampere reading.

Discuss causes of AC motor failures, including faulty bearings.

Introduce the different types of bearings. Describe the characteristics of each type.

Show Transparency 8 (Figure 4).

Provide examples of different types of bearings for trainees to examine.

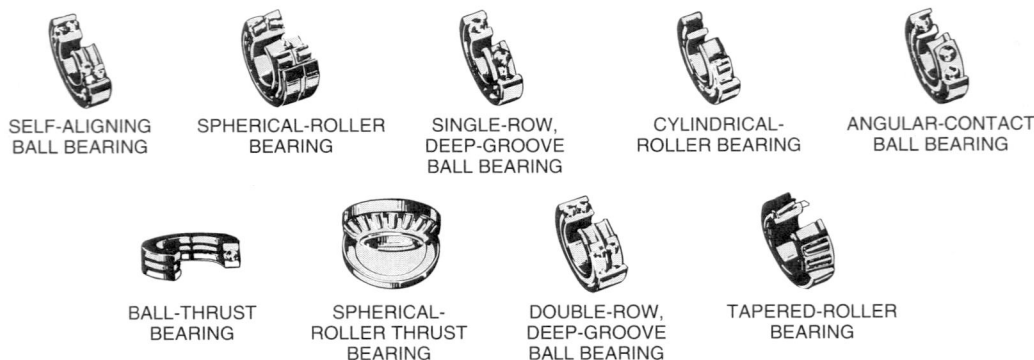

SELF-ALIGNING BALL BEARING

SPHERICAL-ROLLER BEARING

SINGLE-ROW, DEEP-GROOVE BALL BEARING

CYLINDRICAL-ROLLER BEARING

ANGULAR-CONTACT BALL BEARING

BALL-THRUST BEARING

SPHERICAL-ROLLER THRUST BEARING

DOUBLE-ROW, DEEP-GROOVE BALL BEARING

TAPERED-ROLLER BEARING

507F04.EPS

Figure 4 ◆ Various types of bearings.

prevents any exertion of bending influence on the motor shaft—a most important consideration in applications requiring extreme accuracy at high speeds. Self-aligning ball bearings are used for radial loads and moderate thrust loads in either direction.

- *Single-row, deep-groove ball bearing* – The single-row, deep-groove ball bearing will sustain, in addition to radial load, a substantial thrust load in either direction, even at very high speeds. The ability to sustain these loads is made possible by the close-tolerance contact that exists between the roller balls and the continuous groove in each ring. Accurate alignment between the motor shaft and housing is essential in the application of this type of bearing. The single-row, deep-groove bearing is also available with seals and shield, which provide protection from contamination and retain lubricant.

- *Angular-contact ball bearings* – The angular-contact ball bearing can support a substantial thrust load in one direction, combined with a moderate radial load. A steep contact angle assures the highest thrust capacity and axial rigidity. This characteristic is obtained through the addition of a thrust-supporting shoulder on the inner ring, with a similar high shoulder on the opposite side of the ring. These bearings can be mounted singly, or in special applications in tandem, to allow constant thrust in one direction. They can also be mounted in pairs if the sides of the bearings have been ground to a flush finish. This installation provides for a combined load, either face-to-face or back-to-back.

- *Double-row, deep-groove ball bearing* – The double-row, deep-groove ball bearing embodies the same principle of design as the single-row bearing. However, this bearing has a lower axial displacement than occurs in the single-row design, substantial thrust capacity in either direction, and high radial capacity due to the two rows of balls.

- *Spherical-roller bearing* – The exceptional capacity of the spherical-roller bearing can be attributed to the number, size, and shape of the rollers, as well as the accuracy by which they are guided. Since the bearing is inherently self-aligning, angular misalignment between the shaft and housing has no detrimental effect on the application of the bearing. The design and proportion of the bearing are such that both thrust loads and radial loads may be carried in either direction.

- *Cylindrical-roller bearing* – This type of bearing has a high radial capacity, which provides accurate guiding of the rollers, and a close approach to true rolling. The low friction permits operation at high speeds. Cylindrical-roller bearings having flanges on one ring also allow a limited free **axial movement** of the shaft in relation to the housing, and they are easy to dismount even when both rings are mounted within a close tolerance to one another. The double-row type is particularly suitable for machine-tool spindles.

- *Ball-thrust bearing* – The ball-thrust bearing is designed for thrust load in one direction only. The load line through the balls in parallel to the axis of the shaft results in high thrust capacity and minimum axial deflection. Flat bearing seats are essential for heavy loads or for close axial positioning of the shaft.

7.8 MILLWRIGHT ◆ LEVEL FIVE

Instructor's Notes:

_____

_____

_____

- *Spherical-roller thrust bearing* – The spherical-roller thrust bearing is designed to carry heavy thrust loads or combined loads that are predominantly thrust. This bearing has a single row of rollers which roll on a spherical outer race with full self-alignment. The cage, centered by an inner ring sleeve, is constructed so that lubricant is pumped directly against the inner ring's unusually high guide flange. This bearing operates best with relatively heavy oil lubrication.
- *Tapered-roller bearings* – Since the axes of the rollers and raceways of a tapered-roller bearing form an angle with the shaft angle, the tapered-roller bearing is especially suitable for carrying coordinated radial and axial loads. This type of bearing is typically installed adjacent to another bearing capable of carrying thrust loads in the opposite direction. Tapered roller bearings are designed so that their cone (inner ring) and roller/cup assembly (outer ring) are mounted separately.

Recommendations for ball bearing assembly, maintenance, inspection, and lubrication are shown in *Table 4*. Refer to this list often when working with electric motors. Always check the manufacturer's recommendations and local requirements.

Show Transparency 9 (Table 4). Discuss the "Dos" and "Do Nots" of bearing maintenance work.

Assign reading of Sections 6.2.0–9.0.0.

Table 4 Ball Bearing Assembly, Maintenance, and Lubrication Recommendations

DO	DO NOT
DO work with clean tools in clean surroundings.	DO NOT work with poor tools, a cluttered workbench, or dirty surroundings.
DO remove all outside dirt from the housing before exposing the bearing.	DO NOT handle bearings with dirty or moist hands.
DO treat a used bearing as carefully as a new one.	DO NOT spin uncleaned bearings.
DO use clean solvents and flushing oils.	DO NOT spin any bearings using compressed air.
DO lay bearings out on clean paper or cloth.	DO NOT use the same container to clean and rinse the bearings.
DO protect disassembled bearings from dirt and moisture.	DO NOT scratch or nick the bearing surfaces.
DO use clean, lint-free rags to wipe bearings.	DO NOT remove grease or oil from new bearings.
DO keep bearings wrapped in oil-proof paper when not in use.	DO NOT use the incorrect type or amount of lubricant.
DO clean the outside of the housing before replacing the bearings.	DO NOT use a bearing as a measuring tool to check the housing bore or shaft fit.
DO keep bearing lubricants clean when applying and cover containers when not in use.	DO NOT install a bearing on a shaft that shows excessive wear.
DO be sure the shaft size is within the specified tolerances recommended for the bearing.	DO NOT open the bearing carton until the bearing is ready to be installed.
DO store bearings in their original unopened cartons in a dry place.	DO NOT determine the condition of a bearing until it has been properly cleaned.
DO use a clean, short-bristle brush with firmly embedded bristles to remove dirt, scale, or chips.	DO NOT tap directly on a bearing or ring during installation.
DO be certain that, when installed, the bearing is square with and held firmly against the shaft shoulder.	DO NOT overfill during lubrication. Excess oil or grease may enter the motor housing. Too much lubricant will also cause overheating, particularly where bearings operate at high speeds.
DO follow lubricating instructions supplied with the machinery. Use only grease where grease is specified; use only oil where oil is specified. Be sure to use the exact kind of lubricant called for.	DO NOT allow motors to remain idle for long periods of time without rotating their shafts periodically.
DO handle grease with clean paddles or grease guns. Store grease in clean containers. Keep grease containers covered.	

Ensure that you have everything required for teaching this session.

Discuss factors that determine the frequency of motor lubrication.

Describe the lubrication procedure for a ball bearing motor. Stress the need to maintain cleanliness.

Provide motor manufacturers' manuals for trainees to examine. Point out the lubrication recommendations in the manuals.

Discuss methods and procedures for checking bearings.

Explain how to troubleshoot a motor that will not start.

6.2.0 Frequency of Lubrication

The frequency of motor lubrication depends not only on the type of bearing, but also on the motor application. Small- to medium-size motors equipped with ball bearings (except sealed bearings) should be greased every three to six years if the motor duty is normal. On severe applications (high temperature, wet or dirty locations, or corrosive atmospheres), lubrication may be required more often. In severe applications, past experience and condition of the grease are the best guides to the frequency of lubrication.

The lubrication in sleeve bearings should be changed at least once a year or more often when the motor duty is severe or the oil appears dirty.

6.2.1 Lubrication Procedure

Before lubricating a ball bearing motor, the bearing housing, grease gun, and fittings should be cleaned. Care must be exercised to keep out dirt and debris during lubrication. The relief plug should be removed from the bottom of the bearing housing. This is done to prevent excessive pressure from building up inside the bearing housing during lubrication. Install lubrication while the motor is not running. The motor should then be allowed to run for approximately 30 to 40 minutes to expel any excess grease. The relief plug should then be reinstalled and the bearing housing cleaned.

It is important to avoid overlubrication. When excessive grease is forced into a bearing, a churning of the grease may occur, resulting in high bearing temperatures and eventual bearing failure.

On motors that do not have a relief plug, grease should be applied slowly and sparingly. If possible, disassemble the motor and repack the bearing with the proper amount of grease. Always maintain cleanliness during lubrication.

When lubricating sleeve-type bearings, use only the type and amount of lubrication recommended by the manufacturer.

6.3.0 Testing Bearings

Two simple methods commonly used to check bearings during motor operation are touching and listening. If the bearing housing feels unusually hot, or if a growling or grinding sound is being emitted from the area of the bearings, one of the bearings is probably nearing failure. Special stethoscopes are also available for listening to bearings while the motor is running.

> **NOTE**
> Keep in mind that some bearings may safely operate in a higher temperature range than other bearings, even in ranges exceeding 185°C. Check the motor manufacturer's specifications.

Using a feeler gauge, periodically check the air gap on sleeve-bearing motors for bearing wear. Four separate measurements should be taken approximately 90 degrees apart, around the diameter of the rotor. These measurements should be recorded and compared with previous measurements to determine if any deviations are present, which may indicate bearing wear since the last measurements were recorded.

Motors should also be checked for end play, which is the backward and forward movement in the shaft. Ball bearing motors typically will have $\frac{1}{32}$" to $\frac{1}{16}$" of end movement. Sleeve-bearing motors may have up to $\frac{1}{2}$" of end movement.

On larger sleeve bearings, the oil level should be checked periodically and the oil visually inspected for contamination. If safely possible, the oil rings should be checked while the motor is running.

Other inspections may include periodically checking for misalignment or bent shafts, and for excessive belt tension.

7.0.0 ◆ TROUBLESHOOTING MOTORS

If a split-phase motor fails to start, the trouble may be due to one or more of the following faults:

- Tight or frozen bearings
- Worn bearings, allowing the rotor to drag on the stator
- Bent rotor shaft
- One or both bearings out of alignment

Tight or worn bearings may be caused by a failing bearing lubricating system. New bearings sometimes fail if the motor shaft is not kept properly lubricated.

The rotor may not start if the bearings are worn to such an extent that they allow the rotor to drag on the stator. When this condition exists, it can generally be detected by noticeable bright spots on the inside of the stator laminations where they have been rubbed by the dragging rotor.

Instructor's Notes:
_____
_____
_____

A bent rotor shaft will usually cause the rotor to bind in a certain position and then run freely until it returns to that position. An accurate test for a bent shaft can be made by placing the rotor between centers on a lathe and turning the rotor slowly while a tool or marker is held in the tool post close to the surface of the rotor. If the rotor wobbles, it is an indication of a bent shaft.

Uneven tightening of the end shield plates can cause bearings to get out of alignment. When placing end shields or brackets on a motor, the bolts should be tightened alternately, first drawing up two bolts that are directly opposite one another. These two should be drawn up only a few turns and then the others tightened an equal amount all the way around. When the end shields are drawn up as far as possible with the bolts, they should be tapped tightly against the frame with a mallet and the bolts tightened again. Many motor manufacturers specify a bolt-tightening sequence to be applied when reassembling a motor.

8.0.0 ◆ STORING MOTORS

Motors must be stored properly when the project on which they are to be used is not complete. Spare motors also must be stored properly. Industrial installations often keep spare motors as back-ups.

The first consideration when storing motors for any length of time is the location in which they are to be stored. A dry location, and one that does not undergo severe changes in temperature over a 24-hour period, should be selected whenever possible. When the ambient temperature changes frequently, condensation is likely to form on stored motors. Moisture in motor insulation can cause motors to fail upon startup; therefore, guarding against moisture is vital when storing motors of any type.

Large motors commonly have heaters built in. If the motor does have a heater and the motor is to be stored in cold weather, the heater should be wired up by a qualified person to prevent condensation problems.

A means for transporting the motor from the place of storage to the place where it will be used, or for shifting it around in the storage area, is also important. Motors should not be lifted by their rotating shafts. Doing so can damage the alignment of the rotor in relationship to the stator. Even picking up the smaller fractional horsepower motors by the shaft is not recommended. Many workers have received bad cuts from the sharp keyways on motor shafts when they tried to pick them up with their bare hands.

> **WARNING!**
> Never handle a motor by its shaft without proper hand protection. Motor shaft keyways have sharp edges that can cause severe cuts.

When an electric motor is received on a job site, always follow the manufacturer's recommendations for unloading, uncrating, and installing the motor. Failure to follow these recommendations can cause injury to personnel and possible damage to the motor.

Once the motor has been uncrated, check for damage that might have occurred during shipment. Check the motor shaft to verify that it turns freely. This is also an appropriate time to clean the motor of any debris, dust, moisture, or any foreign matter that might have accumulated during shipment.

> **NOTE**
> Motors in storage should have their shafts turned by hand at least once a month or as specified to redistribute the grease in the bearings.

Clean the motor of any debris, dust, moisture, or any foreign matter before putting it into service.

> **WARNING!**
> Never start a wet or damp motor.

Motors are provided with a lug, or lugs, with which to rig them. If there are a number of lugs, it is good practice to use a spreader beam to rig for the lift to keep the motor level. Do not lift by the hood or by the shaft, as you might damage the hood or bend the shaft. You may damage the bearings if you lift by the shaft.

Eyebolts on motors are intended for lifting the motor housing. These lifting devices should never be used when lifting or handling the motor when the motor is attached to other equipment.

The eyebolt lifting capacity is based on a lifting alignment that corresponds to the eyebolt center line. The eyebolt capacity lessens as deviation from this alignment increases.

Discuss considerations and procedures for storing motors.

Demonstrate and/or describe methods for safely moving and transporting a motor.

Explain how to inspect an electric motor when it is received at the job site.

Demonstrate and/or describe how to safely lift a motor using appropriate rigging equipment.

MODULE 15507-09 ◆ INSTALLING ELECTRIC MOTORS 7.11

Review the guidelines for storing motors for a long period of time.

Under your supervision, have trainees practice moving and storing a motor. Note the proficiency of each trainee. This laboratory corresponds to Performance Task 1.

Explain the importance of keeping accurate maintenance records.

Review the list of items that should be documented.

See the Teaching Tips for Sections 1.0.0–9.0.0 at the end of this module.

The following is a list of procedures that should be followed when storing motors for any length of time:

- Make sure motors are kept clean.
- Make sure motors are kept dry.
- Supply supplemental heating in the storage area, if necessary.
- Motors should be stored in an orderly fashion (i.e., grouped by horsepower, etc.).
- Motor shafts should be rotated periodically.
- Lubrication should be checked periodically.
- Protect shafts and keyways during storage and also while transporting motors from one location to another.

9.0.0 ◆ RECORDKEEPING

One of the first steps in establishing a reliable maintenance program is preparing accurate records. As a minimum, records on each motor should include the following:

- A complete description, including age and nameplate data
- Location and application, keeping such notations up to date if motors are transferred to different areas or used for different purposes
- Notations of scheduled preventive maintenance and previous repair work performed
- Location of duplicate or interchangeable motors
- An estimate of the motor's importance in the production process to which it relates

Instructor's Notes:

_____

_____

_____

Review Questions

1. Before prealigning a piece of equipment, the equipment must be _____.
   a. plugged in
   b. inspected
   c. connected to the other pieces
   d. roughly aligned

2. Motors are checked to ensure they are not binding by _____.
   a. plugging them in
   b. looking at the shaft
   c. turning them by hand
   d. turning them with a wrench

3. Rough places in a motor bearing may be caused by _____.
   a. turning them by hand
   b. too much lubricant
   c. improper welding grounds
   d. interference-fit

4. The key should fill _____.
   a. the keyway completely
   b. half of the keyway
   c. the coupling
   d. the cam

5. Punch mark the bolt hole locations through the _____.
   a. baseplate holes
   b. MTBM bolt holes
   c. flange bolt holes
   d. grout

6. Once the STAT has been set, the motor must be _____.
   a. parallel
   b. bolted down
   c. coupled
   d. aligned

7. The motor must be aligned with the crosswise center line.
   a. True
   b. False

8. Move the motor close enough to the STAT to set the _____.
   a. coupling gap
   b. offset
   c. boltholes
   d. shims

9. To find the EMC, push the shaft away toward the couplings until it stops, blue the shaft, energize the motor, and _____.
   a. mark where the coupling touches the shaft
   b. see whether the shaft moves back or forward
   c. scribe the shaft along the edge of the front bearing
   d. try to hold the shaft back

10. If the shaft is not set to the operating EMC before aligning the couplings, the coupling halves may _____.
    a. be offset to one side or the other
    b. show angularity as a consequence
    c. move up and down
    d. bump into each other

11. Set the coupling gap using a _____.
    a. HI-LO gauge
    b. micrometer
    c. thickness gauge
    d. taper gauge

12. Which of the following provides minimum resistance and aligns the motor rotor while turning?
    a. Compensator
    b. Brushes
    c. End bells
    d. Bearings

13. Which of the following is true concerning lubricating motor bearings?
    a. Always add a little more grease than is needed.
    b. Too much grease can be as harmful as insufficient grease.
    c. No grease is better than too much grease.
    d. Too much grease is better than insufficient grease.

14. Which of the following is usually an indication of a bad motor bearing?
    a. Hot bearing housing
    b. Low current draw
    c. Low pull-in torque
    d. Sparking at the brushes

15. The best attachment point for lifting heavy motors is/are the _____.
    a. lugs
    b. shaft
    c. base
    d. end bells

Instructor's Notes:

_____

_____

_____

_____

Summary

This module provided information on the installation of motors, including preparation of the baseplate and prealignment. A description of basic motor inspection, maintenance, and lubrication was provided, as well as information on types of bearings and bearing maintenance. Guidelines for storing and troubleshooting motors were also included.

Notes

Summarize the major concepts presented in the module.

Administer the Module Examination. Be sure to record the results of the Examination on Craft Training Report Form 200, and submit the results to the Training Program Sponsor.

Administer the Performance Test, and fill out Performance Profile Sheets for each trainee. If desired, trainee proficiency noted during laboratory sessions may be used to complete the Performance Test. Record the results on Craft Training Report Form 200, and submit the results to the Training Program Sponsor.

MODULE 15507-09 ◆ INSTALLING ELECTRIC MOTORS   7.15

Axial movement: Movement in the direction of a shaft axis.

Brush: A conductor between the stationary and rotating parts of a machine; usually made of carbon.

Commutator: A device used on electric motors or generators to maintain a unidirectional current.

Instructor's Notes:

_____

_____

_____

Resources & Acknowledgments

Additional Resources

This module is intended to present thorough resources for task training. The following reference works are suggested for further study. These are optional materials for continued education rather than for task training.

R+W America L.P.
www.rw-america.com

Coupling Corporation of America
www.couplingcorp.com

Figure Credits

Baldor Electric Company, 507F02

MODULE 15507-09 — TEACHING TIPS

The following are suggested activities or instructional methods to help you teach the material in this module.

General

When you call on someone to answer a question, the rest of the class relaxes or even tunes out because they expect that the question and answer will take place only between you and the trainee you called on. Instead, use this technique to involve more trainees in answering questions and to keep them on their toes.

1. Ask trainees to define a term or explain a concept.
2. After one trainee has answered, ask a trainee seated nearby if the answer is right. Then ask whether a trainee in the back of the room agrees.
3. Ask trainees to explain why they think an answer is right or wrong.
4. Use the session to clear up incorrect ideas and encourage trainees to learn from their mistakes.

Sections
1.0.0 - 9.0.0

Trade Terms

This Quick Quiz will familiarize trainees with terms commonly associated with electric motors. You will need photocopies of the quiz provided on the following page. Trainees will need pencils. If you allow trainees to use the Trainee Guide, decrease the amount of time you give them to complete the quiz.

1. Make a photocopy of the quiz for each trainee.
2. Give trainees between 5 and 10 minutes to complete the quiz.
3. Go over the answers to the quiz.
4. Ask trainees if they have questions.

Answers to Quick Quiz

1. b

2. c

3. a

Quick Quiz   *Trade Terms*

For each description listed, identify the term that the text best describes. Write the corresponding letter in the blank provided.

_____ 1. A conductor on the stationary and rotating parts of a motor is a(n) _____.

_____ 2. A device used on electric motors or generators to generate a unidirectional current is a(n) _____.

_____ 3. On a motor shaft, movement along the line of the shaft is called _____.

        a. axial movement
        b. brush
        c. commutator
        d. dosimeter
        e. electromagnetic center
        f. takeups
        g. tensiometer
        h. thrust bearing

Section 2.0.0   *Motor Installation Safety*

This exercise will familiarize trainees with motor safety. Trainees will need pencils and paper. You will need to obtain a safety video or arrange for an OSHA or other safety professional to give a presentation on lockout/tagout and other motor safety topics. Allow 20 to 30 minutes for this exercise.

Many organizations, such as Associated General Contractors, the Petroleum Equipment Institute, university safety offices, local safety councils, or local OSHA offices, may have safety videos that can be borrowed free of charge for presentation to the class. Obtain one of the following or another safety training video:

Alternatively, free online training is available at the following locations:

OSHA Lockout Tagout Interactive Training:
   www.osha.gov/dts/osta/lototraining/index.html

Seton Compliance Resource Center
   www.setonresourcecenter.com/safety/loto/

University of South Carolina Lockout/Tagout Online Training:
   ehs.sc.edu/modules/Lockout%20Tagout/loto_intro.htm

1. Tell trainees that a guest speaker will be presenting information on motor safety.
2. Have trainees brainstorm questions for the guest speaker before the speaker arrives.
3. Introduce the speaker or video. Ask the presenter to speak about lockout/tagout procedures and other safety concerns.
4. Have the trainees take notes and write down questions during the presentation.
5. After the presentation, ask the speaker to spend some time to answer questions. Have the trainees ask their questions.

Sections 3.0.0 - 7.0.0   *Types of Motors*

This exercise will familiarize trainees with various types of electric motors. Trainees will need appropriate personal protective equipment, pencils, and paper. You will need to arrange to tour an industrial facility with different types of motors. Allow 20 to 30 minutes for this exercise.

Alternatively, obtain manufacturers' literature, photographs, or illustrations of various types of electric motors. Discuss each type with trainees. Ask trainees to identify typical applications for each type.

1. Describe various types of electric motors and their applications.
2. Tour an industrial facility. Point out various types of motors and their applications.
3. Answer any questions trainees may have.

MODULE 15507-09 — ANSWERS TO REVIEW QUESTIONS

Answer	Section
1. b	2.0.0
2. c	2.0.0
3. c	2.0.0
4. b	3.0.0
5. b	3.0.0
6. d	3.0.0
7. b	3.0.0
8. a	3.0.0
9. c	3.0.0
10. d	3.0.0
11. c	3.0.0
12. d	4.0.0
13. b	4.0.0
14. a	6.3.0
15. a	8.0.0

The NCCER makes every effort to keep these textbooks up-to-date and free of technical errors. We appreciate your help in this process. If you have an idea for improving this textbook, or if you find an error, a typographical mistake, or an inaccuracy in NCCER's Contren® textbooks, please write us, using this form or a photocopy. Be sure to include the exact module number, page number, a detailed description, and the correction, if applicable. Your input will be brought to the attention of the Technical Review Committee. Thank you for your assistance.

Instructors – If you found that additional materials were necessary in order to teach this module effectively, please let us know so that we may include them in the Equipment/Materials list in the Annotated Instructor's Guide.

Write: Product Development and Revision
National Center for Construction Education and Research
3600 NW 43rd St., Bldg. G, Gainesville, FL 32606

Fax: 352-334-0932

E-mail: curriculum@nccer.org

Craft	Module Name

Copyright Date	Module Number	Page Number(s)

Description
_____
_____
_____
_____

(Optional) Correction
_____
_____
_____

(Optional) Your Name and Address
_____
_____
_____

Preventive and Predictive Maintenance

NCCER STANDARDIZED CRAFT TRAINING PROGRAM

The National Center for Construction Education and Research (NCCER) provides a standardized national program of accredited craft training. Key features of the program include instructor certification, competency-based training, and performance testing. The program provides trainees, instructors, and companies with a standard form of recognition through a National Craft Training Registry. The program is described in full in the *Guidelines for Accreditation*, published by NCCER. For more information on standardized craft training, contact the NCCER by writing us at 3600 NW 43rd St., Bldg. G, Gainesville, FL 32606; calling 352-334-0911; or emailing info@nccer.org. More information may be found at our website, www.nccer.org.

HOW TO USE THIS ANNOTATED INSTRUCTOR'S GUIDE

Each page presents two sections of information. The larger section displays each page exactly as it appears in the Trainee Module. The narrow column ties suggested trainee and instructor actions to each page and provides icons (detailed below) to call your attention to material, safety, audiovisual, or testing requirements. The bottom of each page includes space for your notes.

The **Audiovisual** icon indicates an appropriate time to show a transparency or other audiovisual aid.

The **Classroom** icon prompts you to define a term, stress a point, ask trainees to explain a concept, or give examples.

The **Demonstration** icon directs you to show trainees how to perform tasks.

The **Examination** icon tells you to administer the written module examination.

The **Homework** icon is placed where you may wish to assign reading for the next class, assign a project, or advise trainees to prepare for an examination.

The **Laboratory** icon is used when trainees are to practice performing tasks.

The **Materials** icon is a reminder for you to gather materials needed for classes, labs, and testing.

The **Performance Testing** icon tells you to administer a performance test or a portion thereof.

The **Safety** icon is used to emphasize safety issues. It is often keyed to *Caution* and *Warning!* statements in the Trainee Module.

The **Teaching Tip** icon indicates additional guidance is available, such as how to conduct an exercise, get the most educational value from a field trip, or encourage class participation. Teaching Tips may expand on a feature (*Think About It, Did You Know?*) or provide *Quick Quizzes* or similar exercises. You will be referred to the Teaching Tips section at the back of the module if there is additional material.

The **Combination** icon indicates that the laboratory listed corresponds with a performance task. If desired, you can note the proficiency of the trainees during the laboratory, and use it to satisfy performance testing requirements.

PREPARATION

Before teaching this module, you should review the Objectives, Performance Tasks, Materials and Equipment List, and Module Outline. Be sure to allow ample time to prepare your own training or lesson plan and gather all required materials and equipment.

MODULE OVERVIEW

This module provides an overview of the preventive and predictive maintenance processes. Information about nondestructive testing is also included.

PREREQUISITES

Prior to training with this module, it is recommended that the trainee shall have successfully completed *Core Curriculum; Millwright Level One; Millwright Level Two; Millwright Level Three; Millwright Level Four;* and *Millwright Level Five*, Modules 15501-09 through 15507-09.

OBJECTIVES

Upon completion of this module, the trainee will be able to do the following:

1. Explain preventive and predictive maintenance.
2. Explain nondestructive testing.
3. Explain visual and optical inspection.
4. Explain liquid penetrant inspection.
5. Explain magnetic particle inspection.
6. Explain infrared testing.

PERFORMANCE TASKS

This is a knowledge-based module; there are no performance tasks.

MATERIALS AND EQUIPMENT LIST

Overhead projector and screen

Transparencies

Blank acetate sheets

Transparency pens

Whiteboard/chalkboard

Markers/chalk

Pencils and scratch paper

Appropriate personal protective equipment

Examples of flawed welds, stress cracks on parts, etc.

NDT equipment, including:
  Ultrasonic tester
  Pyrometer
  Eddy current tester
  Borescope
  Liquid penetrant kit
  Magnetic particle yoke

Copies of the Quick Quizzes*

Module Examinations**

*Located at the back of this module.

**Located in the Test Booklet.

SAFETY CONSIDERATIONS

Ensure that the trainees are equipped with appropriate personal protective equipment and know how to use it properly.

ADDITIONAL RESOURCES

This module is intended to present thorough resources for task training. The following reference works are suggested for both instructors and motivated trainees interested in further study. These are optional materials for continued education rather than for task training.

An Introduction to Predictive Maintenance, 2002. R. Keith Mobley. Woburn, MA: Butterworth-Heinsmann.

Encyclopedia of Materials Science and Engineering – Supplementary, Vol. 1, 1989. Michael B. Bever and Robert W. Cahn, ed. Cambridge, MA: The MIT Press.

Encyclopedia of Materials Science and Engineering – Supplementary, Vol. 2, 1990. Robert W. Cahn, ed. Cambridge, MA: The MIT Press.

Nondestructive Evaluation and Quality Control Metals Handbook, Vol. 17, 9th Ed., 1989. Materials Park, OH: ASM International.

TEACHING TIME FOR THIS MODULE

An outline for use in developing your lesson plan is presented below. Note that each Roman numeral in the outline equates to one session of instruction. Each session has a suggested time period of 2½ hours. This includes 10 minutes at the beginning of each session for administrative tasks and one 10-minute break during the session. Approximately 10 hours are suggested to cover *Preventive and Predictive Maintenance*. You will need to adjust the time required for hands-on activity and testing based on your class size and resources.

Topic	Planned Time
Session I. Introduction; Preventive Maintenance; Predictive Maintenance	
A. Introduction	_____
B. Preventive Maintenance	_____
1. Program Benefits	_____
C. Predictive Maintenance	_____
1. Requirements and Priorities	_____
2. Documentation	_____
Session II. Nondestructive Testing and Evaluation, Part One	
A. Introduction	_____
B. Ultrasonics	_____
C. Radiography	_____
D. Eddy Current Inspection	_____
E. Visual and Optical Inspection	_____
Session III. Nondestructive Testing and Evaluation, Part Two	
A. Liquid Penetrant Inspection	_____
B. Magnetic Particle Inspection	_____
C. Acoustic Emission Testing	_____
D. Infrared Testing	_____
E. Vibration Analysis	_____
F. Tribology	_____

Session IV. Review and Testing

A. Trade Terms and Quick Quizzes

B. Review

C. Module Examination

    1. Trainees must score 70% or higher to receive recognition from NCCER.

    2. Record the testing results on Craft Training Report Form 200, and submit the results to the Training Program Sponsor.

Millwright Level Five

15508-09

Preventive and Predictive Maintenance

Assign reading of Module 15508-09.

15508-09
Preventive and Predictive Maintenance

Topics to be presented in this module include:

Overview

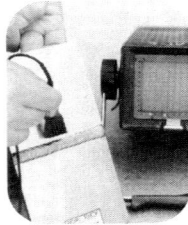

Machinery must be properly maintained in order to operate well for any period of time. In this module, the millwright will be introduced to systematic maintenance procedures. Preventive maintenance includes procedures that help to keep machines running as they should, before they start to fail. Predictive maintenance includes the consideration of problems that are likely to arise, and the scheduling of preventive maintenance to deal with these problems before damage occurs. Predictive maintenance includes nondestructive evaluation and testing methods such as ultrasonics, radiography, and vibration analysis.

Instructor's Notes:

Objectives

When you have completed this module, you will be able to do the following:

1. Explain preventive and predictive maintenance.
2. Explain nondestructive testing.
3. Explain visual and optical inspection.
4. Explain liquid penetrant inspection.
5. Explain magnetic particle inspection.
6. Explain infrared testing.

Trade Terms

Amplitude
Eddy current
Frequency
Hertz
Hysteresis

Resonance
Sensory inspection
Stress
Velocity

Required Trainee Materials

1. Pencil and paper
2. Appropriate personal protective equipment

Prerequisites

Before you begin this module, it is recommended that you successfully complete *Core Curriculum*; *Millwright Level One*; *Millwright Level Two*; *Millwright Level Three*; *Millwright Level Four*; and *Millwright Level Five*, Modules 15501-09 through 15507-09.

This course map shows all of the modules in the fifth level of the Millwright curriculum. The suggested training order begins at the bottom and proceeds up. Skill levels increase as you advance on the course map. The local Training Program Sponsor may adjust the training order.

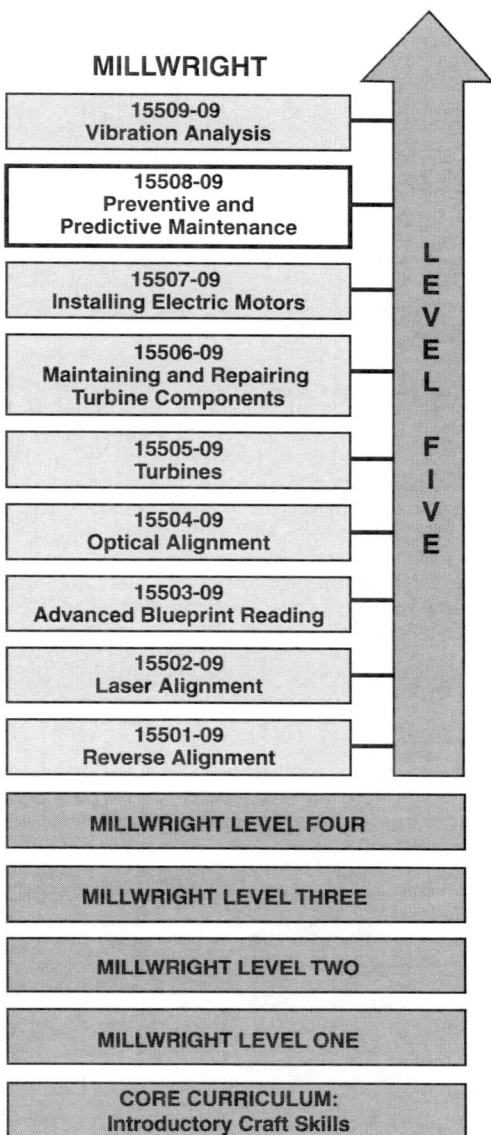

MILLWRIGHT

LEVEL FIVE
15509-09 Vibration Analysis
15508-09 Preventive and Predictive Maintenance
15507-09 Installing Electric Motors
15506-09 Maintaining and Repairing Turbine Components
15505-09 Turbines
15504-09 Optical Alignment
15503-09 Advanced Blueprint Reading
15502-09 Laser Alignment
15501-09 Reverse Alignment

MILLWRIGHT LEVEL FOUR

MILLWRIGHT LEVEL THREE

MILLWRIGHT LEVEL TWO

MILLWRIGHT LEVEL ONE

CORE CURRICULUM: Introductory Craft Skills

508CMAP.EPS

Ensure that you have everything required to teach the course. Check the Materials and Equipment List at the front of this module.

See the general Teaching Tip at the end of this module.

Explain that terms shown in bold are defined in the Glossary at the back of this module.

Show Transparency 1, Objectives. Review the goals of the module, and explain what will be expected of the trainees.

Review the modules covered in Level Five, and explain how this module fits in.

1.0.0 ◆ INTRODUCTION

The purpose of preventive and predictive maintenance is to maximize equipment reliability, which saves money over time. Equipment reliability is maximized by analyzing each situation, accumulating data through testing, compiling repair histories, and applying maintenance programs that use the most straightforward approach. Such programs not only prevent downtime, but they avoid unnecessary maintenance. This allows millwrights to work ahead of anticipated problems instead of falling behind schedule, impacting production.

A critical part of preventive and predictive maintenance is nondestructive testing. Testing attempts to determine what is causing a failure. Charts can be made of changing conditions and equipment reactions to those conditions. This allows the person who schedules the predictive maintenance program to plan ahead logically and productively.

In many companies, these tests are performed by specialists rather than millwrights. However, the millwright should understand the processes and the technologies involved in order to check the specialist's work. The millwright should also be able to set up the area for testing.

2.0.0 ◆ PREVENTIVE MAINTENANCE

Preventive maintenance (PM) is the periodic inspection and maintenance of plant assets and equipment to uncover conditions that may lead to equipment failure or shortened life. The purpose of a PM program is to prevent unscheduled repairs. It is important to set up a system of priorities for determining critical and less-critical items. The duties involved in PM are generally grouped into the following activities:

- Inspecting
- Lubricating
- Replacing components
- Cleaning
- Adjusting
- Documenting
- Reporting

All good PM programs begin with a detailed analysis of past equipment failures, including their costs, causes, remedial actions taken, and changing demands on the equipment. A detailed maintenance history of all equipment within the system must be developed. After this information

has been gathered, it is organized and prioritized according to its effect—ranging from maximum to minimum—on safety factors and product, overhead, and equipment replacement costs. The most costly items should be targeted as first and second priority. Repairable items, which are allowed to fail before they are replaced, are given lower priority.

Primary PM needs are scheduled according to extreme need. The most important PM activities cover items that are extremely important to safety and economics. If a failure could cause loss of life, injury, or environmental damage, it is first priority. Items that could cause maximum economic loss are second priority, although they may be interrelated with first priority items because of the complexity of manufacturing systems. Each component in the system and each system as a whole must be analyzed and prioritized, and projections must be made about present and future needs. At each step in the process, the most dangerous and costly problems can be identified and the program organized so that scheduling becomes more efficient.

Also important is the route the PM path will take through the plant to efficiently use time and equipment to cover priorities. Spare part needs must be studied to balance the desire to have every replaceable item on hand with budget constraints that do not allow all items to be stocked. Some manufacturers believe that it is sometimes possible not to stock spares, and rely completely on vendors to stock and deliver critical items on demand.

Another important facet of a solid PM program is observing equipment conditions while performing scheduled PM. **Sensory inspection**, which involves using sight, sound, smell, and touch, is used to determine how components are performing. Finding clues that help identify problems while a machine is down for PM allows repairs to be scheduled and helps prevent lost time and production.

Some maintenance activities that are related to PM are often mistakenly put into a PM program. The following maintenance activities should not be included in a PM program:

- Improvement or engineering maintenance
- Corrective maintenance
- Scheduled overhauls

Improvement maintenance includes redesign, modifications, and new equipment installation. Although these improvements may enhance the

Instructor's Notes:

_____

_____

_____

maintainability of the equipment and should be planned and scheduled, they should not be included in a PM program. A well-established PM program provides data that leads to these improvements and documents their effectiveness once they have been performed.

Corrective maintenance includes many activities that may be scheduled, but it is not part of a PM program. Corrective maintenance is generally thought of as repairs and modifications that allow production to continue without a complete shutdown. If a PM program is efficient, any interruptions should be minimal and, in most cases, caused by abnormal circumstances that cannot be anticipated. An example would be repairing structural or electronic damage caused by a severe storm.

Scheduled overhauls or periodic production shutdowns are also not considered part of the PM program, although these activities may be driven by the PM program. Documented problems charted by the PM program can be valuable in directing the most important activities during the shutdown period.

2.1.0 PM Program Benefits

Since a PM program involves an investment of time and labor, the expenditure must be justified. A good PM program provides a number of benefits that help to do the following:

- Reduce reactive maintenance
- Increase equipment availability
- Extend the productive life of the equipment
- Improve product quality
- Minimize spare parts inventory
- Reduce total overtime hours

2.1.1 Reduce Reactive Maintenance

Many maintenance activities involve reacting to problems as they occur and performing emergency repairs. Some of this is inevitable, but a good PM program can eliminate most problems before they occur, freeing personnel and resources to perform planned maintenance.

2.1.2 Increase Equipment Availability

PM can be planned and scheduled to meet production needs so that potential repair problems can be detected before they shut down production. Scheduled PM activities can be planned when resources and personnel are most available. This ensures that there is little loss of product and that the equipment remains in use.

2.1.3 Extend the Productive Life of the Equipment

Proper cleaning, lubricating, and adjusting adds to the efficiency and life of the equipment. Regular PM inspections locate potential problems, detect production abuses, and identify operating parameters that have changed. Sometimes a decreased operational demand on the equipment requires less maintenance attention, and increased use requires a more aggressive inspection and repair program.

2.1.4 Improve Product Quality

To be competitive, most manufacturers are on an ever-increasing quality improvement program. Maintaining equipment at the optimum performance level ensures higher product quality. It also eliminates some of the variables that need to be considered when product quality does not meet specification.

2.1.5 Minimize Spare Parts Inventory

The costs involved in stocking spare parts for production cuts directly into profits. With proper PM, the spare parts inventory can be held to minimal emergency spares. Major rehabilitation components can then be purchased as needed instead of maintaining a large inventory of spare parts. Although minimizing the spare parts inventory is a major cost reduction, it can work against you if the PM intensity lags or the follow-through is careless. Minimizing spare parts inventory requires collecting historical data and using common sense, but the net gain is worthwhile.

2.1.6 Reduce Total Overtime Hours

The cost for overtime repairs also cuts directly into profits because it is not factored into the price of the product. Any reduction or elimination of overtime repairs immediately impacts a company's profits, allowing steady employment instead of steady layoffs. Again, planning ahead for maintenance, examining equipment for future problems, and paying close attention to the PM schedule and priorities can head off most equipment-generated emergencies.

Describe each of the benefits of a PM program.

3.0.0 ◆ PREDICTIVE MAINTENANCE

Predictive maintenance (PDM) is the sensing, measuring, and tracking of the physical characteristics or operating performance of equipment. PDM measures unavoidable wear and deterioration and predicts problems before failures occur. PDM is condition-driven, while PM is based on average life or a time schedule. PDM is an extension and enhancement of a good PM program, but it should always remain separate from PM activities to be effective. PDM is set up using the same priorities as PM, and it has the same objectives. The difference is that PDM can be used to analyze equipment while it is operating and producing a product. Preventive and predictive maintenance programs help to prevent critical failures. They also avoid crisis management situations in which maintenance personnel continually respond to breakdowns. *Table 1* shows preventive and predictive maintenance characteristics.

As with PM, the primary objective of PDM is to predict and prevent component failures in equipment. However, PDM's secondary objectives make it more cost-effective than traditional PM. One of the secondary objectives of PDM is to prevent unnecessary replacement of usable parts by singling out only the bad components that are failing. This results in more effective use of all the equipment components. Another objective calls for PDM personnel to recommend that production and maintenance make certain adjustments, calibrations, or changes in procedures. These modifications would enable a seemingly failing piece of equipment to continue running with minor repairs or adjustments instead of a major repair.

3.1.0 Identifying Maintenance Requirements and Priorities

Because no two plants are exactly alike, each PM or PDM program must be customized to fit specific and changing needs. The equipment and machinery must be defined as to function, need, and criticality within the system. In general, follow these guidelines to produce an efficient and flexible PDM program:

- Develop a plant equipment list that identifies all possible maintenance items.
- Identify all health, safety, and environmental hazards.
- Pinpoint all items critical to continued production.
- Determine the significance of each maintenance item as it relates to the following:
  - Safety and the environment
  - Costs and lost production
  - Production schedules
- Determine the consequence of a particular failure in descending order as critical, important, or nonthreatening as it relates to safety, costs, and production schedules.

PM uses a wide and ever-changing variety of diagnostic tools to locate abnormalities in operating equipment. This allows for minor repairs or adjustments to production equipment without causing lost time. If accurate charts are maintained and analyzed, diagnostic tests can be used to predict equipment life and specific failure points.

Table 1 Differences Between Preventive and Predictive Maintenance

Actions	PM	PDM
Diagnosis and inspection	Uses repair history to extend equipment life. Includes standard PM inspections and services to extend equipment life.	Uses diagnostic tools. Compares current operation to normal operating profile.
Detecting abnormal conditions	Consists of visual or teardown inspections, evaluating equipment history, estimating failure points.	Consists of monitoring during operation. Automatically warns of failure. Predicts failure points and time frames.
Correcting deficiencies	Consists of replacing components, periodicallly overhauling equipment.	Consists of spot-replacing unsound components only.
Cost of repairs reduced	Chance of replacing sound components. Possible overstocking of parts.	Provides a more accurate inventory to reduce machine downtime.

508T01.EPS

Instructor's Notes:

_____

_____

_____

Table 2 shows examples of equipment failures grouped into the following categories: critical, important, and nonthreatening.

In general manufacturing statistics, only 20 percent of all equipment or component failures are considered critical. Critical failures are those that create safety hazards, environmental threats, or severe economic loss. If the top 20 percent are avoided, the remaining failures can be corrected with minimal loss and handled as routine maintenance. This is why a detailed analysis of equipment function, as it relates to the consequences of failure, should be maintained and upgraded on a regular basis. These upper-limit failures, sometimes referred to as the critical few, can affect up to 80 percent of a plant's efficiency, safety, and profitability. The principle that 20 percent of the problems create 80 percent of the damage is called the Pareto principle (*Figure 1*). By consistently using PDM techniques, millwrights can address these critical problems before the equipment fails. Then the remaining noncritical problems can be covered by regular maintenance, or the equipment can be allowed to run until it fails.

Another aspect of this analysis is that as the top 20 percent of the problems are eliminated, another set of problems moves up into the top 20 percent, so that eventually, the triangular chart of problems becomes smaller.

3.2.0 Documenting the Equipment Maintenance History

For a complex, multilayered system to work, there must be accurate and consistent documentation of component failures, test results, and all work performed. Since manufacturing techniques are always evolving, the system should

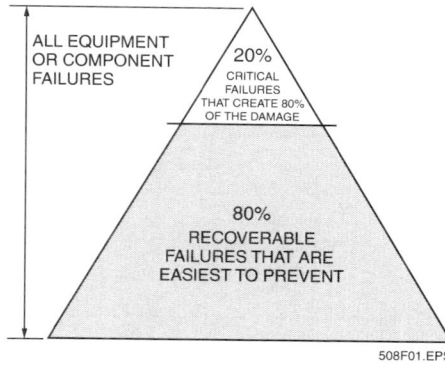

Figure 1 ◆ Pareto principle.

allow for changes and for new testing methods to be added easily. It is better for each plant to generate its own charts and methods in order to keep the system easy to use and relevant to what is happening in the plant.

There are many computer systems and software packages that can be used to track maintenance, but the key to making PDM function properly is consistent monitoring and accurate data. A millwright with the basic instruments and an efficient charting method can provide the same usable results as a computer-driven system. *Figure 2* shows a sample of a typical equipment data sheet.

With the increased use of automation and the constant use of the equipment, the problems faced by millwrights require a high degree of skill and organization. PDM personnel can use the equipment's repair and diagnostic history to predict failure points and avoid multiple or critical equipment failures.

Table 2 Equipment Failure Chart

Function	Functional Failures	Consequences	Severity
Contain liquid (chemicals)	Leaks	Loss of product; possible environmental hazard	Important
Provide flow and pressure at 30 psi	No pressure/flow or low pressure/flow	Reactor mix is off; may cause exothermal reaction	Critical
Input from motor	Bent shaft, misalignment, or bad coupling	Vibration or premature wear on seals and bearings	Nonthreatening

508T02.EPS

Margin notes (right column):

Discuss the categories of equipment failure.

Show Transparency 3 (Table 2).

Show Transparency 4 (Figure 1). Explain the Pareto principle.

Explain the reasons for documenting equipment maintenance.

Show Transparency 5 (Figure 2). Describe a typical equipment data sheet.

See the Teaching Tip for Sections 2.0.0 and 3.0.0 at the end of this module.

Assign reading of Sections 4.0.0–4.4.0.

Ensure that you have everything required for teaching this session.

MODULE 15508-09 ◆ PREVENTIVE AND PREDICTIVE MAINTENANCE 8.5

Describe the use of NDT and NDE and review methods of nondestructive testing.

Show Transparency 6 (Table 3). Discuss the advantages and disadvantages of the NDT methods shown.

Describe the life cycle of a typical machine.

Equipment Data Sheet	
Equipment Number _____	Equipment S/N _____
Equipment Name _____	Model _____
Manufacturer _____	Cost _____
Purchase Date _____	Building _____
Cost Center _____	hp / rpm _____

PDM Priority List	
Environment _____	Safety _____
Avoided Cost _____	Human Relations _____
	The Highest Priority _____

Avoided Cost Calculations

1. DT (Downtime) hr _____ × DT Cost/hr _____ = DT Cost _____
2. Labor _____ + Material _____ = Maintenance Cost _____
3. Defects _____ × Cost/Defect _____ = Defect Cost _____
   Total Avoided Costs _____

PDM Techniques

Technique	Frequency	Test Points	Technique	Frequency	Test Points
VA (Vibration Analysis)			ET (Eddy Current Inspection)		
OA (Oil Analysis)			USL (Ultrasonic Listening)		
IR (Infrared Analysis)			VT (Visual Inspection)		
			Surge Comparison		

508F02.EPS

Figure 2 ◆ Equipment data sheet.

4.0.0 ◆ NONDESTRUCTIVE TESTING AND EVALUATION

Nondestructive testing (NDT) methods are those used to test equipment and components without damaging them. Nondestructive evaluation (NDE) is a system of one or more types of tests used to quantify defects or detect future failures. Most NDT and NDE is done on site under true load conditions, so the equipment does not have to be shut down to be monitored. With NDE, it is possible to detect defects or changes that will lead to premature failure. This information can be used to find the real cause of the change or defect and correct the problem, not just the symptom. NDE is the main tool in a PDM program. *Table 3* lists types of nondestructive testing methods along with the advantages and disadvantages of each method.

NDT focuses on potential critical failures and can also be used to fine-tune equipment and increase productivity with only minor adjustments. In many cases, machinery is not being used to its full capacity. NDE can define the upper limit of a machine's capability while maintaining quality products. Quality control personnel responsible for incoming materials use NDE to locate flaws before they can waste time or compromise safety.

In most cases, the life of a component or piece of equipment follows a predictable failure pattern. When the component is new and initially installed, its break-in period and the learning curve of the operating personnel cause problems and failures. Once the machine matures, it runs smoothly and predictably throughout its expected life with nominal PM. Toward the end of

Instructor's Notes:

_____

_____

_____

Table 3 Types of Nondestructive Testing

NDT	Type of Reading	Advantages	Limitations	Examples of Use
Ultrasonics	Detects changes in acoustic impedance caused by cracks, nonbonds, inclusions, or interfaces.	Can penetrate thick materials; excellent for crack detection; can be automated.	Normally requires coupling either by contact to surface or immersion in a fluid. Orientation, detection, or interpretation of defect can present problems.	Adhesive assemblies for bond integrity. Detection of cracks. Thickness testing.
Radiography	Detects changes in material density from voids, inclusions, material variations, and placement of internal parts. Porosity.	Can be used to inspect a wide range of materials and thicknesses; is versatile; film provides a record of inspection.	Radiation safety requires precautions, is expensive, and detection of cracks can be difficult.	Pipeline welds for penetration, inclusions, and voids. Verification of parts in assemblies.
Eddy currents	Detects changes in electrical conductivity or magnetic permeability caused by material variations, cracks, voids, or inclusions.	Is readily automated, and the cost is moderate.	Is limited to electrically conducting materials and has limited penetration depth. Interpretation of defect signals can be difficult.	Heat exchanger tubes for wall thinning and cracks. Verification of material heat treatment.
Visual-optical	Detects surface characteristics, such as finish, scratches, cracks, or color, and strain in transparent materials.	Is often convenient, and can be automated.	Can be applied only to surfaces, through surface openings, or to transparent material.	Paper, wood, or metal for surface finish and uniformity.
Liquid penetrant	Detects surface openings due to cracks, porosity, seams, or folds.	Is inexpensive, easy to use, readily portable, and sensitive to small surface flaws.	Flaw must be open to the surface. Is not useful for porous materials.	Turbine blades for surface cracks or porosity.
Magnetic particle	Detects leakage magnetic flux caused by surface or near-surface cracks, voids, inclusions, and material or geometry changes.	Is inexpensive and sensitive to both surface and near-surface flaws.	Is limited to ferromagnetic material; surface preparation and postinspection demagnetization may be required.	Railroad wheels for cracks. Detection of weld defects.

508T03.EPS

Show Transparency 7 (Figure 3).

Discuss the factors that cause failures during the life of a machine.

the machine's service life, long-term wear sets in, component failures increase, and rebuilding or replacement becomes necessary. This progression is charted by a bathtub curve (*Figure 3*).

Too much maintenance or unnecessary rebuilds can waste time and money and create problems during the mature life cycle of a machine. Most problems occur either when the machine is new, or much later when components are worn out. Factors that cause failures during the life of a machine include the following:

- Abuse
- Design deficiencies
- Material problems
- Wear or age

Abuse of equipment includes pushing machinery beyond the limits of its physical capabilities and causing purposeful damage through careless or malicious actions. Design deficiencies are also a factor to be dealt with after the machine is in production. NDE can locate equipment abuse and design problems and confirm whether any corrective action taken has been effective. Material problems are flaws and weaknesses that are hidden until the machine is in active use. Long-term use of any machine produces wear, and early detection of flaws and weaknesses is very important.

Failure analysis through NDE can pinpoint problems early and avoid catastrophic failures later. The types of testing equipment used and

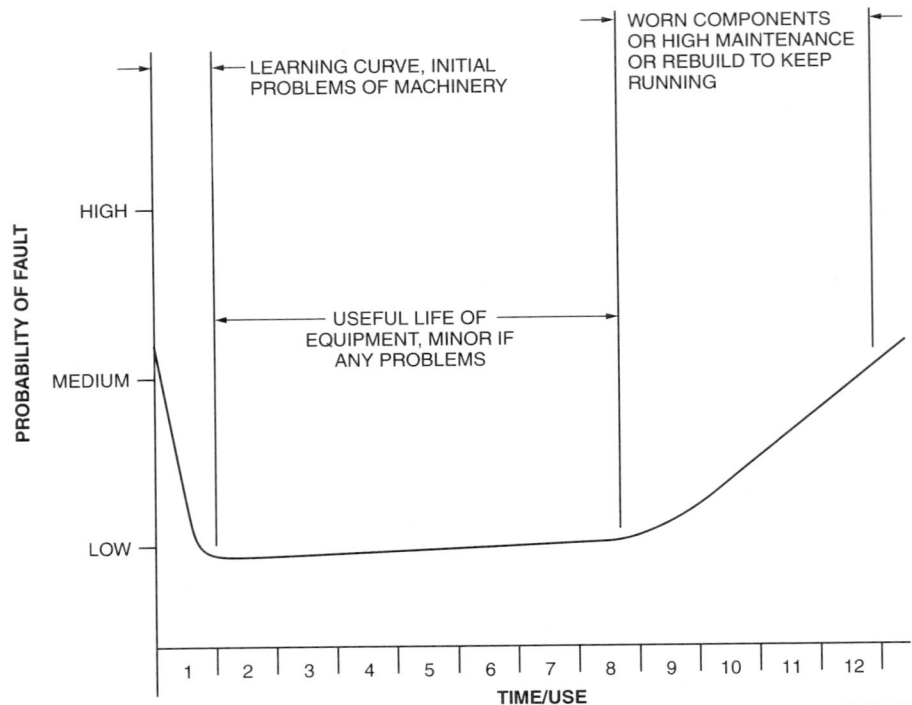

Provide an overview of NDE testing methods.

Explain how ultrasonic waves are used to detect flaws.

Show Transparency 8 (Figure 4). Describe how an ultrasonic flaw detector works.

Note the different types of ultrasonic techniques.

Figure 3 ◆ Bathtub curve of machine life.

the methods employed are varied and constantly changing. Continued awareness of new methods keeps a PDM program efficient and cost effective. The following are some of the more commonly used NDE testing methods:

- Ultrasonics
- Radiography
- Eddy current inspection
- Visual and optical inspection
- Liquid penetrant inspection
- Magnetic particle inspection
- Acoustic emission testing
- Infrared testing
- Vibration analysis
- Tribology

4.1.0 Ultrasonics

Ultrasonics uses high-frequency sound waves to detect the noise generated by plant equipment. Ultrasonic instruments measure the thickness of materials or listen for turbulence in pipes to detect internal leaks and malfunctions. Other ultrasonic testing instruments send focused, amplified sound waves into solid materials, and then compare them with the return wave to find flaws below the surface. Flaws can include corrosion, bad welds, or casting flaws. The **frequency** range for ultrasonic monitoring is from 20,000 **Hertz** (Hz) to 100 kiloHertz (kHz).

In many ways, a beam of ultrasound is similar to a beam of light. Both are waves that travel at a constant **velocity** in a consistent medium. The velocity of the beam depends on the medium it encounters, not on the properties of the wave form. Like a beam of light, ultrasonic waves are bent, reflected from surfaces, refracted when they cross boundaries of different surfaces, and diffracted at the edges of surfaces or around obstacles. When the beam is reflected back, it creates a signature that is interpreted as an image by the ultrasonic monitor.

Ultrasonic monitors (*Figure 4*) can deeply penetrate materials with high sensitivity and accuracy. The output is readily digitized, and operation of the monitor is completely nonhazardous. Various types of ultrasonic techniques are available, including pulse-echo, pulse-transmission,

8.8 MILLWRIGHT ◆ LEVEL FIVE

Instructor's Notes:

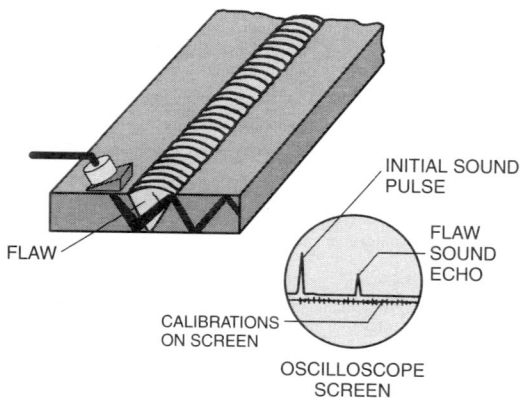

FLAW

INITIAL SOUND PULSE

FLAW SOUND ECHO

CALIBRATIONS ON SCREEN

OSCILLOSCOPE SCREEN

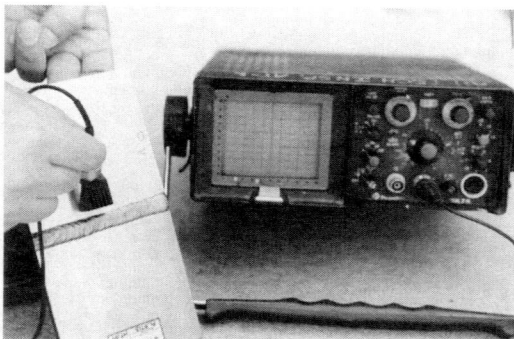

508F04.EPS

Figure 4 ◆ Ultrasonic flaw detector.

ultrasonic attenuation, continuous wave **resonance**, and ultrasonic spectroscopy. One advantage of ultrasonic monitors is that they are very portable. A disadvantage is that they must be monitored by experienced technicians.

Reading an ultrasonic flaw detector requires practice because the screen image shows the peaks of back reflection (*Figure 5*).

4.2.0 Radiography

Radiography uses a source of penetrating radiation, most commonly Cobalt 60, to shoot through an object and project an image onto a detector or sheet of film. The image is formed when the different thicknesses and features of the object create shadows of different densities. The image displays an internal picture of intricate machine parts, castings, weld integrity, and other features. This technology was developed from medical technology, and it includes computer-aided tomography (CAT scans) and magnetic resonance imaging (MRI).

Radiography, which can be applied to any material, produces a sharp internal image throughout the specimen. The advantages of radiography include the ability to use it on a variety of materials, and the depth of detail it displays. The limitations of radiography include its cost and space requirements as well as the technical skill needed to operate the machine and interpret the image. Another disadvantage is that a crack that runs in the same direction as the beam is almost impossible to see. Also, delaminations in layered objects are impossible to detect because there is no difference in absorption. Because these instruments emit radioactivity and are difficult to operate, they should only be used by trained technicians, and all safety procedures must be strictly followed.

When a joint is radiographed (*Figure 6*), the radiation source is placed on one side of the weld and the film on the other. The joint is then exposed to the radiation source. The radiation penetrates the metal and produces an image on the film. The film, called a radiograph, provides a permanent record of the weld quality. Radiography must only be used and interpreted by trained, qualified personnel.

Radiographic inspection can produce a visible image of weld discontinuities, both surface and subsurface, when they are different in density from the base metal and different in thickness parallel to the radiation. Surface discontinuities are better identified by visual, penetrant, or magnetic particle examination.

Show Transparency 9 (Figure 5). Explain how to read flaw detector screen images.

Introduce radiography as an inspection method.

Point out that radiography technologies include CAT scans and MRIs.

Discuss the advantages and disadvantages of radiography.

Explain how radiography applies to the inspection of weld joints.

OSCILLOSCOPE SCREEN

BACK REFLECTION

TRANSMITTING SEARCH UNIT

SOUND BEAM

RECEIVING SEARCH UNIT

DISCONTINUITY ECHO

DISCONTINUITY

CASTING

FRONT AND BACK SURFACE PARALLEL

OSCILLOSCOPE SCREEN

NO BACK REFLECTION

TRANSMITTING SEARCH UNIT

RECEIVING SEARCH UNIT

DISCONTINUITY ECHO

DISCONTINUITY

SOUND BEAM

CASTING

BACK SURFACE NOT PARALLEL TO FRONT SURFACE

RECEIVING SEARCH UNIT

BACK REFLECTION

OSCILLOSCOPE SCREEN

TRANSMITTING SEARCH UNIT

SOUND BEAM

CORED HOLE

ECHO FROM CORED HOLE

DISCONTINUITY ECHO

DISCONTINUITY

CASTING

HOLE AND NEARBY DISCONTINUITY

OVERLAPPING DISCONTINUITY ECHO AND BACK REFLECTIONS

OSCILLOSCOPE SCREEN

TRANSMITTING SEARCH UNIT

SOUND BEAM

RECEIVING SEARCH UNIT

FILLET (TYP)

CASTING

DISCONTINUITY

DISCONTINUITY AT INTERSECTION

508F05.EPS

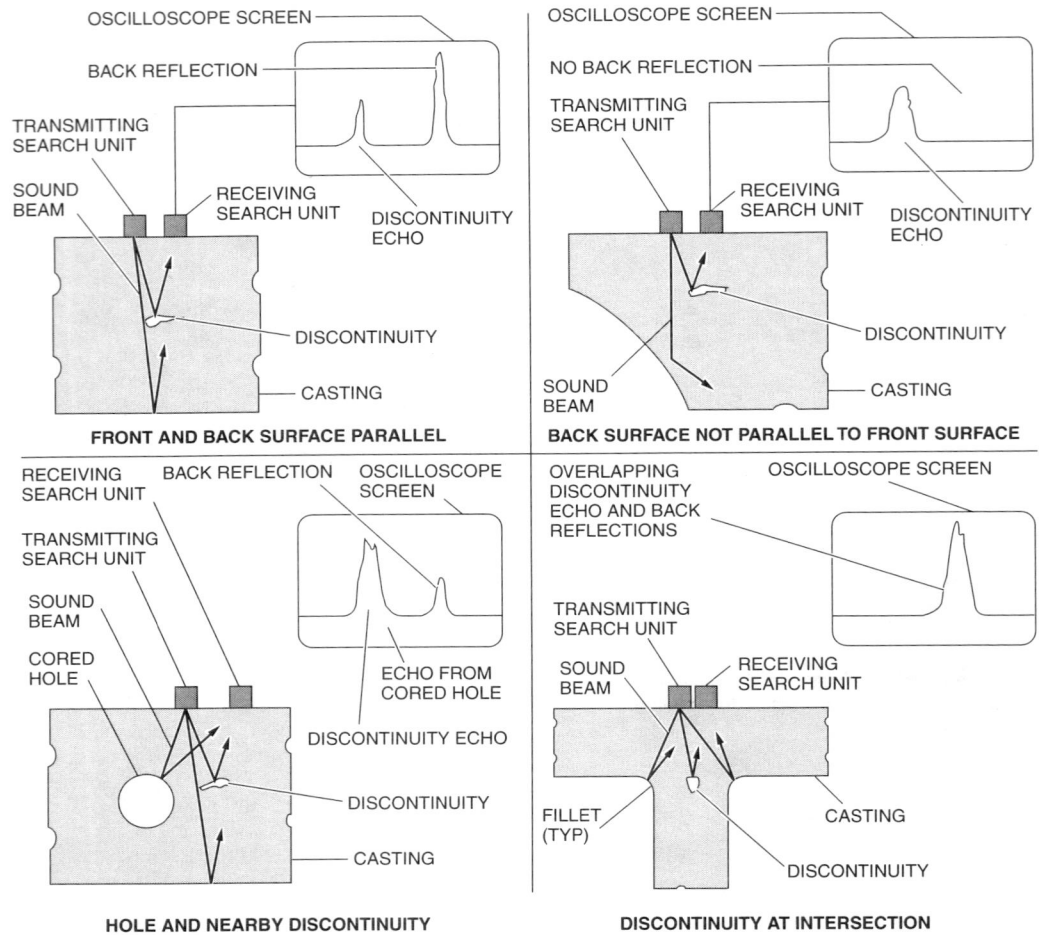

Figure 5 ◆ Flaw detector images.

RADIATION SOURCE

FILM PACK

FILM PLATE

508F06.EPS

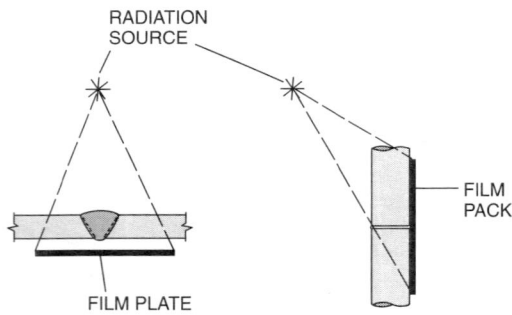

Figure 6 ◆ Radiography examination.

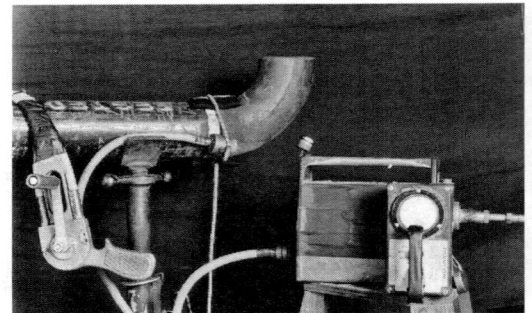

Instructor's Notes:

_____

_____

_____

The X-ray method is used to detect internal defects and to check for proper alignment of assembled parts. X-rays are also used extensively to examine welds, check wall thickness, and examine the quality of casting and forging (*Figure 7*).

WARNING!

Only certified technicians are permitted to perform X-ray activities. **Never** cross a radiation barrier unless instructed to do so by a qualified person.

The following procedure will give you a general understanding of how to X-ray welds. Follow these steps when assisting a technician to X-ray welds:

Step 1 Obtain the X-ray procedure, and review it to ensure that the correct weld identification is listed. The X-ray procedure must be signed and have a detailed explanation of the surface area to be tested.

Step 2 Clean the weld and surface area using a wire brush or half-round bastard file to remove any scale, rust, paint, arc strikes, or gouges.

Explain how X-rays are used to examine welds.

Show Transparency 11 (Figure 7).

Describe the procedure for weld X-rays. Point out that this procedure must be performed by a certified technician.

Warn trainees of the hazards of radiation exposure.

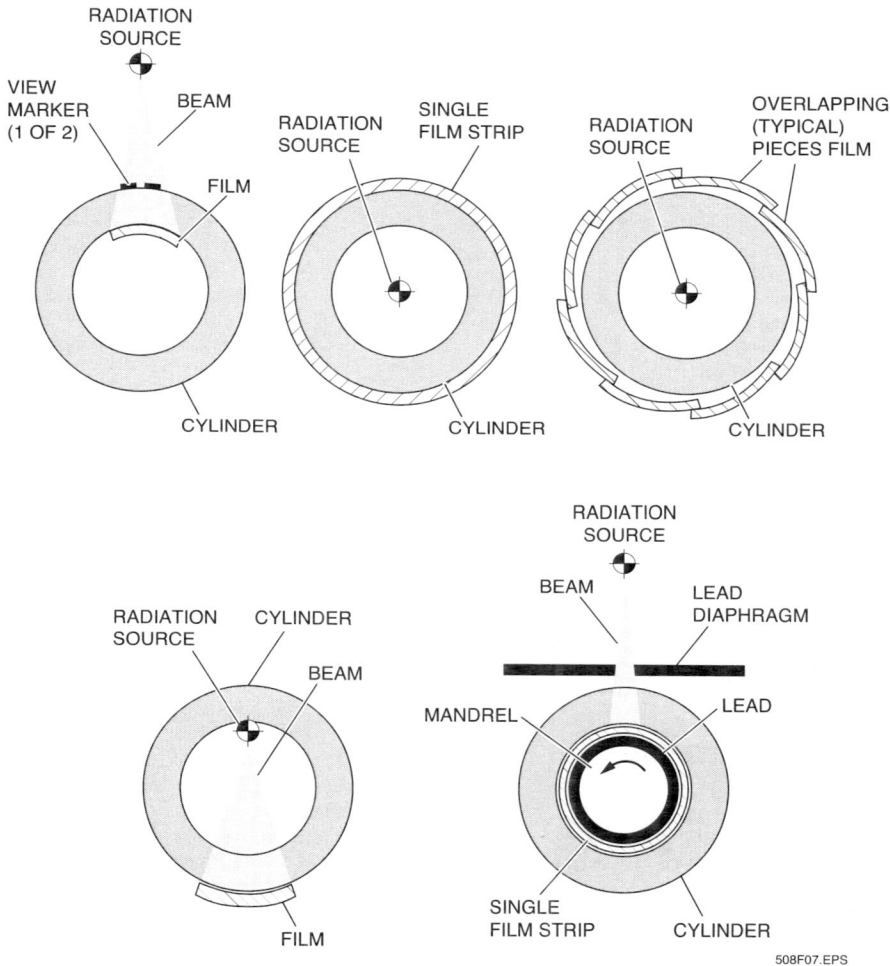

Figure 7 ◆ X-ray using gamma rays.

Step 3 Use a permanent marker to mark off the area where the X-ray film will be placed. Refer to the procedure to make sure that the X-ray film is placed in the correct position.

Step 4 Rope off the area using radiation tape and appropriate signs. The area must be roped off above, below, and along the sides of the target area.

Step 5 Place the approved radiation source at the test area. The radiation source must be located at the approved distance from the weld to be tested.

Step 6 Place the target on the back side of the weld to be tested.

Step 7 Shoot the weld using the radiation source.

Step 8 Remove the X-ray film from the weld.

Step 9 Have the X-ray film processed.

4.3.0 Eddy Current Inspection

Eddy current inspection (*Figure 8*) observes the interaction between electromagnetic fields and metals. The inspection can be used to measure electrical conductivity, magnetic permeability, heat treatment condition, and physical dimensions. Eddy currents can also detect seams, cracks, and voids or measure the thickness of nonconducting coatings on a conductive material.

Eddy current inspection does not require direct electrical contact with the specimen. Its main disadvantage is its sensitivity; it may mask critical variables or lead to misinterpretation.

In an eddy current inspection, electrical currents are induced to flow in the test object by a coil of wire that carries an alternating current. When the probe is placed on the test object, elec-

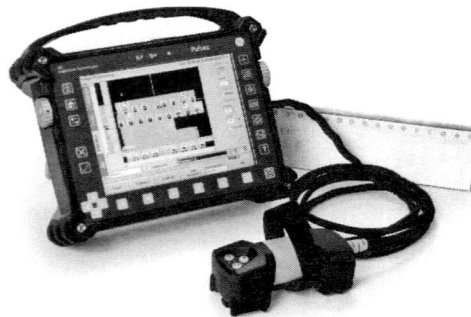

508F08.EPS

Figure 8 ◆ Eddy current inspection system.

tromagnetic energy produced by the coils is partly absorbed and converted into heat by the effects of resistivity and **hysteresis**. Part of the remaining energy is reflected back to the test coil, which has had its electrical signal changed by the test object. The feedback provided by the reflected energy is analyzed and compared to the electrical signature of a reference specimen. Since eddy current inspection penetrates only ¼ inch, it is best used to test for surface cracks, hardness, plating thickness, or other surface defects.

4.4.0 Visual and Optical Inspection

Visual and optical inspection of equipment is the easiest to perform, but is the most likely to be overlooked as an inspection technique. There should be a systematic approach to visual inspection so that no details or events are missed. It is best to make a detailed observation while the equipment is running. Follow the path of the product, carefully watching at every step and noting any abnormalities. With the equipment off or in a safe position, look for any external detail that could provide a clue to potential trouble. Excessive corrosion, bent shafts, leaking seals, and any sign of abnormality should be documented in detail.

Parts of the equipment that are hard to reach, internal, or hot can be inspected with optical borescopes and fiberscopes. The optical borescope (*Figure 9*) reaches deep inside a machine, magnifies like a microscope, and has its own light source. Because the fiberscope is flexible and has a fiber-optic cable, it can provide a close look into areas that can normally be seen only when the equipment is disassembled.

4.5.0 Liquid Penetrant Inspection

Liquid penetrant inspection (*Figure 10*) is used on the surface of smooth or nonporous samples to highlight cracks and gaps that are not visible to the naked eye. The liquid penetrant seeps into surface cracks, pores, delaminations, or shrinkage areas by capillary action. It is also used to highlight wear patterns and to measure **stress** cracks on the surface of materials. Liquid penetrant is supplied as a kit that includes cleaner, dye, and developer.

The inspection involves wetting the cleaned surface of a test object with liquid penetrant. The liquid penetrant flows over the surface to form a continuous, uniform coating that migrates into cracks and surface defects. The liquid is then cleaned off the surface and becomes replaced by a developer. As the developer draws the penetrant out of the defects, it becomes stained by the

8.12 MILLWRIGHT ◆ LEVEL FIVE

Instructor's Notes:

_____

_____

_____

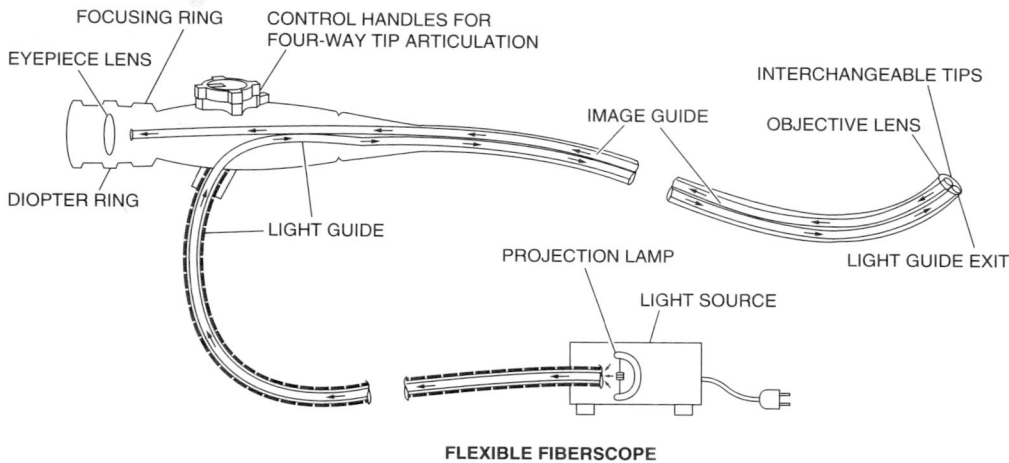

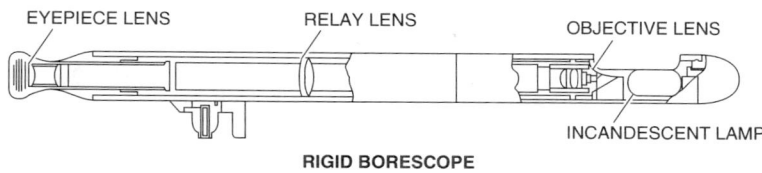

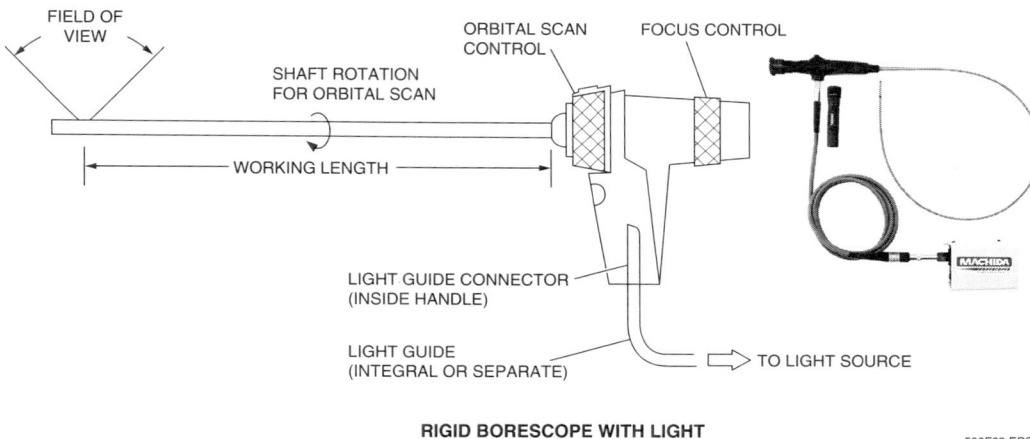

Figure 9 ◆ Fiberscope and borescopes.

penetrant. The stained developer highlights the cracks and surface irregularities and forms a visual image of the flaw.

Liquid dye penetrants reveal two types of indications: true and false. True indications are caused by the penetrant bleeding out from actual discontinuities in the metal. The standard defects indicated include stress or fatigue cracks, underbead cracks, pits, and porosities. Large stress cracks are indicated by wide lines that become apparent quickly after the developer is applied. Underbead cracks are undersurface cracks that bleed through to the surface. Underbead cracks are indicated by a line of dots that appears a few minutes after the developer is applied. Porosity is indicated by dots that come to the surface almost immediately.

MODULE 15508-09 ◆ PREVENTIVE AND PREDICTIVE MAINTENANCE   8.13

Emphasize that the procedure for liquid penetrant testing must be performed carefully to obtain valid results.

Describe the procedure for inspecting welds using a liquid dye penetrant.

Using a liquid penetrant kit, show trainees how to perform the inspection procedure on flawed welds, stress cracks, etc.

Discuss the two major types of developers. Explain that the dry developer must be applied to a dry surface.

Explain that magnetic particle inspection may be used on materials that are magnetized.

Describe types of inductors used for magnetic particle inspection.

The main reasons for false indications are failure to apply the penetrant correctly, or rough, irregular surfaces on the test metal. The procedure for liquid dye penetrant testing must be carefully performed to achieve the proper results. Follow these steps to inspect welds using a liquid dye penetrant:

Step 1 Remove any dirt, grease, scale, acids, chromates, or other contaminants from the surface of the metal using a grinder with a flapper wheel, a wire brush, or a bastard file.

Step 2 Clean the surface with the liquid cleaner supplied with the kit. Apply the liquid penetrant to the test area. The liquid dye penetrant can be applied by dipping the part being tested into the penetrant or by brushing or spraying the penetrant onto the surface.

Step 3 Allow the liquid dye penetrant to remain on the test area for the required waiting period. The length of time will depend on the type of penetrant used and the kind of metal being tested. Follow the penetrant manufacturer's recommendations.

Step 4 Remove any excess penetrant using the recommended solution for the type being used. Do not allow the excess penetrant to dry. If it dries, the metal must be recleaned and the penetrant reapplied. Be careful not to remove the penetrant from possible defects in the metal.

Step 5 Apply the developer to the test area.

> **NOTE**
>
> There are two major types of developers: wet and dry. The wet developer can be either solvent-based or water-based. Both types are applied with a pressurized spray. The dry developer performs the same functions as the wet developer, but it is applied in powder form. The wet developer can be applied to the surface while it is still damp to the touch, but the dry developer must be applied to a dry surface.

Step 6 Inspect the test area for defects.

Step 7 Clean the test area thoroughly using the solvent supplied with the kit.

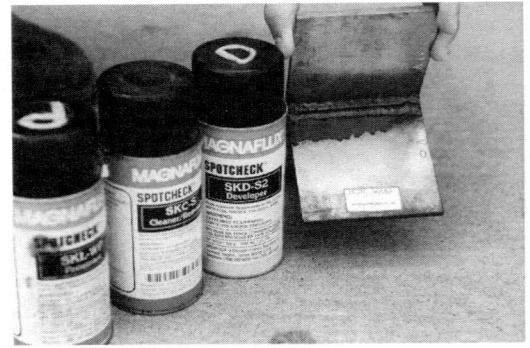

508F10.EPS

Figure 10 ◆ Liquid penetrant materials and inspection.

4.6.0 Magnetic Particle Inspection

When a magnetic field is generated in and around a part made of a ferromagnetic material, such as steel, nickel, or iron alloys, and the lines of magnetic flux are intersected by a defect, such as a crack, magnetic poles are generated on either side of the defect. Magnetic particle inspection works only for materials that can be magnetized.

In magnetic particle inspection, the sample is magnetized with an electrical current, and magnetic particles are applied over the surface. These particles can be applied either dry or in a liquid carrier, such as oil or water. The flaws on or just under the surface of the sample modify the magnetic field. This causes the particles to outline the defect in a pattern which highlights the shape and location of the flaw.

Magnetic particle inspection is very sensitive to minor surface cracks and flaws that are near the surface. Its disadvantages include the limitation to use on magnetic samples and the inability to identify deeper flaws in the sample.

Maximum sensitivity is obtained when the part is oriented perpendicular to the magnetic field so the strength of the field saturates the entire part. The magnetic particle equipment uses either AC or DC current, which can be applied to the part with coils, probes, or flexible conductors that can wrap around a test piece.

Because samples come in various shapes, the inductor for the test must also come in a variety of sizes and shapes. Some sample parts, such as gears, balls, and ring-shaped parts, can be placed in the magnetic field without contact from a probe. Other samples require the use of coils, clamps, central conductors, or other forms of direct contact. For example, a welded joint would use a yoke to induce a magnetic field around the

Instructor's Notes:

_____

_____

_____

_____

weld seam. This would show any discontinuities as extra particles scattered outside the weld seam (*Figure 11*).

4.7.0 Acoustic Emission Testing

Acoustic emission (AE) testing involves releasing a burst of energy into a part and detecting the resulting vibration with either an analog or digital transducer. The source of the return energy is the elastic stress field in the material. If there is no stress in the part, there is no emission. The return signal is analyzed, and any flaws or cracks in the part can be seen by the **amplitude**, dwell time, and other parameters.

AE is useful in finding below-the-surface flaws such as delaminations, cracks, voids, or fiber fractures. It is well-suited for metals, plastics, or composites, and is ideal to use while a part is in use or under real load conditions. AE is also used to do the following: monitor in-process welding and tool conditions during CNC machining; detect wear or loss of lubrication in rotating equipment; and monitor chemical reactions or processes.

4.8.0 Infrared Testing

Infrared tests detect infrared energy emitted by a part or machine. The resulting thermal signature is used to locate both internal and external problems. This method works by converting the invisible heat into an image that is graduated into different colors according to temperature. This image can be used to locate failing bearings, abnormal shaft stress, or faulty electrical hardware. Infrared radiation is produced naturally by all materials at temperatures above absolute zero. Infrared emissions vary in intensity, and these variations can be monitored and recorded using infrared detectors.

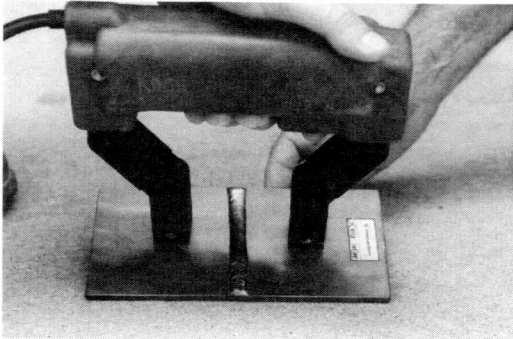

Figure 11 ◆ Electromagnetic yoke on weld sample.

Detector styles include the focused type, for spot checking, or broadfield, for covering a large area.

4.9.0 Vibration Analysis

All machines generate mechanical forces as part of their normal operation. These forces create unique vibrations of various types. Vibration of rotating machinery can be separated into individual vibration cycles. Each cycle (*Figure 12*) begins with the shaft centered in the housing. As the shaft rotates, it bends due to loads, loose bearings, or a bent shaft. As the rpm increases, the shaft moves away from the center line until it reaches a peak. The shaft then moves back through the center line to the peak in the opposite direction. The shaft deflection reverses back to the center line, and this is considered one cycle of vibration. Increases in the load or rpm create a more noticeable wobble, which is measured as increased frequency and amplitude. The vibration is caused by the reversal of movement at each peak of the cycle. The basic components of vibration measurement include frequency, amplitude, displacement, velocity, acceleration, and elapsed time.

In vibration analysis, the equipment used to diagnose vibration is specialized as to function. Vibration meters and monitors are installed on equipment to warn of or measure the amount of vibration. When connected to a chart recorder, a vibration analyzer helps determine the cause and source of vibration by measuring displacement, amplitude, frequency, and timed history. The probes and transducers used are designed for different functions and ranges. To get meaningful results from vibration analysis, it is very important to keep track of the method of recording vibration data, and have a graphic chart of each machine showing where individual readings were taken. *Figure 13* shows a schematic for vibration read points on a steam turbine boiler feed pump.

4.10.0 Tribology

In a PDM program, the three techniques for measuring machine wear through lubrication are lubricant analysis, wear particle analysis, and ferrography. These three methods are part of a field called tribology. Tribology is the process of examining the condition of lubricants in a machine to determine the wear conditions, the state of the lubricants, and the presence and condition of metal particles suspended in the lubricants.

Describe how acoustic emission testing uses vibrations to detect flaws.

Discuss common applications for AE.

Explain how infrared testing uses thermal signatures to locate problems.

Describe vibration analysis and the vibration cycle.

Discuss types of equipment used to diagnose vibration.

Show Transparency 13 (Figure 12).

Show Transparency 14 (Figure 13). Point out the vibration read points on this schematic.

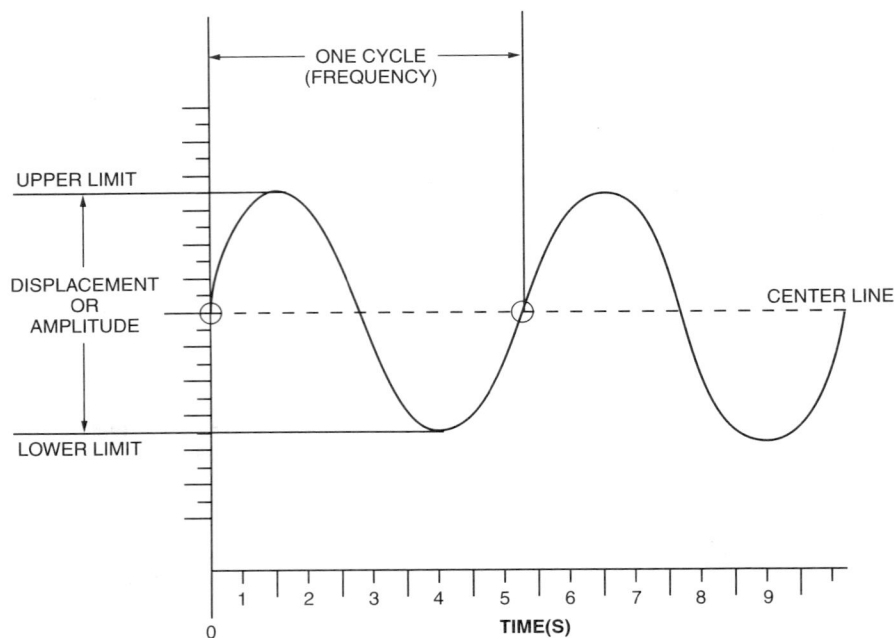

Figure 12 ◆ Vibration cycle.

Lubricant analysis quantifies the condition of a machine's lubricant by analyzing a sample for the individual chemical elements, additives, and contaminants. Over time, consecutive samples of lubricant can chart wear patterns or give early warnings of future failures. Frequent sampling and long-term data storage can save time and money by pointing out exactly when the lubricant should be changed or if different additives are required.

Wear particle analysis furnishes direct information about the wear of mechanical parts. Particles in the lubricant of a machine can provide important information about the internal parts.

Wear particle analysis can predict mechanical failures through the study and history of the size, shape, composition, and quantity of particles in the lubricant. This leads to planned and detailed rebuild or replacement of the machine or components when it is needed.

Ferrography is similar to lubricant and wear particle analysis in the way in which samples are taken, but it has two major differences. Ferrography separates the particles in the lubricant, using a magnetic field. This means that the only particles caught will be ferrous alloys. The other difference is that smaller particles can be captured, providing a better representation of the contamination.

Instructor's Notes:

_____

_____

_____

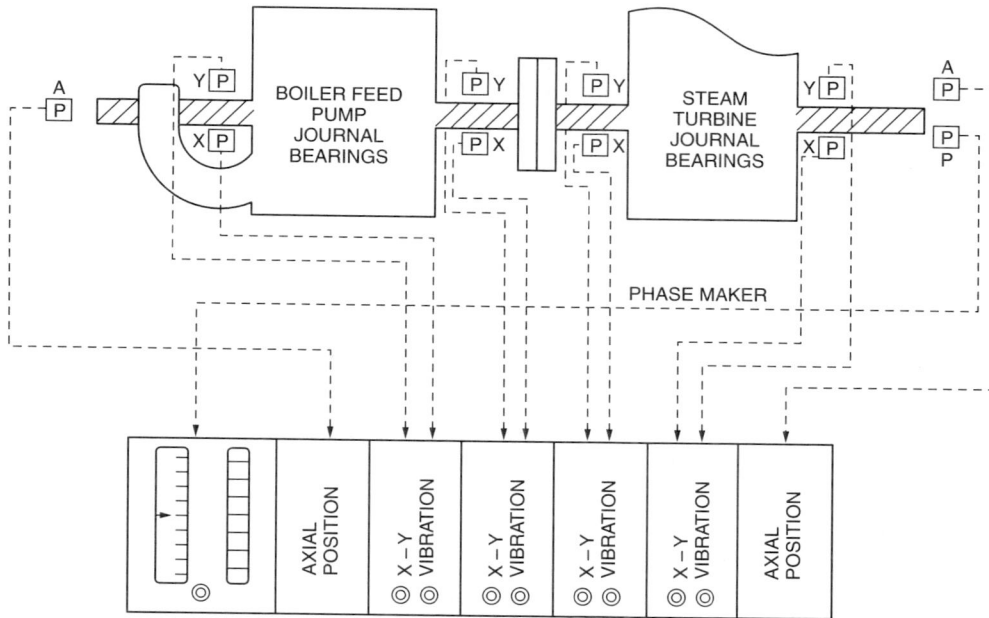

LEGEND:

P	A	PROXIMITY PICKUP FOR AXIAL POSITION
P	X	PROXIMITY PICKUP FOR X RADIAL VIBRATION
P	Y	PROXIMITY PICKUP FOR Y RADIAL VIBRATION
P	P	PROXIMITY PICKUP FOR PHASE REFERENCE

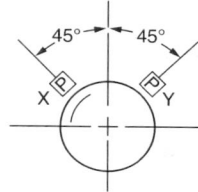

508F13.EPS

Figure 13 ◆ Schematic for vibration read points.

Have trainees complete the Trade Terms Quick Quiz for Sections 1.0.0–4.0.0, and go over the answers prior to administering the Review Questions.

Have trainees complete the Review Questions, and go over the answers prior to administering the Module Examination.

Review Questions

1. Primary preventive maintenance (PM) priorities are established based on economic and _____ factors.
   a. time
   b. convenience
   c. safety
   d. spare part inventory reduction

2. PM does *not* include _____.
   a. sensory inspection
   b. analysis of past equipment failures
   c. corrective maintenance
   d. lubrication

3. A good PM program reduces _____.
   a. documentation
   b. reactive maintenance
   c. cleaning
   d. component replacement

4. Predictive maintenance (PDM) is _____ instead of being based on a time schedule.
   a. crisis-driven
   b. profit-driven
   c. condition-driven
   d. driven by economics only

5. Predictive maintenance is set up using the same priorities and has the same objectives as _____ maintenance.
   a. engineering
   b. corrective
   c. scheduled
   d. preventive

6. One of the objectives of PDM is to prevent unnecessary replacement of useable components.
   a. True
   b. False

7. The main tool in a PDM program is _____.
   a. pareto analysis
   b. statistical process control
   c. nondestructive evaluation
   d. selective observation

8. Acoustic emission testing is useful for finding below-the-surface flaws such as _____.
   a. splatter
   b. delaminations
   c. slag
   d. paint specks

9. Vibration analyzers determine the cause and source of vibration, and vibration meters and monitors are installed to _____ the vibration.
   a. measure
   b. cancel
   c. stop
   d. increase

10. Over time, you can chart wear patterns or predict future failures by taking consecutive lubrication _____.
    a. temperature readings
    b. visual examinations
    c. samples
    d. measurements

Instructor's Notes:

_____

_____

_____

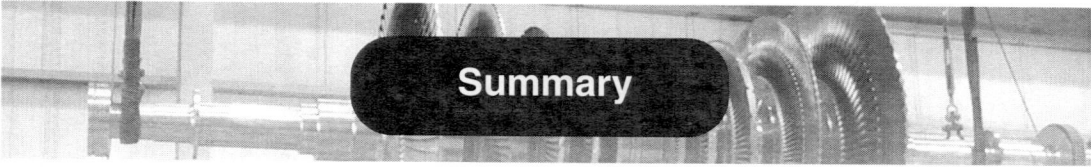

Summary

The complexity of manufacturing equipment and the goal of avoiding any unnecessary downtime make PM and PDM programs an important part of a millwright's job. The primary objective of a preventive or predictive maintenance program is to keep equipment running smoothly, and to maximize safety and minimize downtime and cost. A PDM program uses nondestructive testing and evaluation to confirm what is really happening in the equipment. When changes occur in equipment operation, nondestructive evaluation helps to detect defects or changes that may lead to premature failure.

Summarize the major concepts presented in the module.

Administer the Module Examination. Be sure to record the results of the Examination on Craft Training Report Form 200, and submit the results to the Training Program Sponsor.

Notes

MODULE 15508-09 ◆ PREVENTIVE AND PREDICTIVE MAINTENANCE   8.19

Trade Terms Introduced in This Module

Amplitude: The magnitude of dynamic motion or vibration as expressed in terms of peak-to-peak, zero-to-peak, or root mean square.

Eddy current: An electrical current that is generated and dissipated in a conductive material in the presence of a magnetic field.

Frequency: The repetition rate of a periodic event expressed either in cycles per second (Hertz), revolutions per minute (rpm), or multiples of rotational speed.

Hertz (Hz): The unit of frequency represented by cycles per second.

Hysteresis: The portion of a system response in which a change in input does not produce a change in output; also called deadband.

Resonance: The reinforced vibration of a body caused by a matching frequency of vibration of another body; a common example is a tuning fork.

Sensory inspection: Inspection that includes sight, sound, smell, and touch.

Stress: A force acting on a body per unit area, normally measured in pounds per inch or newtons per square meter.

Velocity: Measures the rate and direction of movement of an object.

Instructor's Notes:

_____

_____

_____

_____

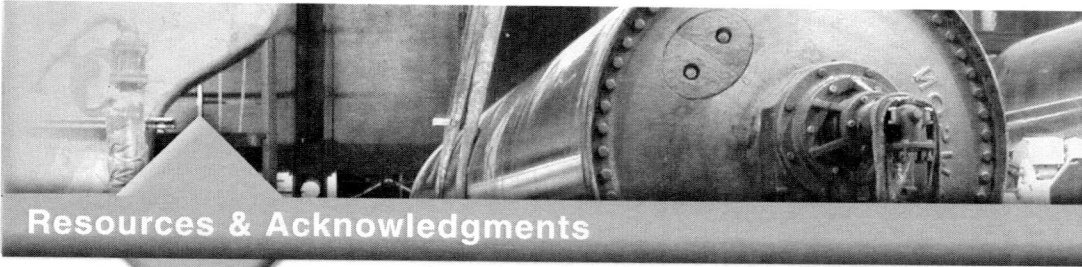

Resources & Acknowledgments

Additional Resources

This module is intended to be a thorough resource for task training. The following reference works are suggested for further study. These are optional materials for continued education rather than for task training.

An Introduction to Predictive Maintenance, 2002. R. Keith Mobley. Woburn, MA: Butterworth-Heinsmann.

Encyclopedia of Materials Science and Engineering – Supplementary, Vol. 1, 1989. Michael B. Bever and Robert W. Cahn, ed. Cambridge, MA: The MIT Press.

Encyclopedia of Materials Science and Engineering – Supplementary, Vol. 2, 1990. Robert W. Cahn, ed. Cambridge, MA: The MIT Press.

Nondestructive Evaluation and Quality Control Metals Handbook, Vol. 17, 9th Ed., 1989. Materials Park, OH: ASM International.

Figure Credits

Fluke Corporation, 508F04 (middle photo), reproduced with permission.

Topaz Publications, Inc., 508F04 (bottom photo), 508F06 (photo), 508F10, 508F11

GE Sensing & Inspection Technologies, 508F08

Machida Borescopes, 508F09 (photo)

The following are suggested activities or instructional methods to help you teach the material in this module.

General

When you call on someone to answer a question, the rest of the class relaxes or even tunes out because they expect that the question and answer will take place only between you and the trainee you called on. Instead, use this technique to involve more trainees in answering questions and to keep them on their toes.

1. Ask the trainees to define a term or explain a concept.
2. After one trainee has answered, ask a trainee seated nearby if the answer is right. Then ask whether a trainee in the back of the room agrees.
3. Ask the trainees to explain why they think an answer is right or wrong.
4. Use the session to clear up incorrect ideas, and encourage the trainees to learn from their mistakes.

Sections
1.0.0 – 4.0.0 *Trade Terms*

This Quick Quiz will familiarize trainees with trade terms commonly used in relation to predictive and preventive maintenance. You will need photocopies of the quiz provided on the following page. Trainees will need pencils. If you allow trainees to use the Trainee Guide, decrease the amount of time you give them to complete the quiz, and remind them to bring their books to class.

1. Make a photocopy of the quiz for each trainee.
2. Give trainees between 5 and 10 minutes to complete the quiz.
3. Go over the answers to the quiz.
4. Ask trainees if they have questions.

Answers to Quick Quiz

1. i
2. d
3. a
4. b
5. g
6. f
7. h
8. e
9. c

Quick Quiz   *Trade Terms*

For each description listed, identify the term that the text best describes. Write the corresponding letter in the blank provided.

_____ 1. A measure of the rate and direction of movement of an object is the _____.

_____ 2. The unit of frequency represented by cycles per second is the _____.

_____ 3. The term used to describe the magnitude of dynamic motion or vibration expressed as peak-to-peak, zero-to-peak, or root mean square is the _____.

_____ 4. An electrical current that is generated and dissipated in a conductive material in the presence of a magnetic field is a(n) _____.

_____ 5. Inspection that includes sight, sound, and touch is called _____.

_____ 6. The reinforced vibration of a body caused by the matching frequency of vibration of another body is called _____.

_____ 7. A force acting on a body, normally measured in pounds per inch or newtons per square meter, is _____.

_____ 8. The portion of a system response in which a change in input does not produce a change in output is called deadband, or _____.

_____ 9. The repetition rate of a periodic event expressed either in cycles per second, revolutions per minute, or multiples of rotational speed is referred to as the _____.

a. amplitude
b. eddy current
c. frequency
d. Hertz
e. hysteresis
f. resonance
g. sensory inspection
h. stress
i. velocity

Sections
2.0.0 and 3.0.0 *Preventive and Predictive Maintenance*

This Quick Quiz will familiarize trainees with preventive and predictive maintenance procedures and characteristics. You will need photocopies of the quiz provided on the following page. Trainees will need pencils. If you allow trainees to use the Trainee Guide, decrease the amount of time you give them to complete the quiz, and remind them to bring their books to class.

1. Make a photocopy of the quiz for each trainee.
2. Give trainees between 5 and 10 minutes to complete the quiz.
3. Go over the answers to the quiz.
4. Ask trainees if they have questions.

Answers to Quick Quiz

1. PM
2. PDM
3. PDM
4. PDM
5. PM
6. PDM
7. PM
8. PDM

Quick Quiz   *Preventive and Predictive Maintenance*

Identify the characteristics of the maintenance system. Write in the corresponding letters of either PM (for preventive maintenance) or PDM (for predictive maintenance) in the blank provided.

_____ 1. Uses repair history to extend equipment life

_____ 2. Uses NDT diagnosis tools

_____ 3. Compares current operation to normal profile

_____ 4. Consists of monitoring during operation

_____ 5. Consists of periodically overhauling equipment and replacing components

_____ 6. Only replaces identified unsound components

_____ 7. Consists of teardown inspections, estimating failure points

_____ 8. Automatically warns of failure

Section 4.0.0 *Nondestructive Examination*

This Quick Quiz will familiarize trainees with nondestructive examination procedures. You will need photocopies of the quiz provided on the following page. Trainees will need pencils. If you allow trainees to use the Trainee Guide, decrease the amount of time you give them to complete the quiz, and remind them to bring their books to class.

1. Make a photocopy of the quiz for each trainee.
2. Give trainees between 5 and 10 minutes to complete the quiz.
3. Go over the answers to the quiz.
4. Ask trainees if they have questions.

Answers to Quick Quiz

1. a
2. g
3. b
4. c
5. d
6. h
7. e
8. i
9. f
10. e
11. f
12. e
13. f
14. d
15. d

Quick Quiz *Nondestructive Examination Methods*

Identify the nondestructive examination (NDE) process described from the list of terms below. Write the corresponding letter in the blank provided. Some terms may be used more than once.

_____ 1. Changes in acoustic impedance are detected using _____.

_____ 2. Heat is converted into an image through _____.

_____ 3. Inclusions or voids are detected in pipe welds using _____.

_____ 4. Changes in electrical conductivity or magnetic permeability are detected through the method of _____.

_____ 5. Surface characteristics, such as finish, are detected via _____.

_____ 6. Frequency, amplitude, displacement, velocity, acceleration, and elapsed time are components included in _____.

_____ 7. Turbine blades are checked for surface cracks or porosity using _____.

_____ 8. Lubricant analysis, wear particle analysis, and ferrography are parts of the field called _____.

_____ 9. Railroad wheels are checked for cracks through _____.

_____ 10. Surface openings due to cracks, porosity, seams, or folds are detected using _____.

_____ 11. An inexpensive NDE method that is sensitive to surface and near-surface flaws is _____.

_____ 12. A method that is not useful for porous materials is _____.

_____ 13. The method that discovers defects using lines of magnetic flux is _____.

_____ 14. The NDE method used on paper, wood, or metal to check for uniformity and surface finish characteristics is _____.

_____ 15. The method that can only be applied to the surface, through surface openings, or to transparent materials is _____.

a. ultrasonics
b. radiography
c. eddy current inspection
d. visual and optical inspection
e. liquid penetrant inspection
f. magnetic particle inspection
g. infrared testing
h. vibration analysis
i. tribology

Answer		Section
1.	c	2.0.0
2.	c	2.0.0
3.	b	2.1.0
4.	c	3.0.0
5.	d	3.0.0
6.	a	3.0.0
7.	c	4.0.0
8.	b	4.7.0
9.	a	4.9.0
10.	c	4.10.0

CONTREN® LEARNING SERIES — USER UPDATE

The NCCER makes every effort to keep these textbooks up-to-date and free of technical errors. We appreciate your help in this process. If you have an idea for improving this textbook, or if you find an error, a typographical mistake, or an inaccuracy in NCCER's Contren® textbooks, please write us, using this form or a photocopy. Be sure to include the exact module number, page number, a detailed description, and the correction, if applicable. Your input will be brought to the attention of the Technical Review Committee. Thank you for your assistance.

Instructors – If you found that additional materials were necessary in order to teach this module effectively, please let us know so that we may include them in the Equipment/Materials list in the Annotated Instructor's Guide.

Write: Product Development and Revision
National Center for Construction Education and Research
3600 NW 43rd St., Bldg. G, Gainesville, FL 32606

Fax: 352-334-0932

E-mail: curriculum@nccer.org

Craft _____ Module Name _____

Copyright Date _____ Module Number _____ Page Number(s) _____

Description _____

_____

_____

_____

(Optional) Correction _____

_____

_____

(Optional) Your Name and Address _____

_____

_____

Vibration Analysis

NCCER STANDARDIZED CRAFT TRAINING PROGRAM

The National Center for Construction Education and Research (NCCER) provides a standardized national program of accredited craft training. Key features of the program include instructor certification, competency-based training, and performance testing. The program provides trainees, instructors, and companies with a standard form of recognition through a National Craft Training Registry. The program is described in full in the *Guidelines for Accreditation*, published by NCCER. For more information on standardized craft training, contact the NCCER by writing us at 3600 NW 43rd St., Bldg. G, Gainesville, FL 32606; calling 352-334-0911; or emailing info@nccer.org. More information may be found at our website, www.nccer.org.

HOW TO USE THIS ANNOTATED INSTRUCTOR'S GUIDE

Each page presents two sections of information. The larger section displays each page exactly as it appears in the Trainee Module. The narrow column ties suggested trainee and instructor actions to each page and provides icons (detailed below) to call your attention to material, safety, audiovisual, or testing requirements. The bottom of each page includes space for your notes.

The **Audiovisual** icon indicates an appropriate time to show a transparency or other audiovisual aid.

The **Classroom** icon prompts you to define a term, stress a point, ask trainees to explain a concept, or give examples.

The **Demonstration** icon directs you to show trainees how to perform tasks.

The **Examination** icon tells you to administer the written module examination.

The **Homework** icon is placed where you may wish to assign reading for the next class, assign a project, or advise trainees to prepare for an examination.

The **Laboratory** icon is used when trainees are to practice performing tasks.

The **Materials** icon is a reminder for you to gather materials needed for classes, labs, and testing.

The **Performance Testing** icon tells you to administer a performance test or a portion thereof.

The **Safety** icon is used to emphasize safety issues. It is often keyed to *Caution* and *Warning!* statements in the Trainee Module.

The **Teaching Tip** icon indicates additional guidance is available, such as how to conduct an exercise, get the most educational value from a field trip, or encourage class participation. Teaching Tips may expand on a feature (*Think About It, Did You Know?*) or provide *Quick Quizzes* or similar exercises. You will be referred to the Teaching Tips section at the back of the module if there is additional material.

The **Combination** icon indicates that the laboratory listed corresponds with a performance task. If desired, you can note the proficiency of the trainees during the laboratory, and use it to satisfy performance testing requirements.

PREPARATION

Before teaching this module, you should review the Objectives, Performance Tasks, Materials and Equipment List, and Module Outline. Be sure to allow ample time to prepare your own training or lesson plan and gather all required materials and equipment.

MODULE OVERVIEW

This module covers the causes of vibration, vibration analysis and monitoring techniques, vibration test equipment, and field balancing of machines.

PREREQUISITES

Please refer to the Course Map in the Trainee Module. Prior to training with this module, it is recommended that the trainee shall have successfully completed the following: *Core Curriculum; Millwright Level One; Millwright Level Two; Millwright Level Three; Millwright Level Four;* and *Millwright Level Five*, Modules 15501-09 through 15508-09.

OBJECTIVES

Upon completion of this module, the trainee will be able to do the following:

1. List four causes of vibration.
2. Identify characteristics of a vibration cycle.
3. Identify and explain the different kinds of basic vibration test equipment.
4. Explain vibration monitoring.
5. Explain field balancing of machines.

PERFORMANCE TASKS

This is a knowledge-based module; there are no Performance Tasks.

MATERIALS AND EQUIPMENT LIST

Overhead projector and screen

Transparencies

Blank acetate sheets

Transparency pens

Whiteboard/chalkboard

Markers/chalk

Pencils and scratch paper

Appropriate personal protective equipment

Examples of equipment with the following causes of vibration:
  Unbalance
  Misalignment
  Bent drive shafts
  Loose mounting bolts
  Worn or damaged bearings
  Improper gear meshing

Loose drive belts
Insufficient lubrication
Electrical problems

Examples of the following vibration test equipment, as available:
  Transducers
  Vibration meter
  Oscilloscope
  Spectrum analyzer
  Electronic filters
  Stroboscope
  Strip chart recorders
  Data collectors

Balancing machine

Copies of the Quick Quizzes*

Module Examinations**

* Located at the back of this module.

**Located in the Test Booklet.

SAFETY CONSIDERATIONS

Ensure that the trainees are equipped with appropriate personal protective equipment and know how to use it properly.

ADDITIONAL RESOURCES

This module is intended to present thorough resources for task training. The following reference works are suggested for both instructors and motivated trainees interested in further study. These are optional materials for continued education rather than for task training.

http://www.reliabilityweb.com/fa/vibration.htm (for vibration analysis testing resources and links).

http://www.plant-maintenance.com/maintenance_articles_vibration.shtml

TEACHING TIME FOR THIS MODULE

An outline for use in developing your lesson plan is presented below. Note that each Roman numeral in the outline equates to one session of instruction. Each session has a suggested time period of 2½ hours. This includes 10 minutes at the beginning of each session for administrative tasks and one 10-minute break during the session. Approximately 5 hours are suggested to cover *Vibration Analysis*. You will need to adjust the time required for hands-on activity and testing based on your class size and resources.

Topic	Planned Time
Session I. Introduction; Causes of Vibration; Vibration Analysis, Test Equipment, and Monitoring	
A. Introduction	_____
B. Causes of Vibration	_____
1. Unbalance, Misalignment, Bent Drive Shafts, Loose Mounting Bolts	_____
2. Worn/Damaged Bearings, Improper Gear Meshing, Loose Drive Belts	_____
3. Insufficient Lubrication, Electrical Problems, Destructive Resonant Frequencies	_____
C. Vibration Analysis	_____
1. Frequency	_____
2. Velocity	_____
3. Acceleration	_____
4. Displacement	_____
D. Vibration Test Equipment	_____
1. Transducers	_____
2. Vibration Analysis Equipment	_____
3. Vibration Recording Instruments	_____
E. Vibration Monitoring	_____
1. Identifying Equipment to be Monitored	_____
2. Establishing Schedules and Determining Monitoring Point Locations	_____
3. Setting Up Record Keeping and Continuous Monitoring Systems	_____

Session II. Field Balancing of Machines; Review and Testing

   A. Field Balancing of Machines   _____

      1. Determining Causes of Unbalance   _____

      2. Calculating Unbalance Force   _____

      3. Determining Corrective Action   _____

   B. Trade Terms Quick Quiz   _____

   C. Module Review   _____

   D. Module Examination   _____

      1. Trainees must score 70% or higher to receive recognition from NCCER.

      2. Record the testing results on Craft Training Report Form 200, and submit the results to the Training Program Sponsor.

15509-09

Vibration Analysis

Assign reading of Module 15509-09.

15509-09
Vibration Analysis

Topics to be presented in this module include:

Overview

Each type of defect in a piece of equipment has its own vibration signature. The frequency, velocity, acceleration, and displacement of a vibration cycle determine the location and cause of the vibration. Vibration analysis enables the millwright to pinpoint the problem in the machinery to make the necessary repairs. The use of computers allows substantial amounts of data to be compared, providing realtime answers to problems. Failures can be predicted before they become inevitable. A good vibration monitoring and analysis program prolongs the life of the equipment and increases productivity by reducing downtime.

Instructor's Notes:

_____

_____

_____

Objectives

When you have completed this module, you will be able to do the following:

1. List four causes of vibration.
2. Identify characteristics of a vibration cycle.
3. Identify and explain the different kinds of basic vibration test equipment.
4. Explain vibration monitoring.
5. Explain field balancing of machines.

Trade Terms

Cycle
G force
Pickup coil
Sheave

Transducer
Velocity
Vibration signature

Required Trainee Materials

1. Pencil and paper
2. Appropriate personal protective equipment

Prerequisites

Before you begin this module, it is recommended that you successfully complete *Core Curriculum; Millwright Level One; Millwright Level Two; Millwright Level Three; Millwright Level Four;* and *Millwright Level Five,* Modules 15501-09 through 15508-09.

This course map shows all of the modules in the fifth level of the Millwright curriculum. The suggested training order begins at the bottom and proceeds up. Skill levels increase as you advance on the course map. The local Training Program Sponsor may adjust the training order.

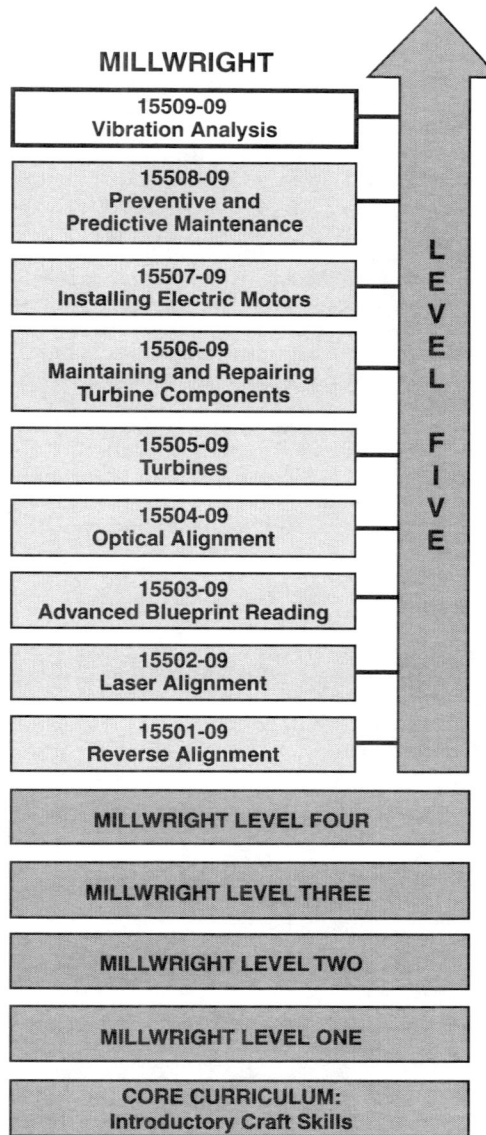

MILLWRIGHT

LEVEL FIVE
15509-09 Vibration Analysis
15508-09 Preventive and Predictive Maintenance
15507-09 Installing Electric Motors
15506-09 Maintaining and Repairing Turbine Components
15505-09 Turbines
15504-09 Optical Alignment
15503-09 Advanced Blueprint Reading
15502-09 Laser Alignment
15501-09 Reverse Alignment

MILLWRIGHT LEVEL FOUR

MILLWRIGHT LEVEL THREE

MILLWRIGHT LEVEL TWO

MILLWRIGHT LEVEL ONE

CORE CURRICULUM: Introductory Craft Skills

509CMAP.EPS

Ensure that you have everything required to teach the course. Check the Materials and Equipment List at the front of this module.

See the general Teaching Tip at the end of this module.

Explain that terms shown in bold are defined in the Glossary at the back of this module.

Show Transparency 1, Objectives. Review the goals of the module, and explain what will be expected of the trainees.

Review the modules covered in Level Five, and explain how this module fits in.

MODULE 15509-09 ◆ VIBRATION ANALYSIS 9.1

1.0.0 ◆ INTRODUCTION

Every rotating machine vibrates to some extent, even when it is new. But as machines are used, bearings and other parts are subjected to wear and damage, resulting in an increase in vibration. Excessive vibration can cause shafts to break, pulleys and gears to come loose, and motor mounts to break, which may send steel fragments or equipment flying through the air and cause personal injury. Vibration monitoring and analysis allow you to pinpoint problems before they cause damage to the equipment or become safety hazards. This module explains the causes of vibration, vibration analysis and monitoring techniques, vibration test equipment, and field balancing. Each of the vibration analysis and monitoring techniques explained in this module is a form of nondestructive evaluation, or NDE. NDE is a method of examining an object or event without destroying it in the process. These procedures are performed by specially trained and certified personnel.

2.0.0 ◆ CAUSES OF VIBRATION

Understanding the causes of vibration is the first step toward controlling vibration in rotating machinery. Early detection and control of vibration can greatly extend the life of the machinery. There are many possible causes of vibration in rotating machinery, including the following:

- Unbalance
- Misalignment
- Bent drive shafts
- Loose mounting bolts
- Worn or damaged bearings
- Improper gear meshing
- Loose drive belts
- Insufficient lubrication
- Electrical problems
- Destructive resonant frequencies

2.1.0 Unbalance

Unbalance is the most common cause of vibration. It is caused by unequal weight or stress at two opposite points around the axis or shaft of a rotating machine. Unbalance can be caused by a manufacturing problem, such as a hidden void in a casting, or by an operational problem, such as the full or partial loss of a fan blade. *Figure 1* shows these two causes of unbalance.

2.2.0 Misalignment

Misaligned couplings, bearings, or V-belt pulleys can cause severe vibration. Misalignment can be offset, angular, or a combination of the two (*Figure 2*).

2.3.0 Bent Drive Shafts

Bent drive shafts can also cause severe vibration. If not detected and corrected early, a bent drive shaft can damage bearings, equipment mountings, and gears.

2.4.0 Loose Mounting Bolts

Loose mounting bolts result in a pounding action, caused by the machine rocking, that is detected as vibration. The severity of the vibration depends on how loose the bolts are. As vibration continues,

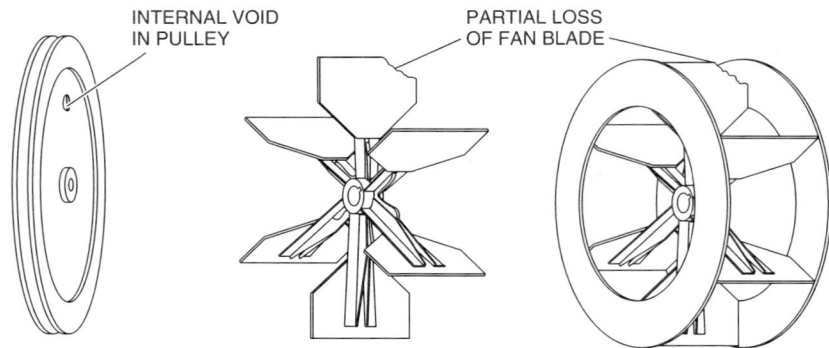

INTERNAL VOID IN PULLEY

PARTIAL LOSS OF FAN BLADE

509F01.EPS

Figure 1 ◆ Two causes of unbalance.

Instructor's Notes:
_____
_____
_____
_____

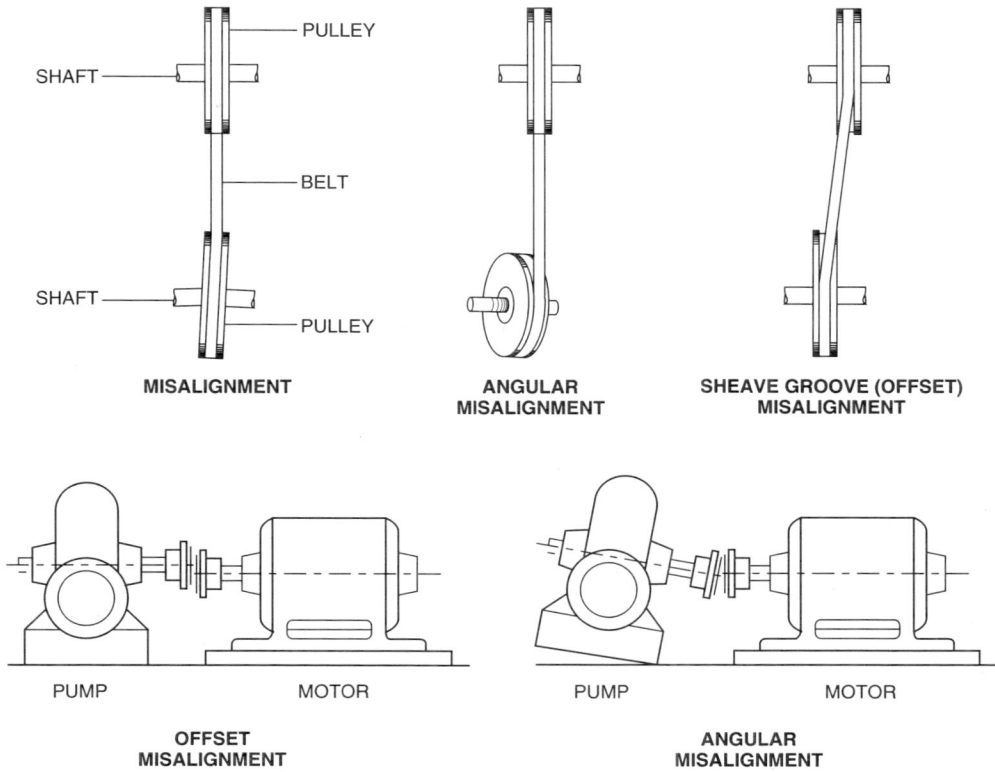

Figure 2 ◆ Misalignment.

the bolts become progressively looser. Complete destruction of related equipment can result if the problem is not corrected.

2.5.0 Worn or Damaged Bearings

Worn or damaged antifriction bearings create a high-frequency vibration. The high frequency results from the size differences between the balls or rollers and the larger inner and outer races. Bearing vibration can sometimes be detected by a high-pitched noise.

Vibration in sleeve bearings can be caused by incorrect clearances or oil whirl. Too little clearance causes low-amplitude, high-frequency vibration. Too much clearance resulting from wear causes high-amplitude, low-frequency vibration.

Oil whirl can occur in loose-fitting, pressure-lubricated bearings in high-speed machinery. If any outside influence, such as imbalance or a sudden shock, displaces the shaft from its normal position in the bearing, pressurized oil rushes in to fill the gap left by the displaced shaft. This fur-

ther displaces the shaft, causing the shaft to whirl in the direction of rotation (*Figure 3*).

2.6.0 Improper Gear Meshing

Improper gear meshing causes vibration between gears and can be caused by several different factors. Improper gear alignment, broken or severely worn gear teeth, and scored gear teeth are the most common causes. Improper gear meshing usually does not result in immediate destruction of the equipment, but the condition slowly worsens if it is not corrected.

2.7.0 Loose Drive Belts

Loose drive belts can usually be easily detected except when one or more belts are loose on a multiple-belt drive. Refer to the equipment or belt manufacturer's instructions for the proper tension of the belt. Use a tension meter to check the tension of the belts, and adjust them as necessary.

MODULE 15509-09 ◆ VIBRATION ANALYSIS   9.3

Discuss the types of problems caused by worn or damaged bearings.

Show Transparency 4 (Figure 3). Discuss the causes of oil whirl and its effects on the shaft.

Discuss the causes and effects of improper gear meshing and loose drive belts.

Emphasize the need to refer to the equipment or belt manufacturer's instructions for the proper tension of a belt.

Show trainees examples of equipment with worn or damaged bearings, improper gear meshing, and loose drive belts.

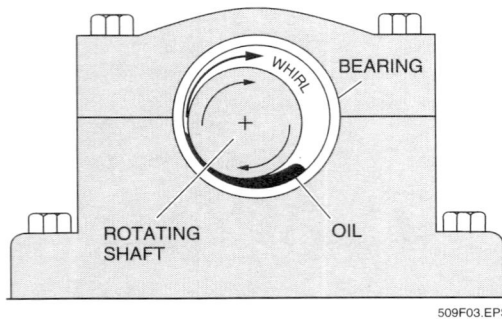

Figure 3 ♦ Oil whirl in a sleeve bearing.

2.8.0 Insufficient Lubrication

Insufficient lubrication between a shaft and a sleeve bearing can cause vibration. Normally, a lubricating grease or heavy oil fills the void between the shaft and the sleeve bearing. If the lubricant is allowed to dry out or becomes used up, any imbalance in the load carried by the shaft will cause the shaft to vibrate in the bearing.

2.9.0 Electrical Problems

Electrically caused vibration in a motor or generator is the result of unequal magnetic forces between the rotor and stator. Electrical vibration can be easily identified. If the vibration stops immediately after removing electrical power, the cause is electrical. If the vibration level decreases slowly as the motor or generator slows to a stop, the problem is mechanical. Usually, the only mechanical problems in a motor are bad bearings, a bent rotor shaft, or an off-balance cooling vane in the back of the motor.

2.10.0 Destructive Resonant Frequencies

When one of a machine's vibration frequencies coincides with a natural frequency of a component or structure within the machine, it creates a resonant frequency. When this vibration matches its frequency with another piece of equipment coupled with this machine, it doubles the amplitude of the resonant frequency and can cause a destructive harmonic frequency that is physically evident. A good example of this is a strong musical note shattering a glass. The natural resonant frequency of the glass matches the note and amplifies itself to the point at which it becomes destructive.

3.0.0 ♦ VIBRATION ANALYSIS

Each vibration has a set of distinctive characteristics known as the **vibration signature**. The process of gathering vibration signatures, determining their causes, and comparing them with stored baseline data is known as vibration analysis.

Vibration of rotating machinery can be separated into individual vibration **cycles** (*Figure 4*). Each cycle begins with the shaft centered in the housing. As the shaft rotates, it bends due to loads, loose bearings, or a bent shaft. As the rpm increases, the shaft moves away from the center line until it reaches a peak. The shaft then moves back through the center line to the peak in the opposite direction. The shaft deflection reverses back to the center line, and this is considered one cycle of vibration. As the load or rpm increases, this creates a more noticeable wobble that is measured as increased frequency and amplitude. The vibration is caused by the reversal of movement at each peak of the cycle.

A vibration cycle consists of four measurable characteristics that make up the vibration signature, or fingerprint. In the vibration analysis process, each characteristic is monitored and recorded. Analyzing the characteristics helps identify the source of the vibration and the defect that caused it. Each type of defect in a rotating piece of equipment has its own specific vibration signature. For example, a bent shaft causes axial vibration at a frequency of one or two times the rotational speed of the shaft. The following are the four characteristics of a vibration cycle:

- Frequency
- **Velocity**
- Acceleration
- Displacement

3.1.0 Frequency

The frequency of a vibration is the number of individual vibration cycles that occur in a given time period. In vibration analysis, frequency is measured in cycles per minute (cpm). If the vibration cycle occurs 60 times in one minute, the frequency is 60 cpm.

3.2.0 Velocity

Vibration velocity is the speed at which the vibrating shaft completes one cycle. The velocity is not constant. When the shaft reaches the maximum distance in each direction from the center line, the velocity is zero because the shaft stops and changes

Instructor's Notes:

_____

_____

_____

direction. The velocity is at its maximum as the shaft travels through the center line. This highest, or peak, velocity is used to analyze velocity and is recorded in inches-per-second peak (*Figure 5*).

3.3.0 Acceleration

In vibration analysis, acceleration is the increasing rate of change in velocity. Acceleration is measured in gravitation constants, or *g*s. One *g* is equal to 9.8 meters per second per second or 32.2 feet per second per second. One *g* is also equal to the acceleration produced by the earth's gravity at sea level. Vibration acceleration refers to the force of *g*s at each cycle peak. The acceleration of a vibration cycle is zero at the center line and reaches peak acceleration at each peak (*Figure 6*).

3.4.0 Displacement

Displacement, also known as amplitude, is the total distance that the shaft travels from one peak to the other peak during one cycle of vibration (*Figure 7*). This distance is measured in mils. One mil is $\frac{1}{1,000}$ inch.

4.0.0 ◆ VIBRATION TEST EQUIPMENT

The frequency, velocity, acceleration, and displacement of a vibration cycle determine the location and cause of the vibration. Vibration analysis allows the technician to pinpoint the problem in the machinery to make the necessary repairs. Vibration test equipment is used to analyze equipment and determine the type of vibration

Show Transparency 6 (Figure 5). Note the points of zero and peak velocity.

Explain how acceleration is measured in *g*s.

Show Transparency 7 (Figure 6). Note the points of zero and peak acceleration.

Show Transparency 8 (Figure 7). Explain that displacement is measured in mils.

Introduce the three types of basic vibration test equipment.

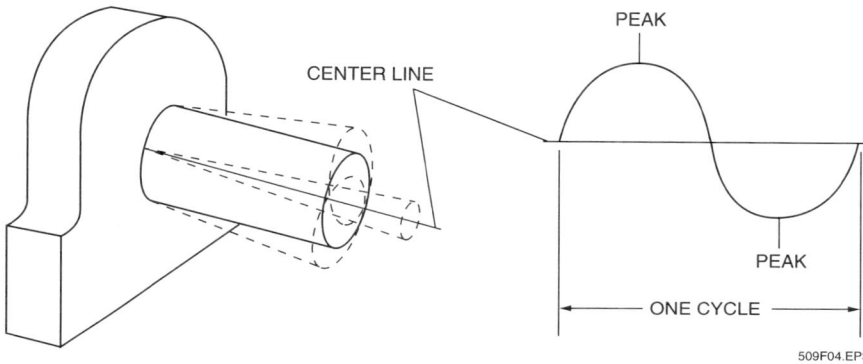

Figure 4 ◆ One cycle of vibration.

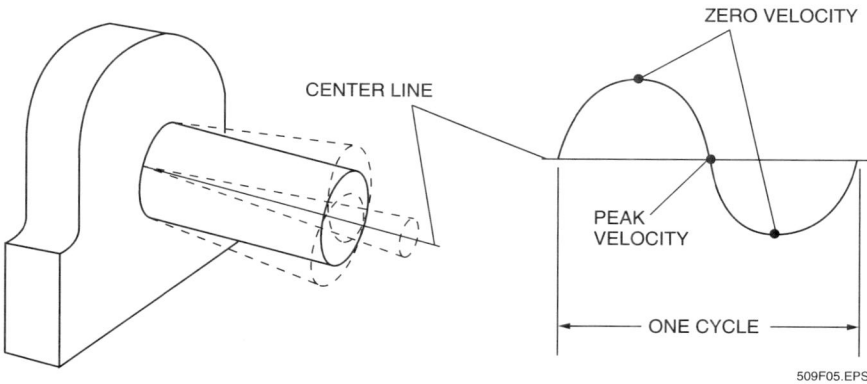

Figure 5 ◆ Zero and peak velocity.

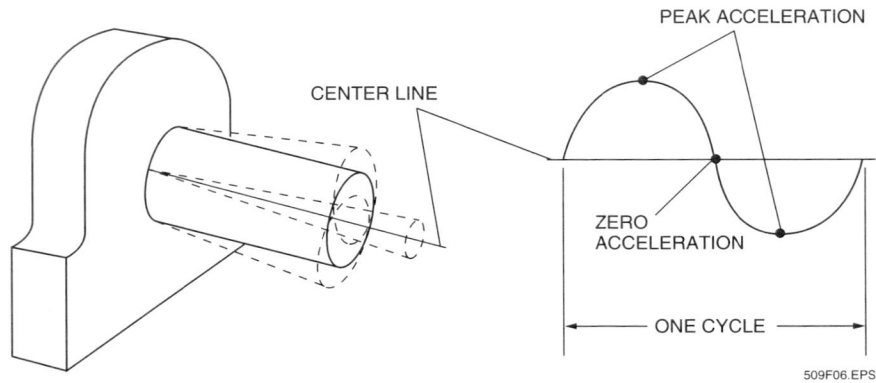

Figure 6 ◆ Zero and peak acceleration.

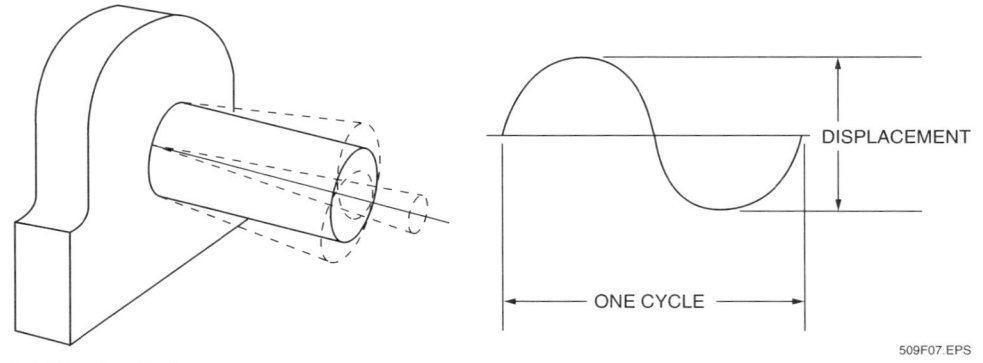

Figure 7 ◆ Vibration displacement.

defect. The basic vibration test equipment
includes **transducers**, vibration analysis equipment, and vibration recording instruments. This
equipment is used to monitor and analyze
machine vibrations.

4.1.0 Transducers

Transducers convert mechanical vibrations into
electrical signals which are reproduced as sounds
or graphic displays. A transducer provides the
link between the equipment being analyzed and
the vibration analysis equipment. The three types
of transducers include the following:

• Velocity transducers
• Accelerometers
• Displacement transducers

4.1.1 Velocity Transducers

Velocity transducers are known as contact transducers because they are mounted directly on the
vibrating surface. These transducers pick up frequency signals in the intermediate range, which is
the frequency range of most industrial equipment.
These transducers can be used with portable,
hand-held test equipment, or they can be permanently mounted to the vibrating surface for use
with a continuous-monitoring system.

A velocity transducer (*Figure 8*) consists of
either a piezoelectric crystal or a coil of fine wire
wrapped around a metal core and restrained by
soft springs. The coil assembly is surrounded by a
magnet attached to the inside of the transducer
case. When the transducer is in contact with the
equipment being analyzed, the case and magnet
move back and forth with the vibrating equipment, while the coil assembly inside the transducer remains in one place. As the transducer
moves back and forth around the coil, the magnetic field surrounding the coil induces voltage in
the coil. This voltage is proportional to the vibration velocity. The faster the transducer case and
magnet vibrate, the higher the output voltage of
the coil.

Instructor's Notes:

_____

_____

_____

4.1.2 Accelerometers

The most accurate measurement of vibration acceleration can be obtained with an accelerometer (*Figure 9*). Accelerometers are contact transducers with wider frequency ranges than velocity transducers. Accelerometers are used to analyze various forces, loads, and stresses that emit low- to high-frequency sounds. Some types of accelerometers require an amplifier to strengthen low-frequency signals. Accelerometers can be permanently mounted to the equipment being monitored. The three ways to mount an accelerometer to machinery are to attach it with a heavy magnet, screw it into a tapped hole, or screw it into an attached adapter.

Accelerometers contain crystal discs that generate a voltage proportional to the vibration acceleration when the crystal is compressed. When the machinery accelerates, **g-forces** cause the crystal discs to compress. The induced voltage is amplified and fed to the analyzer to be displayed as vibration acceleration.

4.1.3 Displacement Transducers

Displacement transducers (*Figure 10*) are noncontact devices that are mounted a specific distance from the rotating shaft of the machinery being analyzed. They are used to detect the small amount of vibration produced by lightweight, high-speed shafts in large, heavy casings. Noncontact displacement transducers are more accurate than contact transducers for this application because they sense the movement of the shaft through a magnetic field.

The displacement transducer sensor is placed by DC voltage at a set gap from the shaft surface and produces a magnetic field that is distorted by the movement of the shaft. As the shaft vibrates, the gap size and the amount of magnetic field distortion change, generating varying voltages. A **pickup coil** in the transducer senses the changes in voltage and transmits the information to the vibration recorder or analyzer. Displacement transducers can provide accurate vibration measurements on turbines, compressors, centrifugal pumps, and similar machinery.

4.2.0 Vibration Analysis Equipment

Vibration analysis equipment is used to determine the cause of vibration. Machinery malfunctions can be identified by comparing the frequency of the vibration with the running speed of the machine. These comparisons are used to determine the sources of the vibration. The electronic equipment used for vibration analysis includes the following:

- Meters
- Oscilloscopes
- Spectrum analyzers
- Electronic filters
- Stroboscopes

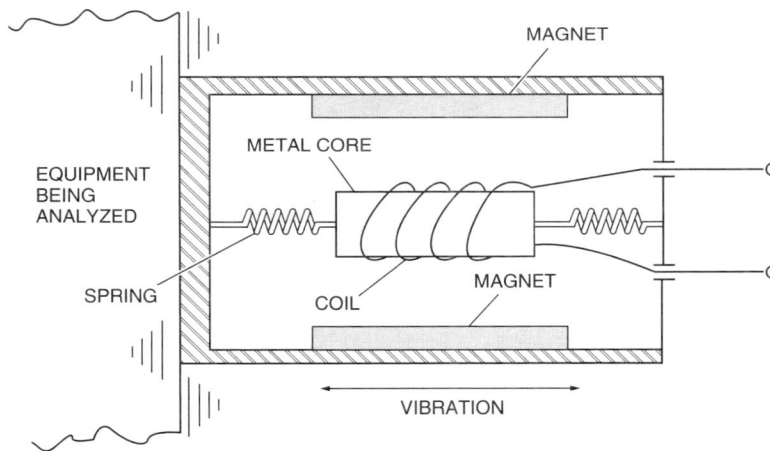

Discuss the use of accelerometers to precisely measure acceleration.

Show Transparency 10 (Figure 9). Describe how voltage is generated by the crystal discs in proportion to vibration acceleration.

Describe the components and operation of displacement transducers.

Explain the functions of the transducer sensor and the pickup coil.

Show Transparency 11 (Figure 10).

Explain how vibration analysis equipment identifies machine malfunctions.

List the types of electronic equipment used for vibration analysis.

Figure 8 ◆ Velocity transducer.

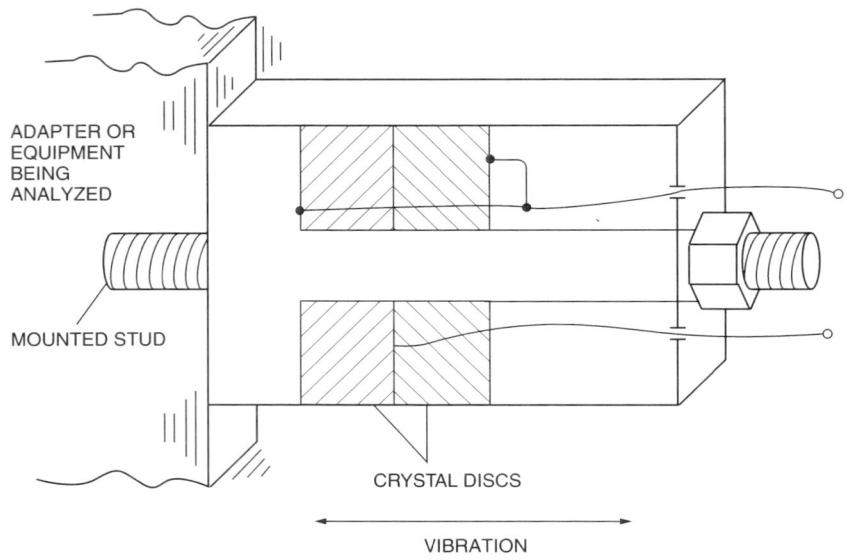

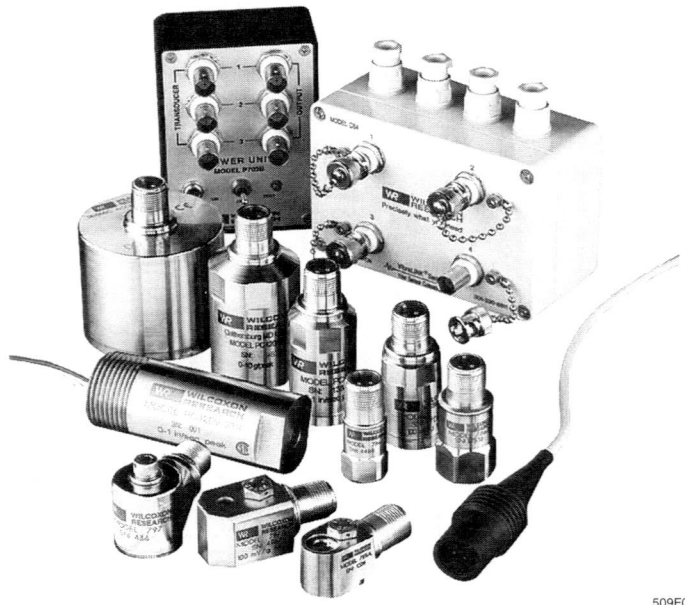

ADAPTER OR
EQUIPMENT
BEING
ANALYZED

MOUNTED STUD

CRYSTAL DISCS

VIBRATION

509F09.EPS

Figure 9 ◆ Accelerometers.

9.8   MILLWRIGHT ◆ LEVEL FIVE

Instructor's Notes:

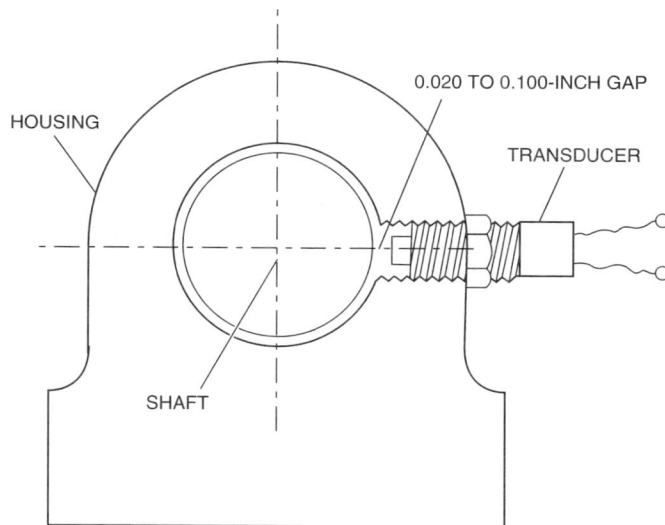

HOUSING

0.020 TO 0.100-INCH GAP

TRANSDUCER

SHAFT

509F10.EPS

Figure 10 ◆ Displacement transducer.

Discuss the operation of vibration meters and the types of vibrations that they measure. Refer to Figure 11.

Discuss the function, operation, and components of oscilloscopes. Refer to Figure 12.

Provide a vibration meter and an oscilloscope for trainees to examine.

.

4.2.1 Meters

Vibration meters (*Figure 11*) are portable instruments used to measure vibrations that occur within a selected frequency. Most portable meters give the amplitude of the vibration displacement in mils, the acceleration of the vibration in *g*s, or the velocity of the vibration in inches per second. Vibration meters are generally battery-powered and use contact transducers.

Some vibration meters are made with the meter and sensor pickup in one housing. However, vibration meters with the sensor connected to a flexible cable are easier to use in a confined space or around obstructions. The three controls on the vibration meter are the displacement/velocity/acceleration selector switch, the amplitude range selector switch, and the battery test/on/off selector switch.

4.2.2 Oscilloscopes

Cathode-ray oscilloscopes are used to test electronic circuits. Oscilloscopes (*Figure 12*) display graphs of rapidly changing voltage and current but are also capable of giving information concerning frequency values, phase differences, and voltage amplitude. The principal components of a basic oscilloscope include a cathode-ray tube (CRT), a sweep oscillator, horizontal and vertical deflection amplifiers, controls, switches, and input terminals. A CRT is a vacuum tube that projects a beam onto a fluorescent screen. The deflection amplifier changes the direction of the beam both vertically and horizontally on the screen. The sweep oscillator, or generator, moves the electronic beam across the screen to form a wave pattern.

MODULE 15509-09 ◆ VIBRATION ANALYSIS   9.9

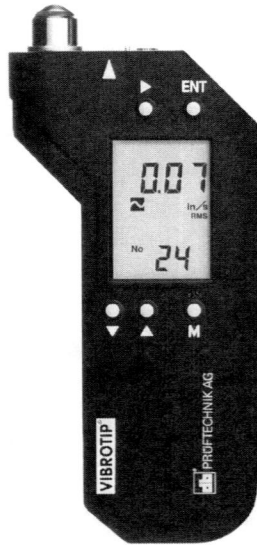

Figure 11 ◆ Vibration meter.

509F11.EPS

The oscilloscope is used as a vibration analysis instrument because it provides a signal that can be displayed on the CRT and analyzed. A signal from the sensor displays a wave pattern on the CRT screen. This signal can be recorded, photographed, or analyzed when it appears on the screen.

Small, portable oscilloscope units are used to monitor signals in the field. Laboratory (bench) oscilloscopes are useful when several people must view the screen simultaneously. A third type of oscilloscope is equipped with a camera adapter and is used to photograph certain signals. *Figure 13* shows a vibration sine wave measurement.

4.2.3 Spectrum Analyzers

Vibration spectrum analyzers (*Figure 14*) measure and compare the amplitude and frequency characteristics of a vibration. Machinery vibration and noise are generally complex, and the analyzer is capable of separating each vibration frequency for individual monitoring.

The analyzer has two amplitude scales. The top scale measures amplitude from zero to one, and the bottom scale measures amplitude from one to three. The severity of the vibration determines which scale to use. A displacement/velocity selector switch allows the technician to measure either vibration displacement in mils or vibration velocity in inches per second.

HANDHELD

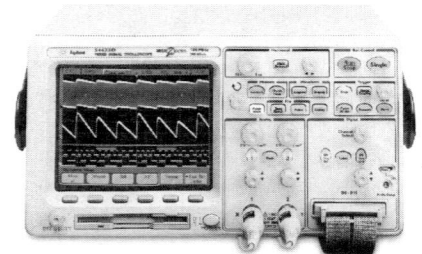

BENCH

BOTTOM PHOTO: © Agilent Technologies, Inc. 2009 Reproduced with permission, courtesy of Agilent Technologies, Inc.

509F12.EPS

Figure 12 ◆ Oscilloscopes.

Instructor's Notes:

_____

_____

_____

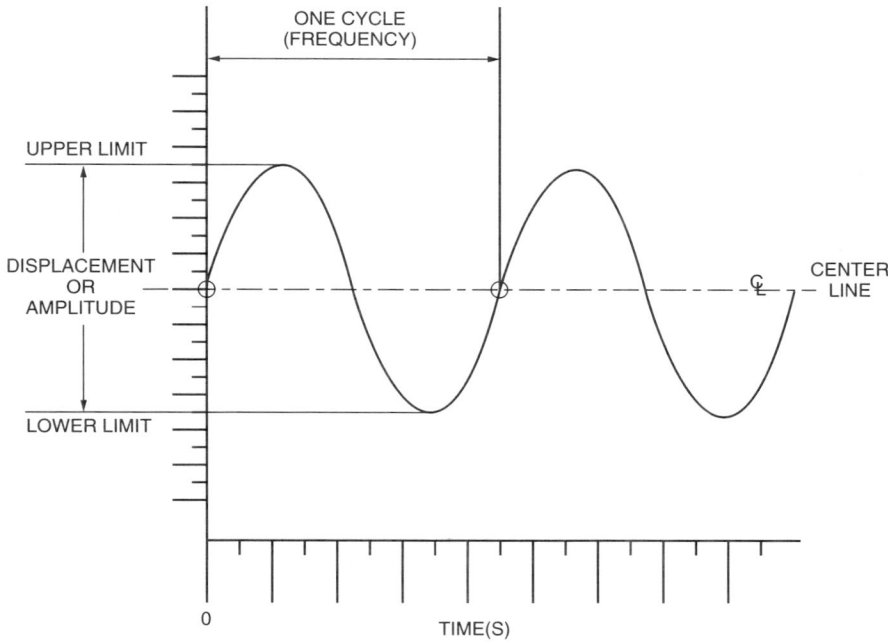

Figure 13 ◆ Vibration sine wave measurement.

ONE CYCLE
(FREQUENCY)

UPPER LIMIT

DISPLACEMENT
OR
AMPLITUDE

CENTER
LINE

LOWER LIMIT

0   TIME(S)

509F13.EPS

A frequency meter measures the frequency of the indicated vibration in cycles per minute (cpm), and a frequency range selector switch allows the technician to select the desired frequency range of the frequency meter.

A turntable filter selector switch allows the technician to select from different filter bandwidths, ranging from sharp to broad. The broad filter bandwidth is used for rapid scanning to quickly determine vibration frequencies present. The sharp filter bandwidth is used to analyze a particular vibration frequency.

4.2.4 Electronic Filters

When a machine vibrates, a signal composed of many frequencies is generated. Basic vibration meters cannot pinpoint specific problems. To determine the frequencies present in the signal, electronic filters are used. They can detect a vibration frequency within a specific narrow band range. Electronic filters can be separate components connected to the vibration analyzer or can be built into the analyzer. Filters can be classified in the following four ways:

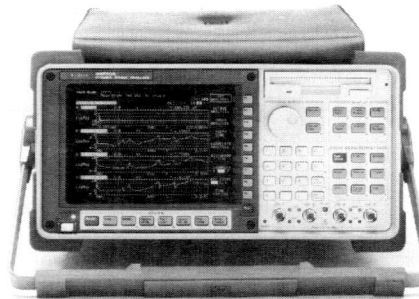

© Agilent Technologies, Inc. 2009 Reproduced with permission, courtesy of Agilent Technologies, Inc.

509F14.EPS

Figure 14 ◆ Vibration spectrum analyzer.

- *Passband frequency* – Passband electronic filters are the most basic types of filters. A passband filter detects a specified band of frequencies and eliminates frequencies above or below the desired passband. High-pass filters detect all frequencies above a specified frequency. Low-pass filters detect frequencies below a specified frequency.

MODULE 15509-09 ◆ VIBRATION ANALYSIS   9.11

- *Method of tuning* – Fixed filters, tunable filters, and automatic filters are used to tune vibration analyzers to specific frequencies. Fixed filters are commonly used for monitoring where known or predictable frequencies occur. Tunable filters allow the filter frequency band to be adjusted so that specific frequencies can be isolated. This type of filter saves time in searching for specific frequencies and can eliminate a cluttered frequency signal. The tuning signal in an automatic, or tracking, filter is generated by and synchronized with the part of the machinery being analyzed. The automatic filter is used in shaft balancing applications and for tracking specific frequencies during start-up. It is also used when the system is coasting to a stop.

- *Component type* – Electronic filters can also be classified as active or passive, depending on the type of components used in their construction. Active filters are commonly used in machinery analysis because they are small and stable and provide a voltage increase. These filters use operational amplifiers to strengthen the vibration signal. Passive filters are much larger and heavier than active filters, limiting their use and portability. The vibration signals from passive filters are much sharper because amplifiers are not required.

- *Roll-off characteristics* – An electronic filter receives frequency signals within a certain frequency band. It does not, however, immediately cut off all frequencies outside the passband. The farther the frequencies are outside the passband, the weaker the frequency signals. This effect is called the filter roll-off. The frequency attenuation varies among filters.

4.2.5 Stroboscopes

Stroboscopes (*Figure 15*) shine a short flash of intense light when triggered by a transducer. The transducer is attached to a piece of equipment. A target on the rotating part of the equipment being analyzed is lighted each time the vibration causes the transducer to trigger the stroboscope. The source of a vibration can be determined because the vibrating part does not appear to move. If the rotating part appears to move when lighted, the vibration may be unsteady or may be coming from another source nearby.

4.3.0 Vibration Recording Instruments

Recording instruments provide either a printed or a magnetic tape recording of vibration data for a permanent record of vibration. Recording instruments are also used to transfer vibration data from one analytical device to another. The major types of recording instruments used in vibration analysis are strip chart recorders and data collectors.

4.3.1 Strip Chart Recorders

A strip chart recorder prints the machinery vibration signature on chart paper. The machinery vibration signature is the frequency and amplitude of machine vibration. A pen records the vibration motion on chart paper, moving between two rollers at a set rate of speed. The chart paper is divided at regular intervals and can be used to record the strength and frequency of the vibration. This provides a hard copy of the machinery signature, which can be reviewed and analyzed. *Figure 16* shows a strip chart recorder built into a vibration spectrum analyzer and an example of information recorded on the strip.

4.3.2 Data Collectors

Data collection from most vibration monitoring equipment is done with computers (*Figure 17*), either with a laptop computer or with a link to a computer elsewhere. Vibration analysis software

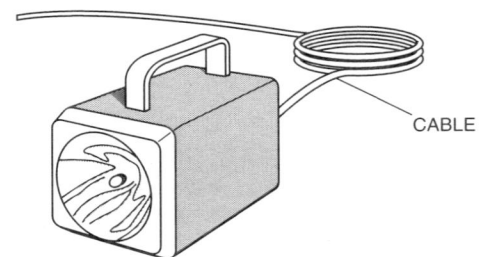

509F15.EPS

Figure 15 ◆ Stroboscope.

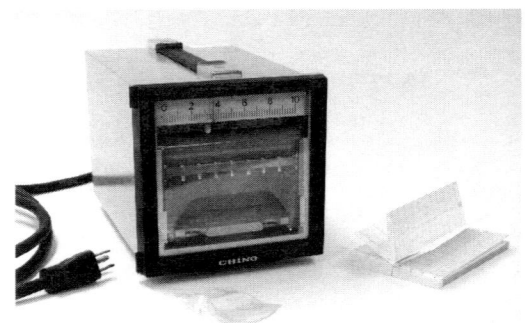

509F16.EPS

Figure 16 ◆ Strip chart recorder and strip example.

Instructor's Notes:
_____
_____
_____
_____

and historical data allow the millwright to compare the vibration patterns at several data points and to observe changes. The software available is capable of determining the probability of damage from vibration or from the causes of vibration.

5.0.0 ◆ VIBRATION MONITORING

Vibration characteristics should be checked and logged when the machine is first installed and then on a regular schedule to monitor increased vibration. Checking vibration regularly and entering the vibration readings on a data sheet or in a computer program is known as vibration monitoring. A vibration monitoring program can greatly reduce production downtime, equipment failure, and safety hazards in a plant. There are several purposes for implementing a vibration monitoring program. They are as follows:

- To identify the equipment to be monitored
- To establish a schedule for monitoring each piece of equipment
- To determine the locations of monitoring points on each piece of equipment

- To develop and maintain vibration monitoring records for each piece of equipment being monitored
- To develop a continuous monitoring system

5.1.0 Identifying Equipment to be Monitored

The first step in setting up a vibration monitoring program is to identify the equipment to be included in the program. The factors to consider include the following:

- Cost of replacement or repair
- Importance of equipment to plant operations
- Possibility of safety hazard
- Lead time for replacement parts

5.1.1 Cost of Replacement or Repair

Equipment is often selected for monitoring on the basis of replacement or repair costs. Monitoring vibration can detect problems early and prevent damage to the equipment, thus reducing costs.

Describe the key reasons for implementing a vibration monitoring program.

List the factors to consider when identifying equipment to be monitored.

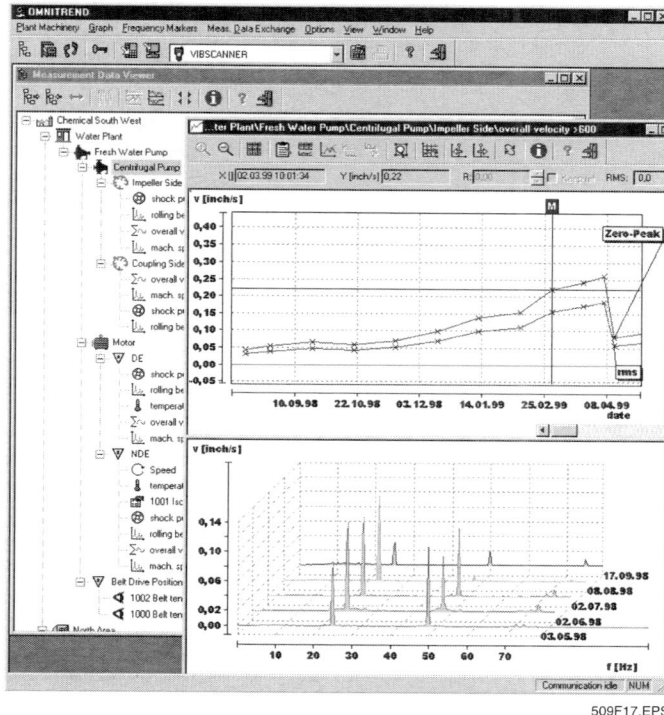

Figure 17 ◆ Printout from a data collector.

509F17.EPS

For this reason, expensive equipment is often included in a vibration monitoring program.

5.1.2 Importance of Machine to Plant Operations

Failure of some large machines can shut down a complete line and even affect the whole plant. A carefully performed vibration monitoring program can often prevent a disruption by predicting an impending failure. Replacement of the part or parts causing the increased vibration can then be performed during routine shutdown.

5.1.3 Possibility of Safety Hazard

Bearing failure is one example of a possible safety hazard. Bearing seizure in a large, high-speed machine can cause disintegration of the shaft and anything connected to it. Pieces of flying metal are a safety hazard to anyone in the area. Vibration monitoring can detect the early stages of bearing failure and considerably reduce the safety hazard.

5.1.4 Lead Time for Replacement Parts

It can take weeks or months to receive machine replacement parts after the order is placed. A good vibration monitoring program can help predict when a part will need to be replaced. Early placement of the order helps ensure that the part is on hand when replacement is needed.

5.2.0 Establishing a Schedule for Monitoring

Many factors must be considered in establishing a schedule for vibration monitoring, including machine speed, number of bearing points, reliability history of similar equipment, and the importance of the machinery. In a vibration monitoring program, the maximum interval for checking important machinery should be three months.

5.3.0 Determining Locations of Monitoring Points

Vibration should be checked at all critical points on the equipment. To make the observation more useful, the vibration of each critical point, such as bearings and shaft couplings, should be measured in three axes: horizontal, vertical, and axial (*Figure 18*).

5.4.0 Developing and Maintaining Vibration Monitoring Records

When vibration is monitored manually, all vibration readings are entered on a data sheet or computer. The recorded data is then compared to the machine baseline vibration data. Baseline data should be established during initial startup or after a machine has been repaired. This data is used as a basis for analyzing vibration signatures gathered periodically throughout the life of a machine. *Figure 19* shows a monitoring form.

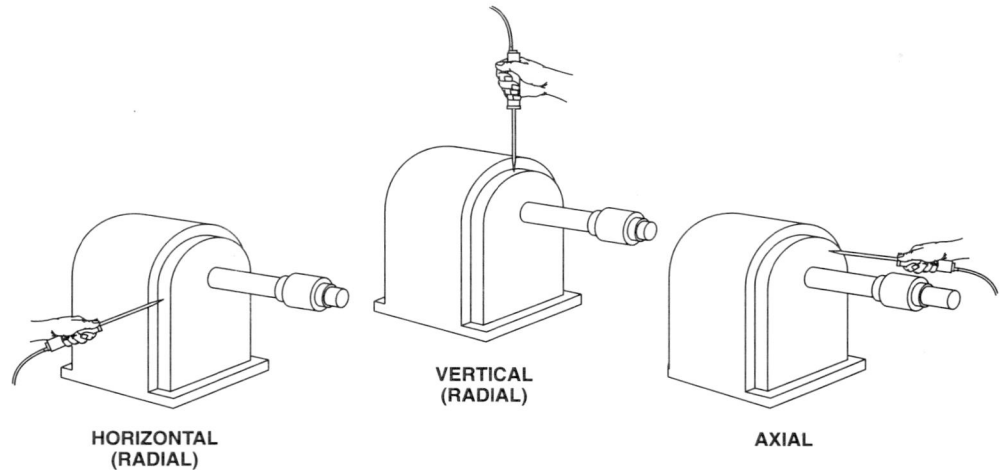

HORIZONTAL
(RADIAL)

VERTICAL
(RADIAL)

AXIAL

509F18.EPS

Figure 18 ◆ Vibration check in three axes.

Instructor's Notes:

_____

_____

_____

5.5.0 Establishing a Continuous Monitoring System

Continuous monitoring uses an instrumentation system that monitors equipment operation at all times. Data collectors, computers with vibration-analysis software packages, and vibration analyzers are used to gather data from the monitors so that it can be more thoroughly analyzed. *Figure 20* shows a typical monitoring system.

The three classes of continuous vibration monitors are standard monitors for general-purpose applications, monitors with basic pre-engineered modifications, and fully engineered monitors. The first two types are used in most continuous monitoring systems. Standard monitors are available as single-channel monitors or multiple interchangeable-monitor modules. Single-channel monitors record vibrations from a single, fixed transducer. Multiple-monitor modules are attached to several fixed transducers and record several types of vibrations simultaneously. Each channel monitor is able to perform the following functions:

- Monitor vibration
- Monitor axial shaft position
- Monitor thrust-bearing wear
- Incorporate warning and shutdown alarms
- Incorporate time delays and test alarms

The most common approach is to link all the information to a computer, where software renders it usable.

6.0.0 ◆ FIELD BALANCING OF MACHINES

Unbalance in a rotating machine is usually caused by a concentration of weight in one spot. Two heavy spots of the same weight exactly opposite each other and equally distanced from the shaft cancel out unbalance. Balancing a machine involves the following:

- Determining the causes of unbalance
- Calculating the unbalance force
- Determining the required corrective action

6.1.0 Determining the Causes of Unbalance

Unbalance can be immediate or time-related. Immediate causes of unbalance are related to manufacturing problems and result in unbalance when the machine or equipment is first started up. The time-related causes develop over a period of time as the equipment is used.

6.1.1 Immediate Causes of Unbalance

Flaws in casting, machining, and assembly show up as unbalance immediately upon startup. These flaws include casting blowholes, eccentricity, keys and keyways, and clearance tolerance buildup. Casting blowholes, or sand traps, in castings are hidden flaws that are undetectable by visual inspection. The lack of metal caused by a hole in a casting results in a weight loss in that area. This weight loss causes a heavy spot directly across the shaft from the blowhole. The result is a significant unbalance (*Figure 21*). For static balancing, place the piece of equipment on a stand, roll it 90 degrees, and see whether it moves from side to side or up and down. To balance the machine, place weights inside at correct points to achieve balance.

The displacement of a shaft from true center is called eccentricity (*Figure 22*). No matter how perfectly round and balanced a **sheave**, rotor, fan blade, or other component, it rotates unbalanced if its shaft is not located exactly at the center of rotation.

Keys and keyways can also cause an immediate unbalance. Motor manufacturers usually balance motors with a half key in the keyway. Sheave manufacturers usually check sheave balance with a half key in the keyway as well. When the motor and sheave are mated and a full key is used, unbalance can result. The key fills half of the keyway on the shaft remaining outside the machine.

The buildup of clearance tolerances from assembling a machine can also result in unbalance (*Figure 23*). For example, the center hole in a sheave or other item has to have a larger inside diameter than the outside diameter of its mating shaft. This difference is required to permit a slip fit. When a set screw is tightened to secure the item to the shaft, the center line of the item is displaced from the shaft center line, causing unbalance.

6.1.2 Time-Related Causes of Unbalance

Rotating equipment, such as blowers, compressor rotors, pump impellers, and other equipment used in material handling, is subject to corrosion or abrasion over a period of time. If the item does not wear evenly, unbalance results.

Another time-related cause of unbalance is the buildup of deposits on rotating equipment. Although deposits tend to form evenly, some unbalance still results. The problem becomes worse as pieces of the deposits begin to break off, increasing the unbalance and causing more pieces to break off. A serious unbalance usually results.

Describe the components and operation of a continuous vibration monitoring system.

Show Transparency 16 (Figure 20).

Discuss the three classes of continuous vibration monitors and their applications.

Assign reading of Sections 6.0.0–6.3.0.

Ensure that you have everything required for teaching this session.

List the three activities involved in balancing a machine.

Define immediate and time-related unbalance.

Describe flaws that manifest as unbalance immediately upon startup.

Show Transparencies 17-19 (Figures 21-23).

Discuss time-related causes of unbalance.

VIBRATION MONITORING DATA SHEET

☐ NOISE ☒ VIBRATION

FOR: _____

_____

EQUIPMENT USED:

_____

_____

PERFORMED BY: _____

LEGEND:
→ PICKUP POINT
X PLAIN BEARING
☒ ANTIFRICTION BEARING
⊣⊢ COUPLING

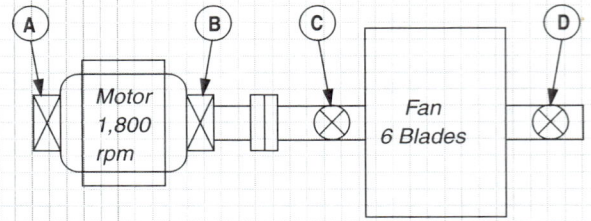

Motor 1,800 rpm

Fan 6 Blades

Bearings "C" & "D"
12 Balls per Bearing

MEAS.		FILTER OUT VIBRATION						SPIKE ENERGY	FILTERED VIBRATION					
		DISP.		VEL.		ACCEL.								
POINTS	POS.	MIL.	CPM	IN/SEC	CPM	G	CPM	G/SE						
	H													
	V													
	A													
	H													
	V													
	A													
	H													
	V													
	A													
	H													
	V													
	A													
	H													
	V													
	A													
	H													
	V													
	A													

UCC   DATA SHEET

509F19.EPS

Figure 19 ◆ Standard vibration monitoring form.

Instructor's Notes:

_____

_____

_____

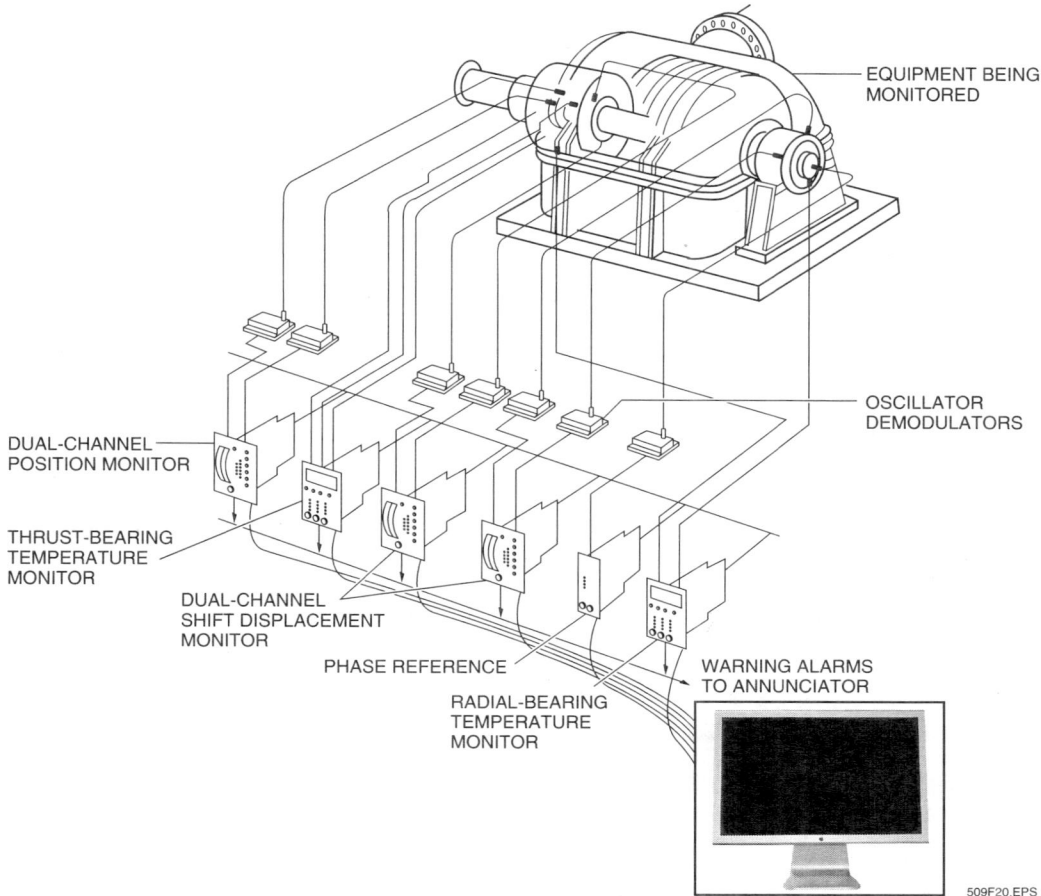

Figure 20 ◆ Continuous vibration monitoring system.

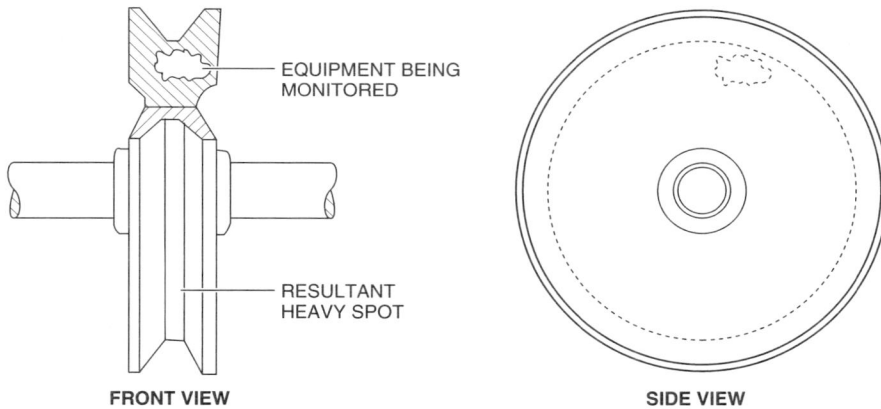

Figure 21 ◆ Blowhole affecting balance in cast sheave.

MODULE 15509-09 ◆ VIBRATION ANALYSIS   9.17

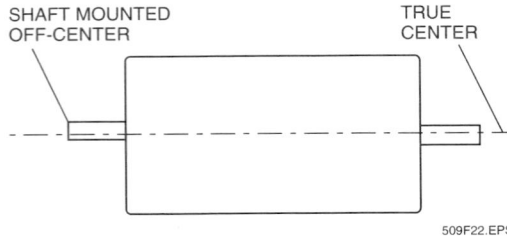

509F22.EPS

Figure 22 ◆ Eccentric shaft.

6.2.0 Calculating Unbalance Force

Unbalance in a rotating object generates a force that causes the object to vibrate. The amount of unbalance is the weight of the heavy spot multiplied by the distance of the heavy spot from the center of rotation. For example, a 2-ounce heavy spot located 3 inches from the center of rotation causes an unbalance of 6 ounce-inches (oz-in). The unbalance force depends on the amount of unbalance and the speed of rotation. The unbalance force for a small piece of equipment can be calculated by using the following formula:

$$F = 1.77 \times (rpm/1,000)^2 \times oz\text{-}in$$

Where:

F	=	unbalance force in pounds
rpm	=	rotating speed of the part
oz-in	=	the amount of unbalance in ounce-inches
1.77	=	the constant required to make the formula dimensionally correct

6.3.0 Determining Required Corrective Action

Whenever possible, equipment balancing is performed without dismantling the machine. When the part requiring balancing must be removed, it is placed on a balancing machine capable of supporting it and driving it at its normal speed (*Figure 24*). A balancing machine consists of four basic components: a stand with two bearing pedestals, two support assemblies with bearings or V-blocks to support the rotor, a method of rotating the rotor, and a vibration instrument to determine the unbalance. A strobe light is often used to detect the position of the heavy spot, and a vibration tester displays the vibration amplitude. From this data, the technician can calculate how much weight must be added to or removed from a given spot to correct the unbalance. Other instruments are available that automatically compute the unbalance of a rotor and that provide a numerical display for the technician to use in balancing the rotor. With these types of instruments, the technician does not need to perform any calculations because balance weights can be placed at the locations indicated by the test instrument.

Point out the factors that are involved in calculating the unbalance force.

Present the formula for calculating the unbalance force for a small piece of equipment.

Provide example unbalance force problems for trainees to solve.

Describe the components, operation, and function of a balancing machine.

Show Transparency 20 (Figure 24).

Provide a balancing machine for trainees to examine.

See the Teaching Tip for Section 6.0.0 at the end of this module.

Instructor's Notes:

_____

_____

_____

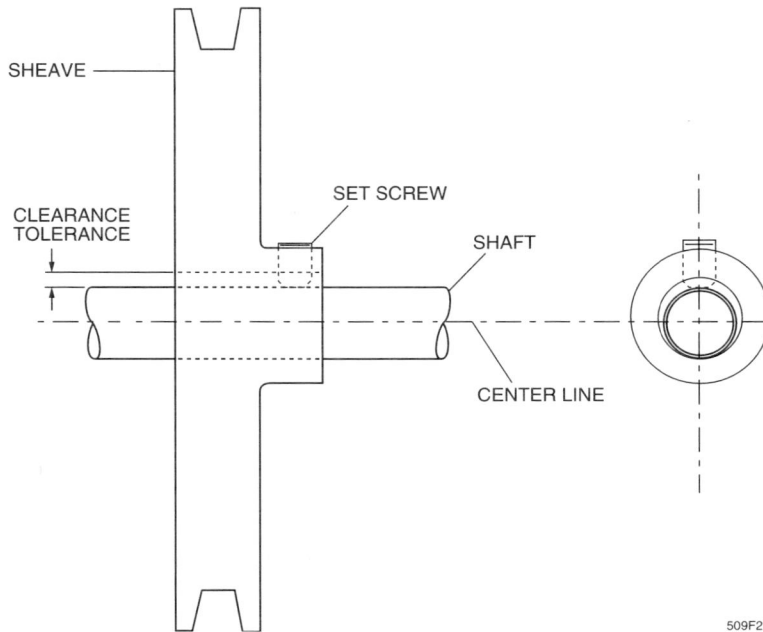

Figure 23 ◆ Clearance tolerance unbalance.

SHEAVE

CLEARANCE
TOLERANCE

SET SCREW

SHAFT

CENTER LINE

509F23.EPS

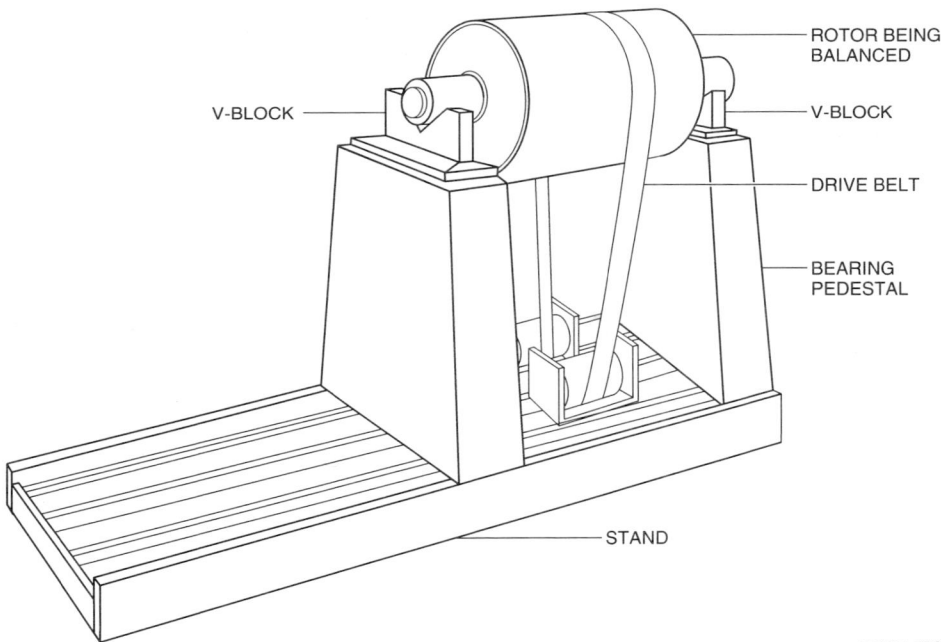

Figure 24 ◆ Balancing machine.

V-BLOCK

ROTOR BEING
BALANCED

V-BLOCK

DRIVE BELT

BEARING
PEDESTAL

STAND

509F24.EPS

Review Questions

1. Misalignment that causes vibration can be offset, angular, or _____.
   a. rotational
   b. circular
   c. a combination of offset and angular
   d. intermittent

2. Vibration is caused by electrical factors if removing electrical power causes the vibration to _____.
   a. increase
   b. decrease
   c. stop immediately
   d. remain the same

3. Vibration cycles consist of _____ measurable characteristics.
   a. two
   b. three
   c. four
   d. five

4. Accelerometers contain _____ discs that generate a voltage proportional to the vibration acceleration.
   a. coil
   b. divided
   c. magnetic
   d. crystal

5. The transducer used to detect small amounts of vibration produced by lightweight, high speed shafts in large, heavy casings is the _____ transducer.
   a. velocity
   b. accelerometer
   c. displacement
   d. frequency

6. Equipment that displays graphs of rapidly changing voltage and current on a CRT screen are known as _____.
   a. meters
   b. strip chart recorders
   c. oscilloscopes
   d. electronic filters

7. A passband filter detects a(n) _____.
   a. amplitude
   b. specified set of frequencies
   c. complete range of frequencies
   d. wavelength

8. The vibration of each critical point should be measured in _____ axes.
   a. one
   b. two
   c. three
   d. four

9. Continuous monitoring uses an instrumentation system that monitors equipment _____.
   a. at all times
   b. automatically at regular intervals
   c. once per shift
   d. once per working day

10. The displacement of a shaft from the true center of the part it supports is called _____.
    a. eccentricity
    b. unbalance
    c. tolerance buildup
    d. roundness

Instructor's Notes:

_____

_____

_____

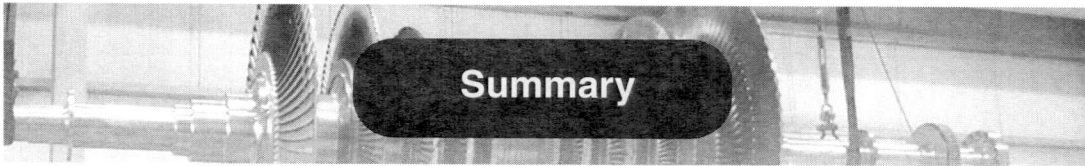

Summary

Each type of defect in a rotating piece of equipment has its own specific vibration signature. The frequency, velocity, acceleration, and displacement of a vibration cycle determine the location and cause of the vibration. Vibration analysis allows the technician to pinpoint the problem in the machinery to make the necessary repairs.

Vibration monitoring provides baseline data and allows the technician to monitor changes in the baseline data and predict failure before it happens. A solid vibration monitoring program prolongs the life of the equipment and increases productivity by reducing downtime.

Summarize the major concepts presented in the module.

Administer the Module Examination. Be sure to record the results of the Examination on Craft Training Report Form 200, and submit the results to the Training Program Sponsor.

Notes

Trade Terms Introduced in This Module

Cycle: One complete sequence of values of alternating quantities.

g-force: A unit of measurement for bodies going through the stress of acceleration.

Pickup coil: A device that converts measurable quantities into corresponding electric signals.

Sheave: A grooved wheel or pulley.

Transducer: Any device or element that converts an input signal into an output signal of a different form.

Velocity: The time rate at which an object changes position.

Vibration signature: The specific characteristics of a vibration measured in frequency, acceleration, displacement, and velocity.

Instructor's Notes:

_____

_____

_____

Additional Resources

This module is intended to present thorough resources for task training. The following reference works are suggested for further study. These are optional materials for continued education rather than for task training.

http://www.reliabilityweb.com/fa/vibration.htm
http://www.plant-maintenance.com/
  maintenance_articles_vibration.shtml

Figure Credits

Wilcoxon Research, Inc., 509F08 (photo), 509F09 (photo)

Photo Courtesy of Grass Technologies, 509F10 An Astro-Med, Inc. Product Group, West Warwick, RI.

Photo courtesy of Ludeca, Inc., 509F11, 509F17 www.ludeca.com

Courtesy of Agilent Technologies, Inc., 509F12 (bench), 509F14, © Agilent Technologies, Inc. 2009 Reproduced with permission,

Fluke Corporation, 509F12 (handheld) Reproduced with permission.

Greenlee Textron, Inc., a subsidiary of Textron Inc., 509F16

Topaz Publications, Inc., 509F20 (photo)

MODULE 15509-09 — TEACHING TIPS

The following are suggested activities or instructional methods to help you teach the material in this module.

General   When you call on someone to answer a question, the rest of the class relaxes or even tunes out because they expect that the question and answer will take place only between you and the trainee you called on. Instead, use this technique to involve more trainees in answering questions and to keep them on their toes.

1. Ask the trainees to define a term or explain a concept.
2. After one trainee has answered, ask a trainee seated nearby if the answer is right. Then ask whether a trainee in the back of the room agrees.
3. Ask the trainees to explain why they think an answer is right or wrong.
4. Use the session to clear up incorrect ideas, and encourage the trainees to learn from their mistakes.

Sections
1.0.0 – 6.0.0   *Trade Terms*

This Quick Quiz will familiarize trainees with trade terms related to vibration analysis. You will need photocopies of the quiz provided on the following page. Trainees will need pencils. If you allow trainees to use the Trainee Guide, decrease the amount of time you give them to complete the quiz, and remind them to bring their books to class.

1. Make a photocopy of the quiz for each trainee.
2. Give trainees between 5 and 10 minutes to complete the quiz.
3. Go over the answers to the quiz.
4. Ask trainees if they have questions.

Answers to Quick Quiz

1. c
2. f
3. a
4. e
5. b
6. d

Quick Quiz

Trade Terms

For each description listed, identify the term that the text best describes. Write the corresponding letter in the blank provided.

_____ 1. One complete sequence of values of alternating quantities is a(n) _____.

_____ 2. The specific characteristics of a vibration measured in frequency, acceleration, displacement, and velocity is the _____.

_____ 3. A device that converts measurable quantities into corresponding electric signals is a(n) _____.

_____ 4. A grooved wheel or pulley is called a(n) _____.

_____ 5. A unit of measurement for bodies going through the stress of acceleration is called _____.

_____ 6. Any device or element that converts an input signal into an output signal of a different form is called a(n) _____.

    a.   pickup coil
    b.   g-force
    c.   cycle
    d.   transducer
    e.   sheave
    f.   vibration signature

Section 2.0.0   *Causes of Vibration*

This Quick Quiz will familiarize trainees with the causes of vibration in rotating machinery. You will need photocopies of the quiz provided on the following page. Trainees will need pencils. If you allow trainees to use the Trainee Guide, decrease the amount of time you give them to complete the quiz, and remind them to bring their books to class.

1. Make a photocopy of the quiz for each trainee.
2. Give trainees between 5 and 10 minutes to complete the quiz.
3. Go over the answers to the quiz.
4. Ask trainees if they have questions.

Answers to Quick Quiz

1. g
2. j
3. i
4. h
5. e
6. c
7. f
8. a
9. d
10. b

Quick Quiz   *Causes of Vibration*

For each description listed, identify the term that the text best describes. Write the corresponding letter in the blank provided.

_____ 1. The most common cause of vibration, which is the result of unequal weight or stress at two opposite points around the axis or shaft of a rotating machine, is _____.

_____ 2. Vibration caused by _____ can sometimes be detected by a high-pitched noise.

_____ 3. If not detected and corrected early, a(n) _____ can damage bearings, equipment mountings, and gears.

_____ 4. Use a tension meter to check the tension of the _____, and adjust them as necessary.

_____ 5. In a motor or generator, _____ is the result of unequal magnetic forces between the rotor and stator.

_____ 6. A cause of vibration that can be described as offset, angular, or a combination of the two is _____.

_____ 7. A cause of vibration in machines that can be compared to a strong musical note shattering a glass is _____.

_____ 8. In a machine that has _____, any imbalance in the load carried by the shaft will cause the shaft to vibrate in the bearing.

_____ 9. When a machine has _____, the result is a pounding action, caused by the machine rocking, that is detected as vibration.

_____ 10. The most common causes of _____ are improper gear alignment, broken or severely worn gear teeth, and scored gear teeth.

a. insufficient lubrication
b. improper gear meshing
c. misalignment
d. loose mounting bolts
e. electrically caused vibration
f. destructive resonant frequencies
g. unbalance
h. loose drive belts
i. bent drive shaft
j. worn or damaged bearings

Section 6.0.0 *Field Balancing of Machines*

This Quick Quiz will familiarize trainees with terms related to the field balancing of machines. You will need photocopies of the quiz provided on the following page. Trainees will need pencils. If you allow trainees to use the Trainee Guide, decrease the amount of time you give them to complete the quiz, and remind them to bring their books to class.

1. Make a photocopy of the quiz for each trainee.
2. Give trainees between 5 and 10 minutes to complete the quiz.
3. Go over the answers to the quiz.
4. Ask trainees if they have questions.

Answers to Quick Quiz

1. g
2. f
3. e
4. j
5. i
6. h
7. c
8. d
9. b
10. a

Quick Quiz *Field Balancing of Machines*

For each description listed, identify the term that the text best describes. Write the corresponding letter in the blank provided.

_____ 1. The amount of unbalance in a rotating object is the weight of the heavy spot multiplied by the _____ of the heavy spot from the center of rotation.

_____ 2. Flaws in casting, machining, and assembly show up as unbalance immediately upon _____.

_____ 3. The unbalance force depends on the amount of unbalance and the _____.

_____ 4. The buildup of _____ from assembling a machine can result in unbalance.

_____ 5. Unbalance in a rotating machine is usually caused by a concentration of _____ in one spot.

_____ 6. The buildup of deposits on rotating equipment is a(n) _____ cause of unbalance.

_____ 7. A casting blowhole causes a(n) _____ directly across the shaft, resulting in a significant unbalance.

_____ 8. The displacement of a shaft from true center is called _____.

_____ 9. Whenever possible, equipment balancing is performed without _____ the machine.

_____ 10. When the part requiring balancing must be removed, it is placed on a(n) _____ capable of supporting it and driving it at its normal speed.

    a. balancing machine
    b. dismantling
    c. heavy spot
    d. eccentricity
    e. speed of rotation
    f. startup
    g. distance
    h. time-related
    i. weight
    j. clearance tolerances

MODULE 15509-09 — ANSWERS TO REVIEW QUESTIONS

Answer		Section
1.	c	2.2.0
2.	c	2.9.0
3.	c	3.0.0
4.	d	4.1.2
5.	c	4.1.3
6.	c	4.2.2
7.	b	4.2.4
8.	c	5.3.0
9.	a	5.5.0
10.	a	6.1.1

CONTREN® LEARNING SERIES — USER UPDATE

The NCCER makes every effort to keep these textbooks up-to-date and free of technical errors. We appreciate your help in this process. If you have an idea for improving this textbook, or if you find an error, a typographical mistake, or an inaccuracy in NCCER's Contren® textbooks, please write us, using this form or a photocopy. Be sure to include the exact module number, page number, a detailed description, and the correction, if applicable. Your input will be brought to the attention of the Technical Review Committee. Thank you for your assistance.

Instructors – If you found that additional materials were necessary in order to teach this module effectively, please let us know so that we may include them in the Equipment/Materials list in the Annotated Instructor's Guide.

Write: Product Development and Revision
National Center for Construction Education and Research
3600 NW 43rd St., Bldg. G, Gainesville, FL 32606

Fax: 352-334-0932

E-mail: curriculum@nccer.org

Craft _____ Module Name _____

Copyright Date _____ Module Number _____ Page Number(s) _____

Description _____

_____

_____

_____

(Optional) Correction _____

_____

_____

(Optional) Your Name and Address _____

_____

_____

Glossary of Trade Terms

Aim: To regulate the direction of a sighting device.

Alidade: The upper part of an instrument that turns in azimuth with the sighting device.

Amplitude: The magnitude of dynamic motion or vibration as expressed in terms of peak-to-peak, zero-to-peak, or root mean square.

Axial movement: Movement in the direction of a shaft axis.

Azimuth: The direction in a horizontal plane.

Azimuth axis: The vertical axis, or the axis of the bearing and spindle that confines rotation to a horizontal plane.

Balloon: In machine drawings, a circle with letters or numbers in it that correlate to the parts in the bill of materials. It usually has a leader and arrowhead pointing to the part or subassembly.

Brush: A conductor between the stationary and rotating parts of a machine; usually made of carbon.

Buck-in: To place an instrument so that the line of sight satisfies two requirements simultaneously, such as aiming at two targets. It is usually accomplished by trial and error.

Carbon ring: A gland shaft packing made of carbon to seal the steam chest.

Circular level: The round level attached to the alidade.

Clamp: The device used to connect the part to be aimed with the stationary part of an instrument so that the tangent screw can operate.

Coherent: Composed of only one color or wavelength and all the waves in phase with each other.

Collimate: To make straight and parallel.

Collinear: Two or more shaft center lines with no offset or angularity between them.

Combustion: The burning of a gas, liquid, or solid in which the fuel is oxidized and emits a great amount of heat.

Combustion chamber: Any enclosed space in which a fuel, such as oil, coal, kerosene, or natural gas, is burned to provide heat.

Commutator: A device used on electric motors or generators to maintain a unidirectional current.

Compressor: A machine used for increasing the pressure of a gas or vapor.

Computer-aided drafting (CAD): Drawings that are used in all aspects of mechanical, civil, electronic, or architectural design.

Computer-aided engineering (CAE): A close interface with CAD that defines design and functionality by the use of finite element analysis, dynamic modeling, and mathematical simulation of all aspects in the use of the end product.

Computer-aided machining (CAM): CAD-generated detail drawings are fed directly into a CNC milling machine, punch press, lathe, or other computer-controlled manufacturing equipment.

Computer-integrated manufacturing (CIM): A total control system in a manufacturing environment that uses computers to link process and machine operations in an operating system.

Condensate: The liquid by-product of cooling steam.

Cycle: One complete sequence of values of alternating quantities.

Datum: A theoretically exact point, axis, or plane from which the location or geometric characteristics or features of a part are established and dimensioned.

Degree: An angle equal to $\frac{1}{360}$ of a full circle.

Dimension: A numerical value indicated on a drawing along with lines, symbols, and notes to define the size or geometric characteristics of a part or feature.

Diopter: The measurement of the power of a lens, equal to the reciprocal of its focal length in meters.

Drawing package: A cohesive group of drawings comprised of assembly and detail drawings and a bill of materials that contain all the information needed to build or assemble a piece of equipment.

Eddy current: An electrical current that is generated and dissipated in a conductive material in the presence of a magnetic field.

Electro-optical instruments: EDMIs that transmit modulated infrared or visible light signals in short wavelengths of about 0.4 micrometers or microns (μm) to 1.2 μm.

Elevation: The direction of a line of sight in a vertical plane.

Glossary of Trade Terms

Elevation axis: The horizontal axis, or the axis of the bearing and the journal of the telescope axle that confines rotation to a vertical plane.

Focus: To move the optical parts so that a sharp image is seen, or the point at which the image is perfectly clear.

Frequency: The repetition rate of a periodic event expressed either in cycles per second (Hertz), revolutions per minute (rpm), or multiples of rotational speed.

g-force: A unit of measurement for bodies going through the stress of acceleration.

Governor: A device used to provide automatic control of speed or power of a prime mover but never as a method for shutting down a turbine.

Hertz (Hz): The unit of frequency represented by cycles per second.

Horizontal: Perpendicular to the direction of gravity.

Hysteresis: The portion of a system response in which a change in input does not produce a change in output; also called deadband.

Infrared light: An electromagnetic signal having frequencies below the visible portion of the frequency spectrum. The infrared light frequencies lie between those of light and radio waves with wavelengths ranging from 0.7 μm to 1.2 μm.

Kinetic energy: The energy of movement.

Laser: A device that amplifies and concentrates coherent light waves in an intense beam of parallel, nonscattering energy. The word *laser* is an acronym for Light Amplification by Stimulated Emission or Radiation.

Machined parts: Parts and components that are designed and made for a specific function.

Machinery train: A group of separate machines connected to each other by a process or physically coupled to transfer power to each other.

Mechanical energy: Energy in the form of mechanical power.

Minute: An angle equal to 1/60 of a degree.

Objective lens: The lens at the front end of a telescope and therefore nearest the object sighted.

Parallax: The change in the apparent relative orientations of objects when viewed from different positions.

Peg test: A procedure used to check for an out-of-adjustment bubble vial on levels and other instruments.

Pickup coil: A device that converts measurable quantities into corresponding electric signals.

Plate level: A fairly sensitive tubular level mounted on the plate of a jig transit, used to place the azimuth axis in the direction of gravity.

Position: The location of an instrument in a horizontal plane about which the instrument turns when it is being leveled. Once a jig transit is leveled, it is the point at which the vertical and horizontal axes intersect.

Purchased parts: Stock parts, components, or hardware that are purchased rather than made.

Reference line: A line of sight from which measurements are made.

Resonance: The reinforced vibration of a body caused by a matching frequency of vibration of another body; a common example is a tuning fork.

Reticle: A series of fine lines that are placed in the focus of the objective lens of an optical instrument to quantify the measurements of angles or distances.

Revision: A change in a part of an engineering drawing that is noted on the drawing.

Second: An angle equal to 1/60 of a minute, or 1/3600 of a degree.

Sensory inspection: Inspection that includes sight, sound, smell, and touch.

Sheave: A grooved wheel or pulley.

Standards: Uprights that support the telescope axle bearings of a jig transit.

Station: The distance given in inches and decimal parts of an inch measured parallel to a chosen center line from a single, chosen point.

Stress: A force acting on a body per unit area, normally measured in pounds per inch or newtons per square meter.

Tangent screw: A hand-operated screw that is used to change the direction of the line of sight either in azimuth or in elevation.

Telescope axle: The horizontal axle that supports the telescope of a jig transit.

Glossary of Trade Terms

Telescope direct: The normal position of a jig transit telescope, as opposed to telescope reversed.

Telescope reversed: The position of a jig transit telescope when it is turned over, or transited, so that it is upside down to its normal position.

Telescopic sight: An optical system that consists of an objective lens and a focusing device and that forms an image on a cross-line reticle which is viewed through an eyepiece that magnifies the image and the cross lines together.

Thermal growth: The expansion of materials with the rise in temperature.

Title block: A section of an engineering drawing blocked off for pertinent information such as the title, drawing number, date, scale, material, draftsperson, and tolerances.

Transducer: Any device or element that converts an input signal into an output signal of a different form.

Turbine: A machine that generates rotary, mechanical power from the energy in a stream of fluid or gas.

Turbine generator: An electric generator driven by a steam, hydraulic, or gas turbine.

Velocity: Measures the rate and direction of movement of an object.

Vertical: In the direction of gravity.

Vibration signature: The specific characteristics of a vibration measured in frequency, acceleration, displacement, and velocity.

Visible light: An electromagnetic signal having frequencies located in the part of the electromagnetic spectrum to which the eye is sensitive. The visible light frequencies have wavelengths ranging from 0.4 μm to 0.7 μm.

Index

Index

TIR. *See* Total indicator reading
Title block, 3.3, 3.8, 3.12, 3.23
Tolerances
    building of clearance, 9.15, 9.19
    in detail drawings, 3.8, 3.9
    motor air gap, 7.6
Tolerance tables, 2.14
Tolerancing, geometric, 3.6–3.7, 3.9, 3.11
Tomography, computer-aided (CAT), 8.9
Torque, 1.9, 2.7, 6.2, 6.6
Total indicator reading (TIR), 1.14–1.16, 1.17
Total station
    capability, 4.21
    controls, 4.25–4.28
    field check for alignment, 4.32–4.37
    history, 4.23–4.24
    initializing, 4.30–4.32
    measurement of errors as ppm, 4.22–4.23
    prisms and reflective targets, 4.24–4.25
    setup and coarse centering, 4.29–4.30
Transducer, 9.6–9.7, 9.8–9.9, 9.22
Transit
    laser, 2.8
    setup, adjustment, and checkout, 4.14–4.17
    trigonometric leveling, 4.37, 4.38
    turning 90-degree angles, 4.17–4.18
Transition pieces, 5.19, 5.21
Travel, 1.11, 1.14, 6.18
Triangle, right, 4.21
Tribology, 8.15–8.16
Tribrach, 4.24, 4.25, 4.26
Trigonometry, 4.21, 4.37, 4.38
Trimble Navigation, 4.21
Trip coil, 5.23
Trip mechanism, overspeed, 6.10, 6.12–6.13, 6.14, 6.18
Tripods, 4.7–4.10, 4.24, 4.29
Trip pin, 6.12, 6.14
Trip speed, 6.15
Troubleshooting, 2.19–2.21, 6.2, 7.10–7.11
Tube, cross-fire, 5.19, 5.20
Turbines
    control and protection system, 5.14
    gas, 5.5–5.7, 5.16–5.26
    hydroelectric (water), 5.2, 5.4, 5.7–5.8
    impulse (constant pressure), 5.2–5.3
    lubrication, 5.10, 5.14
    operating principles, 5.2–5.3
    overview, 5.2, 5.30, 6.2
    reaction, 5.3, 5.4, 5.7
    startup and shutdown, 5.12, 5.16
    steam
        large, 6.18–6.19
        maintenance. *See* Maintenance and repair, steam
            turbine
        parts, 5.8–5.16
        principles of operation, 5.3–5.5

        Westinghouse Type E, 6.2
    troubleshooting, 6.2
    types, 5.3–5.8
U
Ultrasonics, 8.7, 8.8–8.9, 8.10
Ultraviolet light (sunlight), 2.13
Unbalance, 9.2, 9.15, 9.17–9.18
Unbalance force, 9.18
V
Valves
    backseating procedure, 6.16, 6.17
    check, 5.19
    emergency quick-closing, 6.12
    extraction point control, 6.19
    fuel stop, 5.25, 5.26
    governor, 6.7, 6.12, 6.16–6.18
    inlet control, 5.8, 5.9, 5.24
    pilot, 6.7
    pressure relief, 5.16, 6.4
    shutoff, 6.4
    trip, 6.15–6.16
Vanes
    arrangement. *See* Wheel
    gas turbine, 5.16, 5.17
    hydroelectric turbine, 5.7
    steam turbine, 5.4, 5.10, 5.12
Vapor extractor, 5.15
Velocity, 8.8, 8.20, 9.4–9.5, 9.6
Vendor, 3.12, 3.13
Vertex, 4.14
Vertical, 4.5, 4.48
Vertical circle vernier test, 4.17
VFD. *See* Drive, variable frequency
Vibration
    causes, 9.2–9.4
    cycle, 8.16, 9.4, 9.5, 9.11
    from misalignment, 1.8, 1.9, 2.2, 2.6, 2.7
    motor, 7.5, 7.6
    no effect on laser, 2.12
    overview, 9.2
Vibration analysis
    monitoring, 9.7, 9.13–9.15, 9.16–9.17
    overview, 8.15, 8.16, 8.17, 9.4–9.5
    recording instruments, 9.12–9.13
    schematic for read points, 8.17
    test equipment, 8.15, 9.5–9.12
Vibration meter, 9.9, 9.10
Vibration signature, 9.4, 9.22
Views
    in detail drawings, 3.3
    elevation, 2.8, 3.13
    plan, 2.8, 3.13
    section, 3.13
    side, 1.4, 2.8, 2.11
    top, 1.4, 2.8, 2.11
Voltage, transducer, 9.6